Power Systems

For further volumes:
http://www.springer.com/series/4622

Sivaji Chakravorti · Debangshu Dey
Biswendu Chatterjee

Recent Trends in the Condition Monitoring of Transformers

Theory, Implementation and Analysis

Springer

Sivaji Chakravorti
Debangshu Dey
Biswendu Chatterjee
Electrical Engineering Department
Jadavpur University
Kolkata
India

ISSN 1612-1287 ISSN 1860-4676 (electronic)
ISBN 978-1-4471-5549-2 ISBN 978-1-4471-5550-8 (eBook)
DOI 10.1007/978-1-4471-5550-8
Springer London Heidelberg New York Dordrecht

Library of Congress Control Number: 2013949664

Printed on acid-free paper

Springer is part of Springer Science+Business Media (www.springer.com)

Dedicated to
our parents
and also to
Mrs. Manidipa Chakravorti
Ms. Tatiana Chakravorti
Ms. Titir Chakravorti
and
Mrs. Tanaya Dey
who are the strength, support and inspiration
behind each word…

Foreword

Transformers are a major component of electric power systems and are well studied by electric power engineers. The key components of most power transformers are the three phases of electrical copper/aluminum wire windings wound around high magnetic permeability cores surrounded by insulating transformer oil and paper/pressboard insulation. The introduction in Chap. 1 discusses transformer failures and statistics, aging of transformers, diagnosis methods, transformer oil and paper insulation, and remaining life analysis. Chapter 2 analyzes impulse faults in transformers using modern circuit analysis including time-frequency domain methods, wavelet transforms, and fractals. Chapter 3 treats partial discharge measurement and analysis, detection methods, and signal analysis. Chapter 4 treats conventional diagnostic techniques of chemistry, dissolved gas analysis, degree of polymerization measurements, furan in oil analysis, insulation resistance measurements, and dielectric loss-factor measurements. Chapter 5 treats time domain dielectric response measurements, time domain response of transformer insulation, effect of noise on dielectric response measurements in field conditions, modeling of dielectric response, and interpretation of time domain data using a dielectric response model. Chapter 6 on frequency domain spectroscopy treats the basic theory of frequency domain measurements, modeling dielectric response in the frequency domain, and advises on the advantages and disadvantages of dielectric response measurements in time and frequency domains. Chapter 7 on methods of frequency response analysis treats basic measurements, basic features of swept frequency response analysis and measurement connection methods, transfer function features including isolated two-winding transfer functions, layer-type helical winding, continuous disk winding, shielded disk winding, interleaved disk winding, windings placed on outer and middle limbs, and high voltage high power autotransformers. Chapter 7 also treats an assessment of mechanical integrity of a transformer winding, key factors and comparisons of frequency response measurements, and online frequency response analysis. Chapter 8 concludes the book on the topic of remaining life analyses including life estimation based on thermal modeling, cellulose degradation kinetics, and probabilistic approaches. This is a “must have” reference book for anyone working with modern power transformers.

The first author of the book, Prof. S. Chakravorti, was Chairman of the IEEE Kolkata Section in India, and is well experienced to write such a complete book on the important topic of modern power transformers. He is a Professor of Electrical Engineering at Jadavpur University, Kolkata, India, and has been elected to the Fellowship of the National Academy of Sciences India and the Indian National Academy of Engineering. He received his Ph.D. in 1993 from Jadavpur University and became a Lecturer in 1985 working his way to Professor in 2002. His research group, including two co-authors, works on numerical field computation, condition monitoring of transformers, partial discharge analysis, data acquisition and signal conditioning in high voltage systems, computer aided design and optimization of insulation systems, application of artificial intelligence to high voltage systems, and lifelong learning techniques. From 1992 to 2013 he published about 49 papers in refereed international journals and about 26 papers in refereed Indian journals. He has completed 11 sponsored projects and has 2 ongoing projects and has published one book on "Electrical Machines" in 1993. The three authors of this book have one US Patent entitled "Remote Monitoring System" issued on December 25, 2012. From 2005 to 2012 Prof. Chakravorti has given 14 IEEE Power Engineering Society Distinguished Lectures; 15 invited lectures outside India and 11 special invited lectures in India, and 95 other invited lectures within India. This book is a valuable contribution to the teaching and learning of modern power transformers and their applications to modern electric power networks around the world.

July 2013

Prof. Markus Zahn
Department of Electrical Engineering
and Computer Science
Massachusetts Institute of Technology,
Cambridge, MA, USA

Preface

Impulse testing of transformers is being carried out in the High Tension Laboratory of Jadavpur University in Kolkata, India, for more than four decades. The first author of this book is a faculty member associated with this laboratory for the last 28 years and is the In-Charge of this laboratory for the last decade. The insight gained through such testing and also the specific queries of the designers of transformers, which are being tested, gave the impetus to carry out detailed research on impulse fault analysis of transformers. The authors would like to thank the brilliant research scholars, who had taken part in this research program funded by national funding agencies. This research led to the development of accurate tools for impulse fault diagnosis in transformers. However, impulse tests are design acceptance tests that are performed on transformers before they are put in service. In the early parts of the last decade, the first author was having regular discussions with academics as well as engineers from power utilities not only in India but also in Europe, USA, and Australia, wherein the main focus of the discussions mainly veered on development of accurate methodologies for estimating the condition of transformers in service. This is of great importance in view of the large number of transformers that have been installed over the globe in the 1960s and 1970s and are still in service. From the maintenance point of view, these transformers pose a threat to continuity of service, but from the economic point of view these units cannot be replaced overnight. Hence, condition-based maintenance and estimation of remaining life are of paramount importance to power utilities in this deregulated era.

At this juncture, the first author thought of initiating a research program on condition monitoring techniques of transformers. Needless to say, the conventional techniques of transformer condition monitoring have been in place for several decades. But during the last two decades, major works are being carried out at different laboratories around the world for development of reliable and accurate condition monitoring techniques, which are noninvasive in nature. In this context, promising electrical methods were being proposed. So the research at the High Tension Laboratory of Jadavpur University was also focussed in this direction. The greatest difficulty that was faced in this context was the inhibiting cost of diagnostic equipments that were available in the market. The first author did not have enough funds to procure such equipments to carry out the intended research. So there were two alternatives: either to develop the equipments starting from scratch or to abandon the research. At this hour of despair came two outstanding researchers, who not only developed required equipments but also developed accurate analysis tools using state-of-the-art data acquisition and signal processing tools. The efforts of these three have resulted in filing several patents in India as well as in the US, out of which one has already been granted. The news about the work carried out by this research group spread widely within India and quite a few leading power utilities came for cooperative work. One such work has led to successful implementation of the techniques developed by this research group in one of the largest power companies in India. After the completion of the doctoral work of these two bright researchers and successful practical implementation of their techniques in real life, the first author requested them to write this book, wherein they can share their expertise and insight gained through research works over several years.

The authors feel that during the course of these research works and in the preparation of this manuscript they have successfully transcended from academicians involved in laboratory research to ones who know how to implement their research works in real life taking into account all the nuances of practicality. These experiences have enriched their vision of engineering research enormously.

A substantial proportion of the material is original, and has been the outcome of the researches carried out by the authors. However, help has also been taken from the published literature. In writing this book, the authors have assumed that readers are well acquainted with the very basic concepts of transformer design, operation, and testing. The main characteristics of this book are:

(i) it assumes that the reader's goal is to understand the critical balance of cost and quality along with the pros and cons of different condition monitoring techniques;
(ii) it presents a selection of techniques with appropriate technical background to understand their domain of applicability and to consider variations to suit technical and organizational constraints;
(iii) it intends to extend the vision of condition monitoring testing and analysis which is integral to modern maintenance practice; and
(iv) it treats condition monitoring testing and analysis tools together in a coherent framework, as complementary approaches so that adequate quality is achieved at acceptable cost.

By going through this book, students and researchers will gain a proper understanding of the principles and issues in condition monitoring of transformers. Practicing engineers, including quality assurance professionals, will find a variety of techniques with enough discussion on technical issues to support adaptation to the specific requirements of their organization and application domain. Technical managers will find a coherent approach to weaving condition monitoring techniques into the overall maintenance strategy.

The authors would like to express their thanks to many colleagues and professionals who gave constructive criticism and useful suggestions, all of whom could not be mentioned by name here. However, the authors wish to express sincere thanks to the following colleagues, former scholars and professionals, without whom this book would not have materialized: Prof. Sugata Munshi, Dr. Prithwiraj Purkait, Dr. Chiranjib Koley, Mr. P. Rajamani, Mr. Sovan Dalai, Mr. Arijit Baral, and Mr. Monotosh Bhuti. This list of people is necessarily very incomplete. The authors also wish to acknowledge the help and support extended by the Master's students who had at one point of time or other taken part in these research works. This book could not have been completed without the help and support of the families of the three authors. Hence, the authors sincerely thank their families for having the patience and for selflessly cooperating with the authors in taking the challenge, which decreased the amount of time the authors could spend with their families during the course of this book writing.

The authors are indebted to Prof. Markus Zahn for his generosity to find time in his demanding schedule to write the foreword to this book. The contribution of Prof. Zahn in this context is gracefully acknowledged by the authors from the bottom of their hearts.

In addition, the authors would like to acknowledge Jadavpur University and Department of Science and Technology, Govt of India, for providing the financial and infrastructural support necessary for these research and development works.

Finally, the responsibilities for any mistakes and for the ideas expressed in this book are those of the authors' only.

Kolkata, India, July 2013

Sivaji Chakravorti
Debangshu Dey
Biswendu Chatterjee

Contents

Chapter 1
Introduction to Condition Monitoring of Transformer Insulation

1.1 Introduction

Transformer was introduced in electricity network in the late nineteenth century. Today transformers constitute a major portion of the capital equipment of power utilities all over the world and the reliable power supply depend heavily on the fault free operation of transformers. Power rating of the transformers varies from a few tens of kVA to several hundreds of MVA. The cost of replacing the transformers varies from a few hundred dollars to several million dollars. In a large power utility the number of transformers excluding those in the distribution network could be anything between a few hundred and one thousand [1]. Failure of transformer while in service usually lead to significant revenue loss to the utility, potential environmental damage, explosion and fire hazards and expensive repairing or replacement costs. Hence, it is desirable that the transformers should be utilized to the maximum extent consistent with adequate service life.

A "failure'' of transformer is defined as either (a) any forced outage of a transformer due to its failure in service or (b) trouble which requires extensive factory or field repair. Transportation damage and minor troubles which may require an equipment outage are not considered as failures [2]. Transformer failures can be broadly classified as electrical, thermal and mechanical. The failures are also classified in a different manner as internal or external. Failure due to insulation degradation, PD, increased moisture content, overheating, winding resonance etc. fall under the internal category, whereas the faults due to lightning strikes, switching over-voltages, system faults, system overload etc. fall under the external category [3]. So far as the location of failure is concerned, it could be in the main tank, bushings, tap changers or in the transformer accessories.

S. Chakravorti et al., *Recent Trends in the Condition Monitoring of Transformers*, Power Systems, DOI: 10.1007/978-1-4471-5550-8_1, © Springer-Verlag London 2013

1.1.1 Transformer Failure Pattern

Through experience and numerous data gathered by researchers and practitioners, transformer failure rate (FR) has been found to follow the so-called "Bathtub Curve" as shown in Fig. 1.1. The traditional bathtub curve depicts transformer life in three stages. The first stage is known as the infant-mortality period and it has a decreasing FR. The infant mortality is followed by a nearly constant FR period, which is usually long. Finally, the curve ends in the third stage, a period of wear out with an increasing FR [2], which is mostly related to equipment aging. Sometimes the first stage has increasing FR that indicates failures arising from imperfect manufacturing, handling, or control processes. In such cases FR in the first stage peaks quickly and is followed by a period of decreasing FR. In some other cases decreasing FR is followed by an increasing FR in the first stage. The decreasing FR region corresponds to infant mortality of the traditional bathtub curve. The increasing FR which follows is mostly due to problems similar to those mentioned above but which require a longer exposure time for manifestation [2]. The infant-mortality period of the life cycle results from failures in the weak subsamples. Hence, a reasonable definition of the infant mortality period includes all failures prior to the steady state. The percentage of weak subsamples varies with equipment type and manufacturing lot, even for the same manufacturer. Analysis of the time-varying FR is a powerful tool for investigations of failure mechanism and failure causes of transformers.

International experience shows that power transformers are very reliable in service and the global FR is less than 2 percent per year. Studies have also shown that the highest FRs can be observed at power transformers in the upper voltage levels as depicted in Fig. 1.2 [4]. In this context it is to be noted that network and generator transformers considerably differ due to the different loading nature. Generator transformers are normally loaded according to their rated power, whereas network transformers are loaded to 100 % or more in emergency situations only.

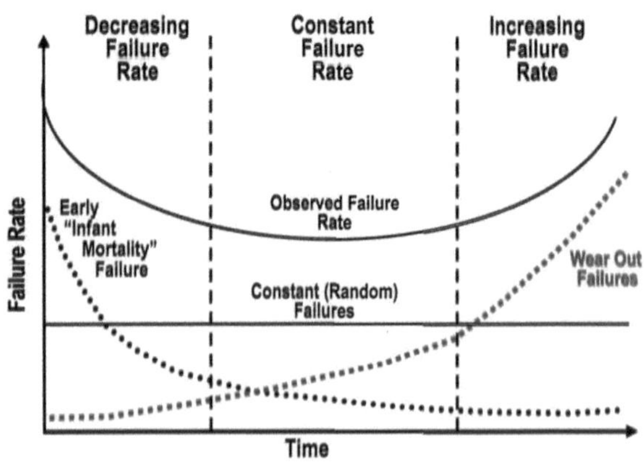

Fig. 1.1 Typical transformer failure pattern

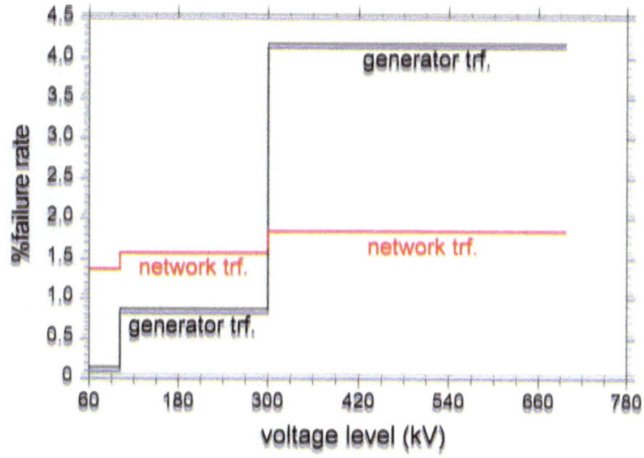

Fig. 1.2 Dependency of failure rate of power transformers on voltage level

1.1.2 Failure Statistics for Power Transformers

CIGRE Working Group 05 carried out a survey on failures in power transformers with and without on-load tap changers [3]. The survey results showed that for transformers with on-load tap changers 40 % failures were due to on-load tap changers while 19 % were due to windings. On the other hand for transformers without on-load tap changers, 26 % failures were due to windings and 33 % were due to terminals. Figure 1.3 shows the survey results for transformers with on-load tap changers [3].

Survey of transformer failure has also been carried out in South Africa and the failure data is reported in [5]. Distribution of each failure mode is shown in Fig. 1.4 [5]. In this study out of 188 failures in transformers, 56 were found to be due to aging of insulation.

Failure statistics for large transformers that had been in service between 15 and 25 years have also been studied in Russia [6] and the results are shown in Fig. 1.5. In another study carried out by IMIA working group 33(03), it has been reported that out of 94 failures in transformers 24 were due to insulation aging [7].

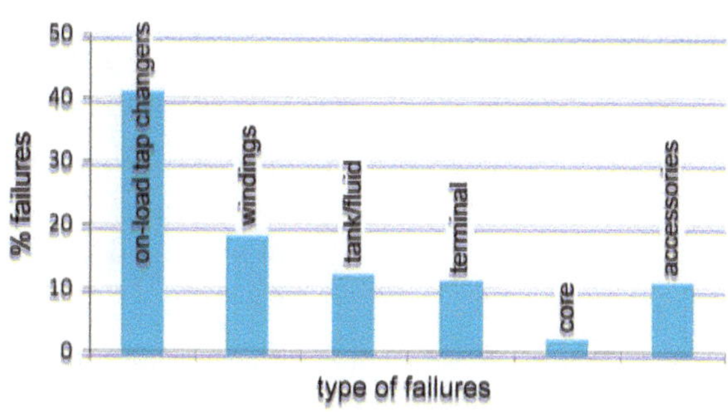

Fig. 1.3 Survey result for failures in transformers with on-load tap changers

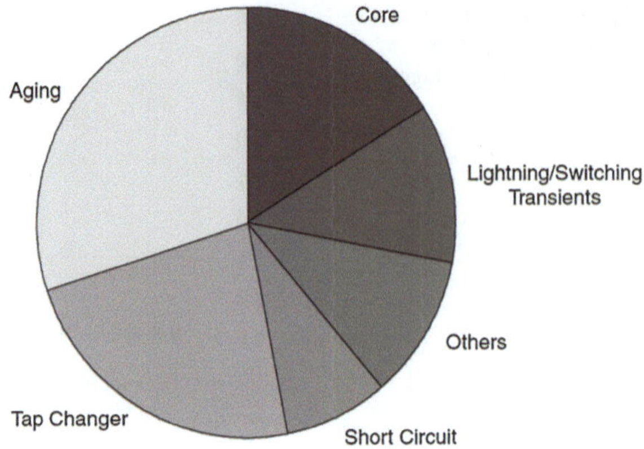

Fig. 1.4 Survey results for failure modes in transformers in South Africa [5]. Reproduced with kind permission ©IET 2012

1.2 Aging Electrical Power Infrastructure

A large growth in base infrastructure industries, especially the electric utilities, took place in the post World War II era as the world went through significant industrial growth. This growth caused the world energy consumption to increase from 1 trillion to 11 trillion kWh in the two decades following the war. It has resulted in installation of large number of transformers between 1950s and 1970s to supply the increasing energy demand. Figure 1.6 depicts the total transformer additions in the U.S.A. in three decades [7]. Australia/New Zealand Transformer Reliability survey report published in 1996 [8] showed that the average age of power transformers in Australia and Nee Zealand power system in 1995 was 28.6 years. The largest transmission system in Australia is the New South Wales Transmission System and in [9] it has been reported that approximately 65 % of installed transformers in that system are over 20 years old with 25 % over 30 years old in 2001. It has also been reported that the majority of transformers on National Grid's UK system were installed between 1955 and 1975 [10]. An important fact is that most of these transformers are free breathing type, which makes their functional age more than their natural age. Most of these transformers

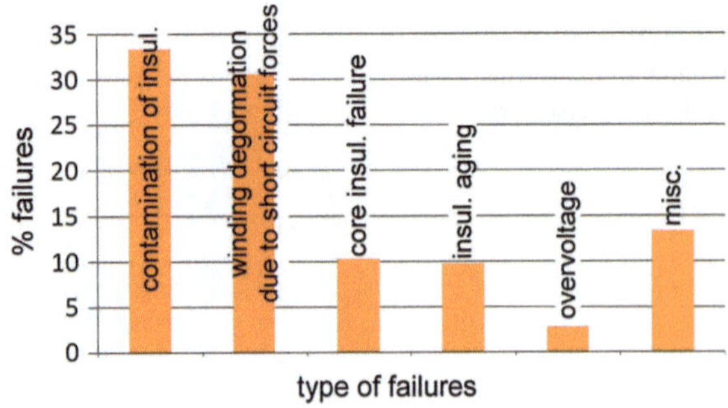

Fig. 1.5 Survey results for failure modes in transformers in Russia

Fig. 1.6 Addition of transformers in the U.S.A

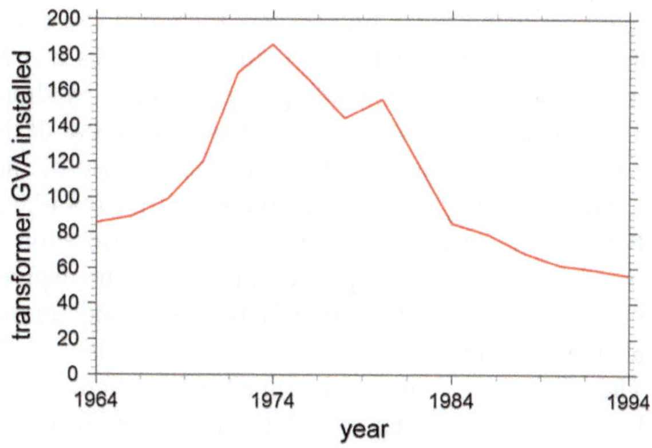

are therefore now in the aging part of its life cycle as represented by the bathtub curve. Consequently they are expected to have an increasing FR in the forthcoming years.

1.2.1 Classification of Visually Detectable Faults in Transformers

The faults that can be identified by visual inspection of transformers after the occurrence of the faults have been classified in IEC 60599 [11]. There are five characteristics faults: (a) PDs which induces small carbonized punctures in paper, (b) Low energy discharges that causes larger punctures in paper, tracking or carbon particles in oil, (c) High energy discharges with power follow through, the evidence of which are extensive carbonization, metal fusion and possible tripping of transformer, (d) Thermal faults below 300 °C in which cases the paper turn brownish and above 300 °C evidenced by paper carbonization and (e) Thermal faults above 700 °C indicated by oil carbonization, metal coloration or fusion.

1.2.2 Transformer End-of-Life

Neither all the factors that affect the life expectancy of transformers are known, nor are all properly understood. However, in general, the following are significant: (a) The initial thermal, electrical and mechanical properties of insulating materials used. It is to be noted here that various materials lose their properties at different rates and during the aging process the same material may lose different properties at different rates, (b) Thermal, electrical and mechanical stresses. Thermal stresses arise due to overload currents, local overheating, leakage fluxes as well as malfunctioning of cooling system. Electrical stresses are caused by system as well as transient over-voltages, winding resonances etc. Mechanical stresses between leads, conductors and windings are produced by short-circuit and inrush currents. Their relative importance is determined by the design to keep the stresses to a minimum commensurate with economical use of the material used, (c) Normal load cycle along with the environmental factors like ambient temperature and (d) Permissible degree of deterioration. Deterioration to complete destruction of transformers is not acceptable in practice. The degree of deterioration that can be permitted is determined largely by safety and service continuity conditions and also by the probability of occurrence of abnormal operating conditions [12]. Therefore there is no well defined point when a transformer should be replaced, but only an increasing probability that the transformer will fail.

Both mechanical and dielectric withstand strength of the transformer is reduced by aging of its insulation system. During the service life of a transformer, it is subjected to faults that result in high radial and compressive forces. With system growth the operating stresses on transformer increase as the load increases. In an aged transformer, typically the conductor insulation is weakened to the point where it can no longer sustain mechanical stresses of a fault. In other words, the insulation becomes so brittle that through fault conditions or even normal vibration may cause major damage. Then dielectric failure of turn to turn insulation occurs, or loosening of winding clamping pressure takes place, which reduces the transformer's ability to withstand future short circuit forces [7].

1.2.3 Maintenance Strategy

Prevention of failure and keeping the transformers in good operational condition is an important issue for power utilities. Traditionally time-based maintenance (TBM) was carried out in which transformers were maintained at regular time intervals irrespective of the necessity of the maintenance. Today, however, power utilities are performing condition-based maintenance (CBM) as opposed to TBM. The roots of this evolution can be found in the restructuring and deregulation of electric power industry. In the changed scenario, the independent power producers, transmission companies, system operators and distribution companies are forced to cut costs in maintenance and operation without endangering steady supply of electrical power. If the actual condition of the equipment is reliably known, then costs can be reduced in CBM by carrying out maintenance only when the condition of transformer requires it. Consequently, reliable diagnostic tools are basic requirements of CBM and there is an increasing need for noninvasive monitoring and diagnostic tools for assessment of internal condition of transformers.

1.2.4 Monitoring and Diagnostics

In general, the term 'monitoring' covers measurement of basic parameters with threshold alarms. In other words, the integration and collection of data, with some local control functions and early warning of incipient failure symptoms can be described as monitoring. It provides the first alert that some form of deterioration of the transformer has exceeded a previously established threshold value. Over the years monitoring has boiled down to utilization of such systems for early warning, alarming and trending [13].

On the other hand, the term 'diagnostics' indicates incorporation of advanced analysis that are capable of performing reliable assessment of equipment condition and of suggesting actions to be taken.

Basic task of a diagnostic system is to take the data and transform it into usable information. Modern diagnostic systems automate the data collected at the transformer or substation into useable transformer information. Present day research in diagnostic system is directed towards developing a system which will not only work with data collected from sensing devices but also take the function related data of the equipment and combine them into a diagnostics engine. Power utilities are seeking intelligent analytical tools to have a greater reliance on automated fault indicators. Preferred functionalities of monitoring and diagnostic systems are detailed in Fig. 1.7 [13]. It is difficult to have all the features shown in Fig. 1.7 in one system. But if two or more features are combined into one system, then it would offer the desired result in most of the cases.

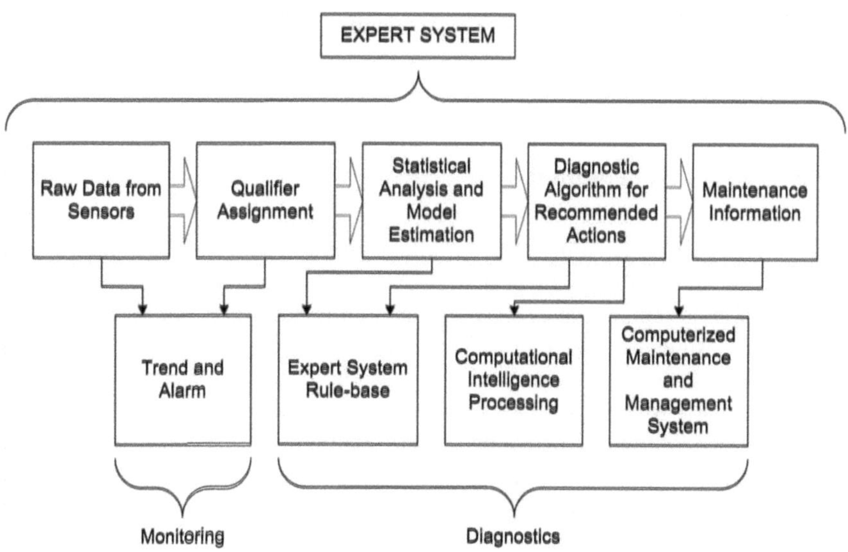

Fig. 1.7 Functionalities of monitoring and diagnostic systems

1.3 Diagnosis Method

An ideal diagnosis test would consist of exposing a large number of full-scale transformers of the design in question to the expected service in the field until failure occurs. But, this procedure would be too expensive, and would require so much time that the results would be of no practical use when the tests were finally complete. Consequently, a certain sacrifice in accuracy is made from practical considerations by the use of a relatively small number of transformers but using a duty cycle sufficiently severe to insure deterioration to a selected level in a reasonably short length of time. Moreover, the selected end point of deterioration should be such that it indicates the probability that the design could continue in normal service but would fail under moderate abnormal conditions. Finally, the procedure should be such that data may be obtained by varying a readily measurable parameter and the data would make it possible to interpret the results obtained by test [12].

1.3.1 Requirements of Diagnosis Method

As each diagnosis method is developed and applied in real-life situations, it is always weighed up against other methods. Methods which have become established over the years satisfy important criteria, some of which are as follows: (a) Sensitivity to important parameters of transformer condition, (b) Reproducibility of results over time and for different testing personnel, (c) Compensation of raw data for significant environmental effects like temperature, (d) Good correlation with other established methods and (e) Availability of valuable information for the time and expense involved [9].

1.3.2 Design Acceptance Test

It has been mentioned earlier that the infant mortality rate depends upon manufacturing defect or inadequacy of material used. In order that infant mortality rate is low, design acceptance tests are performed on transformers before they are put in service. Design acceptance test is defined as a test carried out to determine the adequacy of the design of a particular type, style, or model of equipment and its component parts to meet its assigned ratings and to operate satisfactorily under normal service conditions or under special conditions if specified; and to demonstrate compliance with appropriate industry standards. Lightning impulse voltage withstand test is one of the most important design acceptance test for transformers.

In the lightning impulse voltage withstand test of transformers, impulse voltage sequences are generated in the laboratory and applied to transformers as per IEC 60076, Part 4 [14]. The resulting voltage and current waveforms are recorded and the identification of faults during impulse tests is done based on the recorded current and voltage waveforms. Impulse testing of transformers and fault detection techniques date back to late 1940s [15–17]. The methods reported in those earlier works were based on visual inspection of current oscillograms at the Basic Insulation Level (BIL) and a calibrating level. These initial methods were straightforward and sensitive, but required human experts for interpretation of oscillograms. To eliminate the problem of fault diagnosis in time domain due to deviation in input impulse waveform, the transfer function method was proposed by Malewski and Poulin [18] and Vaessen and Hanique [19]. Problems of the transfer function method have been identified in the course of practical testing and the possible sources leading to error and ambiguity in the transfer function computation were critically analyzed by Pandey and Satish [20]. In the absence of expert human knowledge, a good alternative approach in transformer fault diagnosis is the application of expert system. Purkait and Chakravorti [21] proposed an expert system tool for assisting the testing personnel to obtain unambiguous, reliable and transparent decision in impulse fault diagnosis of power transformers within 33 % of the winding length using both time and frequency domain analyses. Due to complex nature of the composite insulation system of transformers, determination of exact location of faults within transformer windings is a difficult task, particularly identification of minor faults involving a small section of the winding. It has been reported in [22] that the inherent non-stationary pattern of transformer current waveforms during different fault conditions can be effectively classified using the frequency-selective feature of wavelet transform. Other techniques like Fuzzy classification [23] and fractal analysis [24] have also been used for impulse fault identification transformers. In [25] transformer winding responses were decomposed into low and high frequency components using discrete wavelet transform and the time-frequency domain features were extracted out of the decomposed components, which were then analyzed using a pattern classifier, viz. Support Vector Machine (SVM), for fault identification within 10 % of the winding length as it is of higher practical significance. To make such refined fault analysis possible, a new time–frequency domain based modelling technique for transformers has also been proposed in [26]. Refined techniques like SVM aided fractal analysis [27] and rough granular approach using cross-wavelet transform [28] have also been used for transformer impulse fault analysis. However, all the above-mentioned studies were based on static insulation failure, i.e. the faults which do not evolve with the propagation of impulse voltage wave. As a result of the advancement in manufacturing techniques the probability of occurrences of static insulation failure during impulse testing has become lower and majority of the failures that may arise are dynamic in nature, i.e. the faults that arise along the winding due to propagation of impulse voltage wave. In a recent paper [29] wavelet network has been used to identify the dynamic faults in transformers under impulse excitation using the features extracted from cross-correlation sequence of winding currents. Although several methods have been proposed for identification of impulse faults in transformers, accurate identification of fault characteristics is still an open issue owing to the complex structure of the composite insulation system in transformers.

1.3.3 Age Related Failure

When the transformers are at the third stage of the bathtub curve, the failures increase rapidly over time. In the final quartile of transformer life, failure probability can increase by 5 times or more than the normal system FR. Figure 1.8 shows the age distribution vs failure probability of transformers [30]. Figure 1.9 shows the bathtub curves for new as well as rebuilt units which have failed in a non-catastrophic mode [31]. In practice, many utilities have demographic distributions displaying large number of units in the 40, 50 or even older age categories. Figure 1.10 shows demographic data for a population of 50 MVA transformers in which condition data have been assessed and sorted by condition [31]. Figure 1.11 shows the data for failure due to general aging of oil-paper insulation in

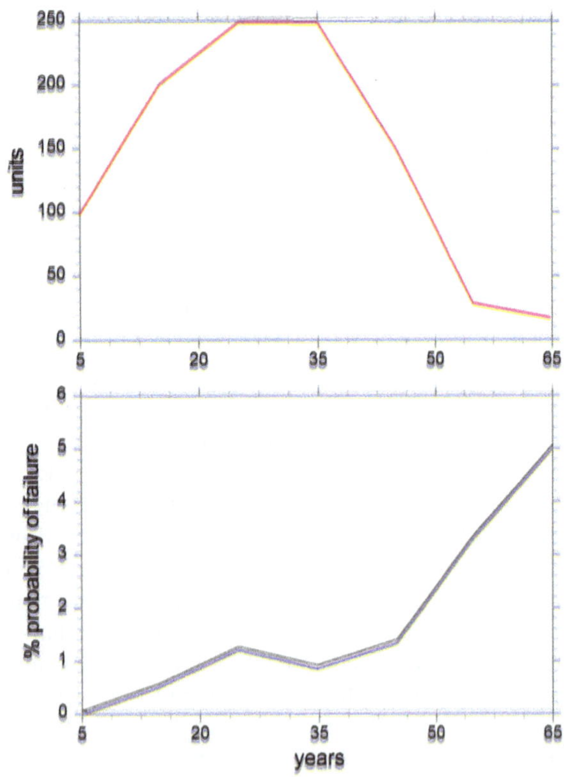

Fig. 1.8 Transformer age distribution vs. failure probability

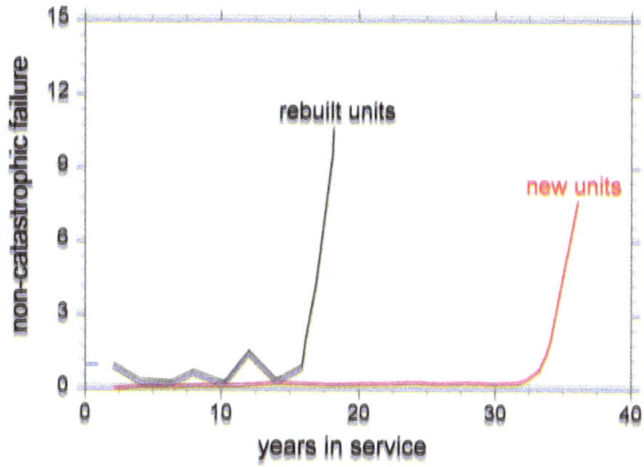

Fig. 1.9 Bathtub curves for new and rebuilt transformers

transformers as a function of the age of transformers as obtained from a survey conducted in South Africa [5]. Table 1.1 presents the distribution of transformer failures by age as obtained from IMIA survey [7]. Figure 1.12 shows the importance of maintaining insulation systems in good condition [31].

There are two fundamental questions to be answered for the development of a technically sound transformer diagnostic system: (i) How does aging affects the failure of a transformer and (ii) What are the observable symptoms of these aging related failures? The answers to these questions require in-depth understanding of transformer design and field experience.

It is necessary to explicitly represent the cause and effect relationships between failures and observable symptoms. The probability of occurrence of each failure mode and the parameters that indicate the strength of causal relationship are needed to be understood. All the major problems that can be detected prior to failure should be identified. Values of measurable parameters in conjunction with diagnostic models are to be used for detecting changes from normal operation. This will in turn reduce the catastrophic failures considerably and enable scheduled repairing during maintenance outages, thereby eliminating downtime associated with forced outages [13].

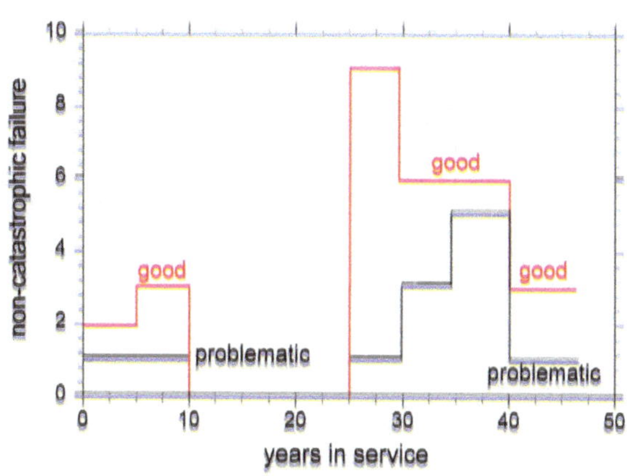

Fig. 1.10 50MVA transformer demographic data

Fig. 1.11 Failure due to aging vs age of transformers in South Africa [5]. Reproduced with kind permission ©IET 2012

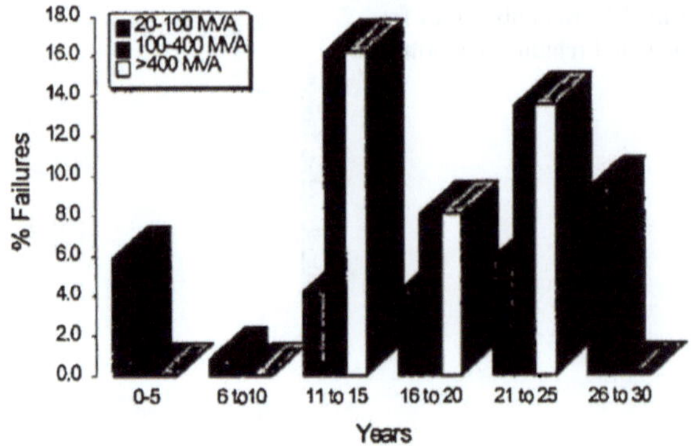

Table 1.1 Distribution of transformer failure by age

Age (Years)	Failures	Age	Failures
0–5	9	21–25 years	10
6–10	6	>25 years	16
11–15	9	Unknown	35
16–20	9	Total	**94**

1.4 Transformer Oil-Paper Insulation System

The significant components of the main transformer insulation are: (a) Insulation between the HV winding and the tank, (b) Insulation between the HV and the LV windings and (c) Inter-phase insulation. These components are the most sensitive to the insulation deterioration as they usually have the smallest margins in the dielectric strength. The monitoring of the solid and liquid insulation in these components is therefore of utmost importance. The margin of dielectric strength is usually significantly higher for the insulation between the LV winding and the core

Fig. 1.12 The impact of insulation system integrity on transformer mortality

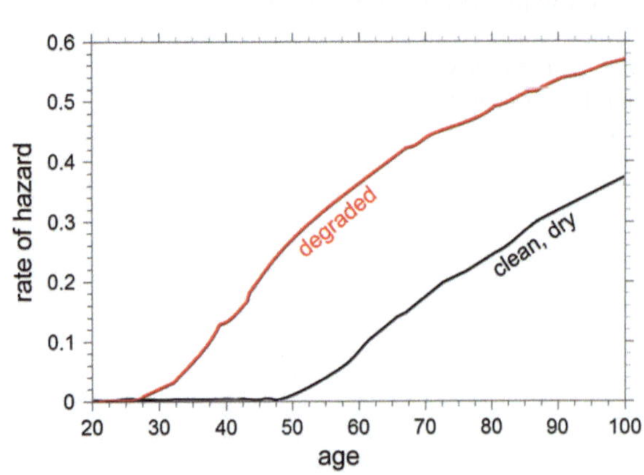

than in the spaces that include HV winding and hence here only a high degree of deterioration would be the cause for concern [32].

Notable defects in transformer major insulation are as follows: (a) Moisture in cellulose insulation, (b) Contamination of oil with water, particles and insulation aging products, (c) Insulation surface contamination, which occurs mainly due to adsorption of polar aging products on a cellulose surface or due to deposition of conducting particles and insoluble aging products and (d) PDs in weaker parts of insulation. The presence of moisture and impurities changes dielectric parameters of deteriorated components, viz. their conductivity, permittivity and dissipation factor, particularly with temperature, which in turn result in related changes in the dielectric characteristics of the whole transformer. Defects in transformer insulation could be categorized as reversible and irreversible. Defects related to excessive moisture, oil contamination or surface contamination usually fall into the category of reversible defects, while the damage created by PD activities is usually irreversible in nature [32].

Defects in minor insulation, e.g. inter turn or coil insulation, such as moisture in cellulose insulation, overheating leading to accelerated aging and insulation surface contamination, have only a small impact on overall dielectric characteristics of the whole transformer, due to relatively high capacitance of turn insulation. Thus detection of a defective condition of minor insulation of a winding is very difficult until a critical PD or noticeable gas generation occurs.

Typical scenarios of insulation failure in transformer are presented in [32] on the basis of analysis of over 200 failures:

- Critical contamination of oil (typically presence of free water) + rapid change of temperature ⇒ PD appearance at rated voltage ⇒ Breakdown.
- Surface contamination + water + rapid change of temperature ⇒ PD appearance ⇒ Flashover.
- Particles contamination + switching surge ⇒ Critical PD ⇒ Breakdown.
- Water + particles contamination (or bubbles present in oil) ⇒ Critical PD ⇒ Creeping discharge ⇒ Breakdown.
- Surface contamination + lightning impulse ⇒ Surface discharge ⇒ Flashover.
- Distortion of winding geometry ⇒ PD appearance ⇒ Creeping discharge ⇒ Breakdown.
- Distortion of winding geometry + Switching surge ⇒ Flashover between coils (sometimes with restoring withstand strength) ⇒ Gas evolution.

The above discussion indicates that there are several ways of diagnosing defects in transformer insulation system:

(i) Detection of PDs
(ii) Detection of distortion of winding geometry
(iii) Detection of aging products by chemical methods
(iv) Detection of changes in dielectric characteristics, which can be performed in time domain and also in frequency domain

The first two tests, i.e. PD and the winding geometry distortion detection, are sensitive to local defects. On the other hand, analyses of oil and paper insulation by chemical methods and by dielectric response measurements provide information on the overall insulation condition.

1.4.1 Partial Discharge Measurement

PD can occur in solids, liquids, and gases in ways which will not affect the breakdown voltage during a high voltage test. Yet PD will eventually cause failure in most systems, possibly years after the system is put to service. PD activity, which is produced by incipient faults in insulation system, is widely regarded as one of the best indicator of insulation degradation and provides an 'early warning' against insulation faults which allows taking corrective action before catastrophic failure occurs.

In general, the following are the main steps in the condition assessment of power transformers by PD measurement. First of all, any PD activity that indicates harmful insulation defects must be detected. After the detection and recording of the PD, the analysis of the PD data is carried out to identify the defect in the transformer. The identification of the defect can be done by finding the location of the insulation defect and by comparing its pattern with other known defects from a reference database. Finally risk assessment for an insulation failure of the transformer is to be done.

PD measurement techniques are based on the macro-physical effects of PD phenomenon, such as optical effect, pressure wave, discharge effect or chemical effect. Consequently, PD detection method could be conventional electrical, high frequency, acoustic, optical or chemical. Conventional electrical method of PD measurement integrates the effect of PD pulses in time domain according to IEC 60270. PD detection by measuring high frequency waves is done by capacitive, inductive or electromagnetic sensors in the frequency range of 3–300 MHz. Presently ultra high frequency sensors for PD detection are available in the frequency range of 300 MHz–3 GHz. Acoustic signal in the frequency spectrum 10 Hz up to 300 kHz are generated as a result of the pressure wave produced by PD. Acoustic PD sensors could be piezo-electric sensor, structure-born sound-resonance type, opto-acoustic-sensor, condenser microphone or accelerometer. Optical detection and localization of PD is possible by low-light amplifier or coronascope when the PD source is accessible from outside. If it is inaccessible, then optical detection could be done by fibre-optic cable, collimator or photodiodes. Chemical method PD measurement is carried out by means of H_2 sensor, gas in oil analysis or ozone analysis.

In all the above-mentioned methods of PD measurement, the following are the main quantities of measurement: PD source quantity, sensor quantity, reading quantity and derived quantity. Each method has its own performance and sensitivity checks.

With the development of PD analyzers which can record very large amount of data, new avenues in PD analysis have been opened by means of studying the temporal transition of PD pulses. This methodology, which is also known as PD pulse train analysis, studies among other parameters temporal transition of pulse sequence, time dependency of number of PD pulses and their heights.

Significant progress in the PD measuring technique has made it possible to effectively reject external interference, detect weak PD signals and diagnose the condition of transformer insulation on-site, similar to well established laboratory tests. Practical experience has confirmed that on-site PD measurement technique provides the sensitivity to PDs in field conditions of about 20 pC in power plants and 50 pC in the 500–750 kV substations [32]. This has led to the development of on-line PD monitoring systems for transformers.

1.4.2 Chemical Methods for Condition Assessment of Transformers

The chemical reactions that occur within transformer insulation due to aging are caused by pyrolysis oxidation and hydrolysis. The factors that accelerate aging are temperature, oxygen and moisture. These age related chemical reactions cause reduction of mechanical strength of cellulose insulation. In the very aged condition the cellulose insulation becomes brittle to the extent that it may crack due to even small mechanical forces and dielectric breakdown occurs due to rupturing of solid insulation. But it is to be noted here that even at that brittle condition paper insulation retains an acceptable level of dielectric strength.

Over the years chemical methods are being used extensively for assessment of the overall condition of transformer. Traditional chemical tests are measurements of dielectric strength of insulating oil, neutralization number and interfacial tension. A reduction in the dielectric strength of insulating oil indicates the presence of moisture, sediment such as sludge and conducting particle in oil. Acids in the insulating oil originate primarily from decomposition/oxidation of oil. These organic acids are detrimental to the entire oil-paper insulation system. Corrosion within the transformer due to acids is more in the presence of moisture. Deterioration of insulation system increases rapidly once the critical acid number is reached. A definite relationship exists between the neutralization number and interfacial tension. Normally an increase in neutralization number is accompanied by a reduction in interfacial tension. Thus interfacial tension is a good back up test for neutralization number. Myers Index Number (MIN) is an excellent way of evaluating oil condition. It is obtained by dividing the interfacial tension by neutralization number. Good quality of oil will have a MIN of around 1500.

A sensitive and reliable technique used for assessing the condition of oil-filled transformers is Dissolved Gas Analysis (DGA). Small quantities of gases are liberated due to decomposition of oil and cellulose insulation under electrical and thermal stresses. The qualitative composition of the liberated gases is dependent upon the type of fault. It is to be mentioned here that a sudden generation of large amount of gas due to a major fault will not cause the generated gases to be dissolved in oil and will cause the Buchholz relay to operate. It has been reported in literature that faults like PD, overheating and arcing in transformer insulation can be detected and distinguished with the help of DGA [33]. In power utilities, often a number of DGA samples are taken over a period of time for trending analysis, which is valuable for preventive maintenance program. DGA helps in providing advance warning of developing faults, monitoring of progression of fault and scheduling of repairing programs conveniently.

The gases that are commonly present as dissolved gases in insulating oil in transformers are Hydrogen (H_2), Methane (CH_4), Ethane (C_2H_6), Ethylene (C_2H_4), Acetylene (C_2H_2), Carbon Dioxide (CO_2), Carbon mono-oxide (CO), Oxygen (O_2) and Nitrogen (N_2). Out of these gases, normally O_2 and N_2 are considered to be non-fault gases. Accuracy of DGA results depend greatly upon the skill with which the gas extraction, separation and identification has been carried out. The following should also be considered while analyzing DGA results: (a) volume of oil in transformer, (b) age and history of the transformer including repairing, degassing, oil treatment of the transformer, (c) type of transformer, e.g. generator or network, sealed or free breathing transformer and (d) the construction of the tap changer.

Interpretation of DGA is not an exact science. There are several methods of interpretation of DGA results, out of which the following are used widely in practice: (a) Total Dissolved Combustible Gases (TDCG), (b) Permissible Concentration of Gases in a Transformer, (c) Key Gas Method, (d) Gas Ratio Methods, viz. Dornenburg Ratio Method, Roger's Ratio Method, IEC Ratio Method and (e) Duval's Triangle.

Depolymerization of cellulose takes place due to various aging mechanisms of paper insulation during its service life. An effective method of evaluation of remaining life of paper insulation is measurement of degree of polymerization (DP). DP value of paper is about 1200 in new condition and it becomes around 1000 after the initial heat treatment of the winding. It is generally suggested that DP values around 200 indicate end-of-life of paper insulation. However, direct measurement of DP is difficult for in-service transformers. A commonly used technique is to lower the oil below the required paper sampling point at the connecting leads and to replace it with new paper as shown in Fig. 1.13 [34]. Figure 1.14 shows the DP profile along the connecting lead from the bottom of the tank, which is expected due to the temperature profile in an in-service transformer [34].

The situation is more complicated for transformer windings due to complex heat and moisture disposition. But a relationship between height and DP similar to connecting leads has been found in [34] for the winding of an in-service transformer as shown in Fig. 1.15. However, it should be noted here that taking paper samples from hot inaccessible parts of transformer winding is only possible during repairing. It has also been reported in [34] that the paper insulation which is in direct contact with bulk oil ages at a higher rate than the layers of paper insulation

Fig. 1.13 Paper sampling from connecting lead for DP measurement

which is shielded by overlying layers of paper. Infrared spectroscopy based study reported in [35] indicated that the oil in the underlying layers of paper insulation was equivalent to new oil even when the bulk oil was oxidized, which could be due to filtering effect of the outer layers of paper insulation. Diminished degradation of paper in the underlying layers is also due to the fact that these layers are not exposed to byproducts of oil oxidation.

DP measurement is an invasive test and unusual steps need to be taken for removing paper sample from an in-service transformer. Hence, it is mostly used for end-of-life evaluation. For regular measures to be taken for preventive

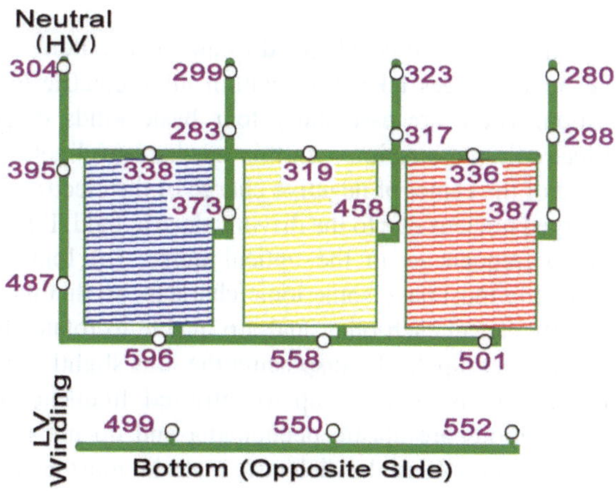

Fig. 1.14 DP profile of connecting lead of an in-service transformer

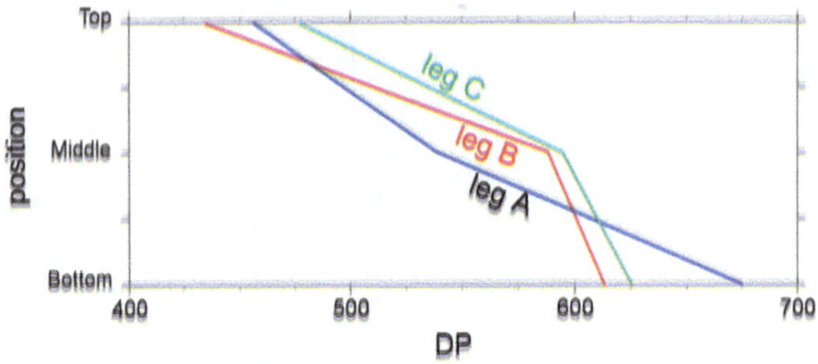

Fig. 1.15 DP profile of HV winding surfaces of an in-service transformer

maintenance programs, it is preferable to have a non-invasive test which could be performed even when the transformer is not taken off the grid. Specific furanic compounds are generated due to paper degradation that remain dissolved in oil. Hence, measurement of furanic concentration by oil testing is a non-invasive and convenient diagnostic technique for CBM of transformers. Out of the different furanic compounds, it has been reported in literature [35] that the concentration of 2-furfuraldehyde in oil, which is the most prominent component of paper degradation, relates well to the DP value of paper insulation. But the paper insulation of a transformer does not age uniformly everywhere due to differences in temperature, moisture distribution, oxygen level and operating condition. Consequently, the furanic concentration indicate the average condition of paper insulation and the paper deterioration at hot-spot will be greater than that indicated by the measured levels of furanic concentration in oil.

1.4.3 Condition Assessment of Transformers by Dielectric Response Measurement

When a dielectric is subjected to an electric field, a relative shift of positive and negative charges take place within the dielectric giving rise to dielectric polarization. There are essentially four basic kinds of polarization mechanisms: (a) Electronic polarization – In this mechanism dipoles are induced by the displacement of the center of negative charge of the electrons with respect to the positively charged nucleus due to the driving electric field. It is extremely fast and is effective up to frequencies in the optical range, (b) Ionic polarization—In this case a dielectric must have some ionic character so that there are internal dipoles, which exactly cancel each other and are unable to rotate. The driving electric field then induces net dipoles by displacing the ions slightly from their position of rest. This mechanism is effective up to infra-red frequencies. Both electronic and ionic polarizations are elastic phenomena and do not involve power loss, (c) Dipole Polarization—Here the dielectric material must have permanent dipoles which can rotate freely. In thermal equilibrium, the dipoles are oriented randomly and thus carry no net polarization. The driving electric field aligns these dipoles to some extent and thus induces a polarization of the dielectric. This mechanism is effective up to ac frequencies of several MHz and (d) Interfacial Polarization—This mechanism takes place in insulation system comprising multi dielectric media, e.g. oil-paper insulation system of transformers. Under the action of the driving electric field, movable positive and negative charges may get deposited at interface of two dissimilar dielectric media forming some kind of dipoles. This mechanism is slow and is effective in power frequency range or lower. Dipole and interfacial polarizations are inelastic phenomena and do involve power loss.

Some or all of these mechanisms may act simultaneously. Electronic polarization, e.g., is always present in any dielectric and thus becomes superimposed on other active mechanism, if any. Electronic, ionic and dipole mechanisms are amenable to basic considerations and calculations. On the other hand, no basic treatment is applicable for interface polarization. There is no general way to calculate the charges on interfaces or their contribution to the total polarization of a dielectric.

Polarization processes in relation to dielectric response have been discussed in details in [36]. Presently there are three main types of dielectric response measurement techniques: (a) Polarization and Depolarization Current Measurement (PDC), (b) Recovery or Return Voltage Measurement (RVM) and (c) Frequency Domain Spectroscopy (FDS). PDC and RVM methods are DC voltage tests in which dielectric response is measured as function of time. As a result these techniques fall under the category of time-domain measurements. On the other hand, FDS is an ac voltage test and in FDS dielectric response parameters are measured as function of frequency. Consequently FDS is categorized as frequency domain measurement.

In PDC, a step voltage of constant magnitude, which is free from ripple, is applied to the test object for a long period of time (e.g. 10,000 s) and the resulting polarization current through the test object is measured, arising from the dc conductivity and the polarization processes having different time constants corresponding to different insulating materials and their different conditions. The measurement of polarization current is stopped when the polarization current becomes either stable or very low. Subsequently, the test object is short-circuited for a long period of time and the consequent depolarization current is measured, which do not have any contribution from the dc conductivity.

The findings reported in [37] indicate that initial parts of polarization and depolarization currents measured for transformer insulation system are influenced greatly by the higher mobility of charge carriers in insulating oil and hence are sensitive to the condition of transformer oil. The steady part at longer times of polarization current measurement is due to the not so mobile charge carriers in paper or pressboard insulation. Hence, the steady part of polarization current is a good indicator of the condition of solid insulation. Thus PDC method is a time domain diagnostic technique by which the condition of paper and oil insulation could be assessed separately with the help of experimental results [38].

In RVM the test object is subjected to a step dc voltage for a certain period of time and then short-circuited for a relatively lesser period of time. Subsequently, the test object is open-circuited. Since the depolarization of all the polarization processes having different time constants will not be complete during the inadequate depolarization duration, the charges present in the incomplete depolarization processes start charging the geometric capacitance in the open circuited condition and a voltage appears across the terminals of the test object, which is known as recovery or return voltage. The shape of recovery voltage measured for an aged transformer is shown in Fig. 1.16 [39]. It has been reported that the initial slope of the recovery voltage is determined by dc conductivity [40]. A lower value of central time constant (CTC), which is the time required to attain the peak value of the recovery voltage, indicate poorer condition of transformer insulation [41]. The foundations of RVM had been reported in [42] and it has been reported that RVM is useful for assessing the uniformity of insulation aging and/or moisture distribution. However, RVM results are found to depend greatly on the geometry of the insulation structure of the transformer under test [43].

Consistency between PDC and RVM results has been highlighted in [38]. It has been shown in literature that PDC [44] and RVM [45] results could be modeled accurately with the help of extended Debye model of transformer insulation under the assumption that the dielectric materials are linear. In other words, the parameters of the extended Debye model could be determined from time domain measurements of dielectric response and these parameters of extended Debye model could then be correlated to the condition of the transformers.

Another excellent non-invasive diagnostic tool for assessing the condition of transformer insulation system is Frequency Domain Spectroscopy (FDS). FDS is an extended version of commonly used method of dielectric dissipation factor measurement at power frequency. But measurement of dielectric dissipation factor at only power frequency is insufficient to detect changes that take place in complex insulation system such as the one in transformers. Moisture content, contamination and other impurities affect the dielectric properties of transformer insulation system at different and specific frequency ranges. Hence, in FDS the magnitude and

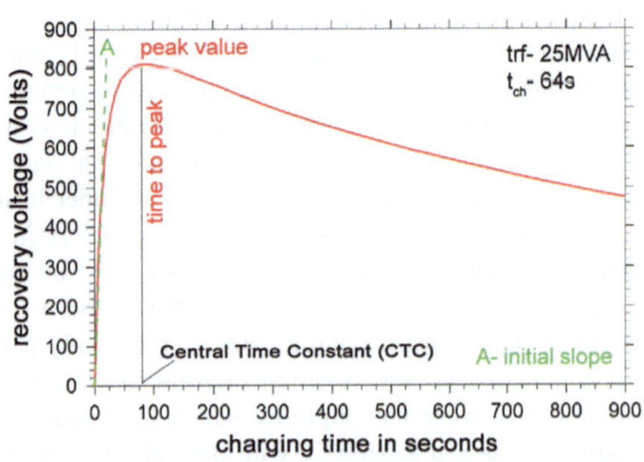

Fig. 1.16 Typical shape of recovery voltage measured for an aged transformer

phase of the current flowing through the transformer insulation system due to sinusoidal excitation voltage is measured at different frequencies over a wide frequency range from 1 mHz to 1 kHz or even wider. From these measured voltage and current, dielectric dissipation factor, complex capacitance and complex permittivity are commonly determined as functions of frequency.

It has been reported that enhanced interfacial polarization increases dielectric dissipation factor mainly in low and very low frequency range [46]. As a result aging and moisture content increase dielectric dissipation factor and capacitance at lower frequencies [47]. FDS results are strongly affected by temperature variation during measurement. Low frequency measurements being very time consuming, it is necessary that temperature of the transformer insulation system should be kept constant during FDS measurements. Otherwise extreme care needs to be taken to interpret the results of FDS in the presence of temperature variations. It has also been reported that significant errors can come into FDS results if proper precautions are not taken particularly with different equipment connected in parallel to the transformer under test [48]. A clear understanding of the physical behaviour of the insulation system is necessary for better understanding and analysis of FDS results. A circuit model based on the principles of linear dielectric response can provide a good fit of the experimentally obtained FDS results and the identified model parameters can be used for condition assessment of transformer insulation system [49].

FDS parameters of transformer insulation system can be computed from time-domain measurements, if the dielectric response has been measured for an adequate period of time. Conversely conversion from frequency domain to time domain is also possible. The key to such conversions is determination of the extended Debye model of transformer insulation with its known branch elements [46].

1.4.4 Monitoring of Transformer Winding Displacement

Short circuit forces appearing during the service life of transformers can cause winding distortion and weakening of winding clamping can cause winding movement. Both winding distortion and winding movement can cause major failure leading to substantial damage to the transformer and unwanted loss of revenue. Traditionally measurement of leakage inductance had been used for detecting changes in the geometry of transformer windings. However, it has been established that only radial deformations can be detected using this method. This is because of the fact that only a small change in the leakage flux is introduced by axial deformations and hence is very difficult to identify by leakage inductance measurement [50]. Consequently, the Frequency Response Analysis (FRA) as suggested by Dick and Erven [51] has become increasing popular over the last two decades for assessing the mechanical integrity of power transformers. FRA, which is a method of system characterization by means of analyzing its frequency responses, has been successfully used for detecting mechanical faults of windings, core as well as clamping structure in not only factory but also in field applications [52].

In FRA frequency dependence of parameters such as open and short-circuit impedances of windings, transfer impedance, voltage transfer ratio between primary and secondary are measured considering the transformer as a "black box". As capacitive and inductive parameters of transformer winding get changed due to mechanical deformations, such deformations cause deviations in the FRA spectrum and hence could be identified through appropriate measurements.

FRA is essentially a comparative method in which frequency dependent parameters are measured over a wide frequency range, typically from 50 Hz up to at least 1 MHz, which are then compared with reference responses. The reference measurement is regarded as the FRA fingerprint of healthy transformer and could be the FRA measurement result obtained from previous tests. But in reality fingerprints are rarely available, particularly for transformers that are in service for quite some time. In such cases other information such as results from tests on a similar transformer or from tests on different phases of the same transformer has to be taken for diagnosis. Given the option, the inter-phase comparison should be the last alternative.

Graphical inspection is the traditional way of comparing the FRA spectra, in which changes in the overall shape of the responses are identified, e.g. appearance of new resonant frequencies, disappearance of existing resonant frequencies, shifts or damping of existing resonances etc. [53]. It should be mentioned here that the frequency response of a power transformer retain typically ill-scaled peaks and valleys that may drastically affect the deviation between two measurements [54]. Moreover, transformer design and manufacturing tolerances, measuring setup and noises can generate differences in FRA spectra that may add complexities to FRA diagnosis [55]. Thus interpretation of FRA results requires an expert. To overcome such limitation, researchers have used several numerical as well as statistical indicators for fault identification. Precise modeling of high frequency behaviour of transformer winding provides improved interpretation of FRA results, as they may assist in identifying the salient features of winding responses, e.g. the effects of localized winding faults causing small changes in circuit parameters [50].

Two test methods are commonly used for FRA, viz. Sweep Frequency Response Analysis (SFRA) and Impulse Frequency Response Analysis (IFRA). In both the methods the excitation voltage is applied to one winding while the response is measured at another winding or terminal. In SFRA the frequency response is obtained in the entire frequency range with sinusoidal excitation voltage. On the other hand, in IFRA the response is obtained with an impulse excitation voltage [56].

In order to identify a spectral deviation, which indicate a mechanical damage inside a transformer, the FRA measurements should be free from other influences. However, it has been reported in literature that FRA measurements are sensitive to insulation aging, layout of measuring leads, core magnetization, insulation contamination to name a few. Studies have also shown that FRA results are sensitive to temperature of transformer winding as well as moisture content of paper insulation [57]. Thus it is of practical importance that the key factors are identified that contribute to FRA measurement results.

1.5 Remaining Life Analysis

Aging assets, rising energy demand, and the need to deliver energy without outage are issues facing utilities all over the world. At the same time, financial constraints demand an increased return on investment over reduced maintenance spending. These apparently contradictory demands can be met through optimized asset management. This, in turn, requires accurate and reliable models for remaining life analysis considering both technical and economical criteria. This is especially relevant now as many power transformers that are installed long back are expected to reach their end-of-life in the coming years.

Mineral oil impregnated paper is the insulation system in most of the aged power transformers in service. The oil-paper insulation system of transformer suffers degradation due to thermal, electrical and mechanical stresses acting on the insulating structure. Thermal stress is considered to be the most important factor that affects the life of oil filled transformers. Empirical formulae have been proposed in literature for estimation of elapsed life based on the hottest spot temperature of the winding [58, 59]. Subsequently, condition based evaluation of transformer insulation system has been suggested based on thermo-chemical reaction kinetics in the IEEE loading guide [60], which has been supplemented by CIGRE document [61]. Standard test procedure for thermal evaluation of oil-filled transformers has been prescribed in another IEEE standard [62].

Generally, life of a transformer is equal to the life of its insulation, to be precise its solid insulation. Thus aging of paper insulation primarily determines the end-of-life of a transformer in service. The end-of-life of oil impregnated paper insulation can be assessed with higher confidence by performing destructive testing, viz. measurement of degree of polymerization [63]. However, degradation of paper generates a class of hydrocarbons known as Furans that exist in oil in dissolved state. It has been reported that results of Furan as well as DGA analysis could be correlated well with DP and thus results of these non-destructive diagnostic techniques based on oil testing could be used for estimation of elapsed life of transformers [63].

In a nutshell, various aspects of remaining life analysis of transformers are as follows: (a) Determination of functional age of insulation from oil test results like DGA and Furan, (b) Measurement of DP of paper samples taken from in situ transformers, if practicable, and (c) Application of appropriate life estimation models for power transformers to determine remaining life. Researchers over the entire world are trying to come up with reliable and accurate model for estimating the remaining life of power transformers based on diagnostic parameters.

References

1. Wang M, Vandermaar AJ, Srivastava KD (2002) Review of condition assessment of power transformers in service. IEEE Electr Insul Mag 18(6):12–25
2. Kogan VI, Fleeman JA, Provanzana JH, Shih CH (1988) Failure analysis of EHV transformers. IEEE Trans Power Delivery 3(2):672–683
3. CIGRE Working Group 05 (1983) An international survey on failures in large power transformers. Electra, No. 88
4. Metwally IA (2011) Failures, monitoring, and new trends of power transformers. IEEE Potentials 30:36–43
5. Minhas MSA, Reynders JP, de Klerk PJ (1999) Failures in power system transformers and appropriate monitoring techniques. In: Proceedings of the 11th ISH London, UK, Paper no. 1.94.S23
6. Grechko ON, Kalacheva I (1996) Current trends in the development of in-service monitoring and diagnostic systems for 110–750 kV power transformers. Appl Energy: Russ J Fuel, Power, Heat Syst 34(5):84–97
7. Bartley WH (2003) Analysis of transformer failures. 36th annual conference of international association of engineering insurers, Stockholm
8. CIGRE Publication (1996) Australia/New Zealand transformer reliability survey report
9. Fetherston F, Finlay B (2001) Power transformer condition assessment—the second century and beyond. Proceedings of AUPEC conference, Perth
10. Jarman P, Wang Z, Zhong Q, Ishak T, (2009) End-of-life modeling for power transformers in aged power system networks. CIGRE 6th Southern Africa regional conference, Paper C105
11. IEC 60599 (1999) Mineral oil-impregnated electrical equipment in service-guide to the interpretation of dissolved and free gases analysis
12. Lockie AM (1955) Functional life expectancy tests for liquid filled distribution transformers. AIEE Trans Power Apparatus Syst Part III 75:977–985
13. Sparling BD (2001) Moving forward from monitoring to diagnostics. IEEE-PES transmission and distribution conference and exposition, Atlanta, pp 960–963
14. IEC 60076, Part 4 (2002), Guide to the lightning impulse and switching impulse testing—power transformers and reactors
15. Aicher LC (1948) Experience with transformer impulse failure detection methods. AIEE Trans 67:1621–1631
16. Beldi F (1950) The impulse testing of transformers. Brown Boveri Rev 37:179–193
17. Hagenguth JH, Meador JR (1952) Impulse testing of power transformer. AIEE Trans 71:697–704
18. Malewski R, Poulin B (1988) Impulse testing of power transformers using the transfer function method. IEEE Trans Power Delivery 3:476–490
19. Vaessen PTM, Hanique E (1992) A new frequency response analysis method for power transformers. IEEE Trans Power Delivery 7:384–390
20. Pandey SK, Satish L (1998) Multi resolution signal decomposition: a new tool for fault detection in power transformers during impulse tests, IEEE Trans Power Delivery 13:1194–1200
21. Purkait P, Chakravorti S (2002) Time and frequency domain analyses based expert system for impulse fault diagnosis in transformers. IEEE Trans Dielectr Electr Insul 9(3):433–445
22. Purkait P, Chakravorti S (2002) Pattern classification of impulse faults in transformers by wavelet analysis. IEEE Trans Dielectr Electr Insul 9(4):555–561
23. Purkait P, Chatterjee A, Chakravorti S, Bhattacharya K (2003) Translationally adaptive Fuzzy classifier for transformer impulse fault diagnosis. IEE Proc Gener Transm Distrib 150(1):33–40
24. Purkait P, Chakravorti S (2003) Impulse fault classification in transformers by fractal analysis. IEEE Trans Dielectr Electr Insul 10(1):109–116

25. Koley C, Purkait P, Chakravorti S (2006) Wavelet aided SVM tool for impulse fault identification in transformers. IEEE Trans Power Delivery 21(3):1283–1290
26. Koley C, Purkait P, Chakravorti S (2006) Time-frequency representation of resistance for modeling of transformer winding under impulse test. IEEE Trans on Power Delivery 21(3):1367–1374
27. Koley C, Purkait P, Chakravorti S (2007) SVM classifier for impulse fault identification in transformers using fractal features. IEEE Trans Dielectr Electr Insul 14(6):1538–1547
28. Dey D, Chatterjee B, Chakravorti S, Munshi S (2008) Rough-granular approach for impulse fault classification of transformers using Cross-Wavelet transform. IEEE Trans Dielectr Electr Insul 15(5):1297–1304
29. Rajamani P, Dey D, Chakravorti S, Cross-correlation aided wavelet network for classification of dynamic insulation failures in transformer winding during impulse test. IEEE Trans Dielectr Electr Insul 18(2):521–532
30. Prevost TA, Woodcock DJ (2007) Transformer fleet health and risk assessment. IEEE PES Transformers Committee Tutorial, Dallas
31. Kurtz C, Ford G, Vainberg M, Lebow M, Ward B (2005) Managing aged transformers. Transm Distrib World 57(7):36–40
32. Sokolov V, Berler Z, Rashkes V (1999) Effective methods of assessment of insulation system conditions in power transformers: a view based on practical experience, Available online at www.ztzservices.com/files/EIC-99Transformers.pdf
33. Duval M (2002) A review of faults detectable by gas-in-oil analysis in transformers. IEEE Electr Insul Mag 18(3):8
34. Allan D, Jones C, Sharp B (1991) Studies of the condition of insulation in aged power transformers. Part-1: insulation condition and remnant life assessments for in-service units. Proceedings of the 3rd international conference on properties and applications of dielectric materials, Tokyo, 8–12 July 1991
35. Shroff DH, Stannett AW (1985) A review of paper aging in power transformers. IEE Proc, Part C 132:312–319
36. Gaefvert U (1996) Condition assessment of insulation systems: analysis of dielectric response methods', Nordic insulation symposium NORD IS 96. Norwegian University of Science and Technology, Trondheim, pp 1–19
37. Frimpong G, Gaefvert U, Fuhr J (1997) Measurement and Modeling of Dielectric Response of Composite Oilrpaper Insulation. IEEE 5th Int Conf Prop Appl Dielectr Mater (ICPADM) 2:86–89
38. Gaefvert U, Frimpong G, Fuhr J (1998) Modelling of dielectric measurements on power transformers. International conference on large HV electric systems, CIGRE, Paris, Paper No. 15–103
39. Saha TK, Yao ZT (2003) Experience with return voltage measurements for assessing insulation conditions in service-aged transformers. IEEE Trans Power Delivery 18(1):128–135
40. Bognar A, Csepes G, Kalocsai L, Kispal I (1991) Spectrum of Polarization Phenomena of Long Time-constant as a Diagnostic Method of Oil-paper Insulating Systems. IEEE Proc 3rd Int Conf Prop Appl Dielectr Mater (ICPADM) 2:723–726
41. Osvath P, Csepes G, Webb M (1992) Polarization spectrum analysis for diagnosis of insulation systems. 6th International Conference on Dielectric Materials, Measurement and Applications, IEE London, UK, pp. 270–273 (Conf Publ No. 363)
42. Csepes G, Kispal I, Fekete J, Romvari Z, Brooks R, Szebeni M, Bognar A, Uri E, Babos S (1998) Correlation between electrical and chemical testing techniques for assessing degradation of oil-paper insulation. International Conference on Large HV Electrical Systems, CIGRE, Paris, France, Paper No. 15–202
43. Saha TK, (2003) Review of modern diagnostic techniques for assessing insulation condition in aged transformers. IEEE Trans Dielectr Electr Insul 10(5):903–917

44. Saha TK, Purkait P, Mueller F (2005) Deriving an equivalent circuit of transformers insulation for understanding the dielectric response measurements. IEEE Trans Power Delivery 20(1):149–157
45. Jota PRS, Islam SM, Jota FG (1999) Modeling the polarization spectrum in composite oil-paper insulation systems. IEEE Trans Dielectr Electr Insul 6:145–151
46. Zaengl WS (2003) Dielectric spectroscopy in time and frequency domain for HV power equipment, Part I: theoretical considerations. IEEE Electr Insul Mag 19(5):5–19
47. Setayeshmehr A, Fofana I, Eichler C, Akbari A, Borsi H, Gockenbach E (2008) Dielectric spectroscopic measurements on transformer oil-paper insulation under controlled laboratory conditions. IEEE Trans Dielectr Electr Insul 15(4):1100–1111
48. Blennow J, Ekanayake C, Walczak K, Garcia B, Gubanski SM (2006) Field experiences with measurements of dielectric response in frequency domain for power transformer diagnostics, IEEE Trans Power Delivery 21:681–688
49. Fofana I, Hemmatjou H, Meghnefi F, Farzaneh M, Setayeshmehr A, Borsi H, Gockenbach E (2010) On the frequency domain dielectric response of oil-paper insulation at low temperatures. IEEE Trans Dielectr Electr Insul 17:799–807
50. Abeywickrama N, Serdyuk YV, Gubanski SM (2008) High-frequency modeling of power transformers for use in frequency response analysis (FRA). IEEE Trans Power Delivery 23(4):2042–2049
51. Dick EP, Erven CC (1978) Transformer diagnostic testing by frequency response analysis. IEEE Trans Power Apparatus Syst 97(6):2144–2150
52. Abeywickrama KGNB, Podoltsev AD, Serdyuk YV, Gubanski SM (2007) Computation of parameters of power transformer windings for use in frequency response analysis. IEEE Trans Magn 43(5):1983–1990
53. Abeywickrama N, Serdyuk YV, Gubanski SM (2008) Effect of core magnetization on frequency response analysis (FRA) of power transformers. IEEE Trans Power Delivery 23(3):1432–1438
54. Kim JW, Park BK, Kim SW, Park PG (2005) Fault diagnosis of a power transformer using an improved frequency response analysis. IEEE Trans Power Delivery 20(1):169–178
55. Wang ZD, Li J, Sofian DM (2009) Interpretation of transformer FRA responses—Part-I: influence of winding structure. IEEE Trans Power Delivery 24(2):703–710
56. Wang M, Vandermaar AJ, Srivastava KD (2004) Transformer winding movement monitoring in service-key factors affecting FRA measurements. IEEE Electr Insul Mag 20(5):5–12
57. Reykherdt AA, Davydov V (2011) Frequency response analysis measurements and power transformer diagnostics. IEEE Electr Insul Mag 27(1):22–30
58. Montsinger VM (1930) Loading the transformers by temperature. AIEE Trans 49:776–792
59. Dakin TW (1947) Electrical insulation deterioration treated as a chemical reaction rate phenomenon. AIEE Trans 66:113–122
60. IEEE loading guide for mineral oil immersed transformer, C57.91–1995
61. CIGRE working group 12.09 (1993) Lifetime evaluation of transformers. Electra pp 39–51
62. IEEE standard test procedures for thermal evaluation of insulation systems for liquid-immersed distribution and power transformers, C57.100–2011
63. Pradhan MK, Ramu TS (2005) On the estimation of elapsed life of oil-immersed power transformers. IEEE Trans Power Delivery 20(3):1962–1969

Chapter 2
Impulse Fault Analysis

2.1 Introduction

Outdoor transformers are often subjected to lightning and switching impulse/surges; these surges propagate along the transmission lines and thereafter into the winding of transformer. Any weakness of insulation may result in failure of the transformer due to such overvoltages. To prevent breakdown of the transformer and thereby interruption of supply, the equipment must be designed to withstand such highvoltage surges, the peak of which may be several times the normal working/rated voltage of the system. So, the impulse withstand test is an acceptance test for transformers.

The impulse test is normally done with impulse voltages, having standard wave shape of 1.2/50 μs as per IEC 60076-4 [1]. These standard wave shapes are impressed on one side of the transformer winding, while keeping other end of the winding and tank to ground through a low resistance (in the case of tank current method). During impulse testing of transformers the oscillographic traces of applied voltage and corresponding current waveforms are investigated for judging the insulation condition of the transformers. This procedure requires significant human expertise and knowledge for proper judgment of insulation condition. Development of high speed processors with machine intelligence techniques makes it possible to analyze the impulse test waveforms by the machine itself without/with minimal requirement of human intervention. Different computer-aided methodologies, such as, artificial neural network (ANN) [2], fuzzy systems [3], and wavelet-based analysis [4] have already been reported for the classification and localization of fault within the transformer winding. Recent studies include information granulation-based approach using rough set analysis for this purpose.

For most of the cases regarding the fault analysis by machine intelligence techniques, faults are simulated either in a software-based model or in an analog model of the transformer. Suitable features are extracted from the response waveforms (fault current waveforms). A rule base is generated on the basis of those features with known fault conditions employing a suitable classification algorithm. This phase may be considered as training phase of a supervised learning. Thereafter the performance of the scheme can be tested upon real-life data.

S. Chakravorti et al., *Recent Trends in the Condition Monitoring of Transformers*, Power Systems, DOI: 10.1007/978-1-4471-5550-8_2, © Springer-Verlag London 2013

2.2 Impulse Test Methods as per IEC

The impulse test is normally done with impulse voltages, having standard wave shape of 1.2/50 μs as per IEC 60076-4, as stated earlier [1].

The routine impulse test normally consists of one reduced and one full-wave impulse, or two full-wave impulses, with the standard wave shape of 1.2/50 μs as per IEC. The full-wave impulses are of a crest value equal to the rated BIL of the terminal being tested. These standard wave shapes are impressed on one side of the transformer winding, while keeping other end of the winding and tank to ground through a low resistance (called *tank current method*). The winding currents are recorded by measuring the voltage across the standard or known low resistance shunt. These currents are analyzed in order to find the faults within the transformer winding.

For Impulse testing of a three-phase winding, the impulse voltage is applied at one terminal at a time. The other two terminals are shorted at that time and grounded through a resistance equal to the surge impedance of the line, as shown in Fig. 2.1 [5]. Different test voltage levels depending on the voltage rating of the transformer winding under test is given in Table 2.1.

2.3 Faults in a Winding and Data Acquisition of Fault Current

Transformer failures during impulse test can be classified into four major classes as listed below:

- Winding-to-ground failure involving major insulation.
- Failure across relatively large portions of the winding but not involving ground.
- Inter winding breakdown, i.e., between high voltage (HV) and low voltage (LV).
- Failure between small portions of windings, such as coils, sections, and turns.

All the above-mentioned kinds of faults may be broadly classified under two heads, namely the *series faults* and the *shunt faults*. Series faults represent faults between turns, layers, and disks, whereas shunt faults stand for winding to ground, winding to tank, and winding to winding faults [1, 5].

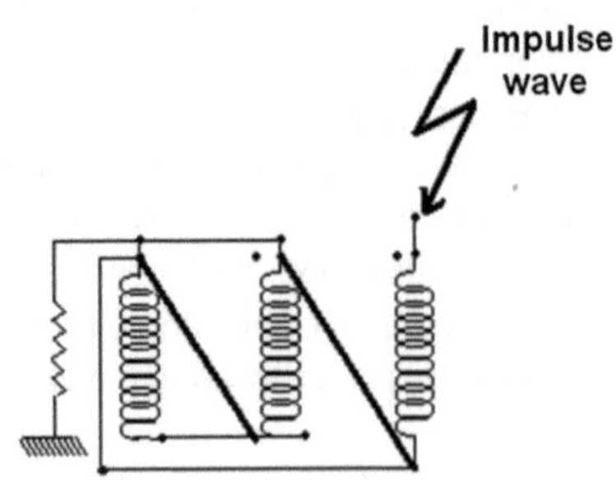

Fig. 2.1 Winding connections for impulse test

Table 2.1 Rated withstand voltages for transformer windings with highest voltage for equipment $U_m < 300$ V

Highest voltage for equipment, U_m	Rated lightning impulse withstand voltage[a]	
kV_{rms}	kV_{peak}	
	List 1	List 2
3.6	20	40
7.2	40	60
12	60	75
17.5	75	95
24	95	125
36	145	170
52	250	
72.5	325	
123	450/550	
145	550/650	
170	550/650/750	
245	750/850/950	

[a] The choice between List 1 and List 2 in this table for $U_m < 52$ kV and the choice between alternative rated withstand voltages for $U_m = 123$, 145, 170, and 245 kV depends on the severity of overvoltage conditions to be expected in the system and on the importance of particular installation

In most of cases the winding responses to faults were investigated by the 'tank current method' of fault detection, which is also called the 'line current method'. In this method impulse is impressed on line end of the winding while the earth end of the winding and the transformer tank is connected to ground through a low resistance. The schematic of the "tank-current" method is given in Fig. 2.2, where the tank and the earth end of the winding are connected to ground through a shunt resistance (R_{sh}). C_g, C_s, L_n, and R_n represent the capacitance to ground, series capacitance, equivalent inductance, and resistance of the winding for each disc, respectively. In 'tank current' method, the current components shunting out through the ground capacitors are also included in the recording of the response. This makes the 'tank current' method more sensitive as compared to the conventional 'neutral current' method.

These waveforms can be transferred to a PC for further analysis using suitable interfaces. A typical scheme for data acquisition is shown in Fig. 2.3 [6].

2.4 Impulse Fault Current Analysis

Detection of transformer insulation failure during impulse test in laboratory has been an important issue over a long period of time. Advancement in instrumentation and computer based signal processing tools have reformed this issue. Various tools and techniques have been used over the years for impulse fault

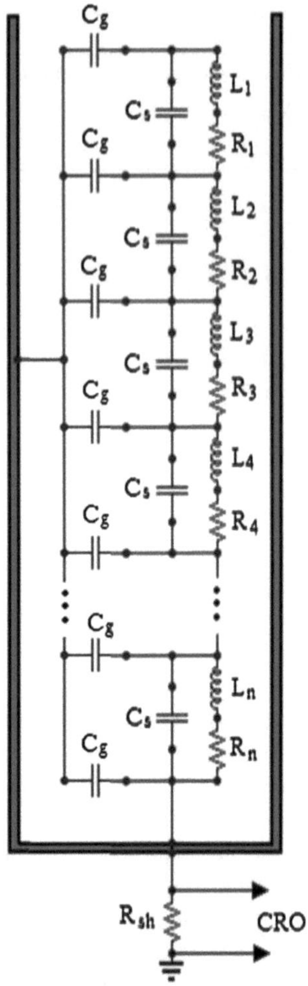

Fig. 2.2 The schematic of tank-current method

diagnosis in transformers, including manual oscillographic analysis [7], frequency domain analysis [8], and modern signal processing techniques like wavelet [9], fractal [4], artificial intelligent tools like ANN [10], and fuzzy techniques [2].

The initial voltage distribution along the winding length in a transformer when a surge voltage is applied, depends upon the ratio of winding-to-ground capacitance (C_g) to interturn capacitance (C_s). These capacitances of the winding depend upon the winding geometry, i.e., the physical disposition of the turns and coils relative to each other and to the grounded parts of the structure. Therefore, different winding styles, e.g., layer and disc windings, have different values of the ratio (C_g/C_s) and as a result they perform differently under the application of a voltage surge. In the present chapter, the schemes described with experimental results for the analysis of impulse faults of transformers, are based on disc-type

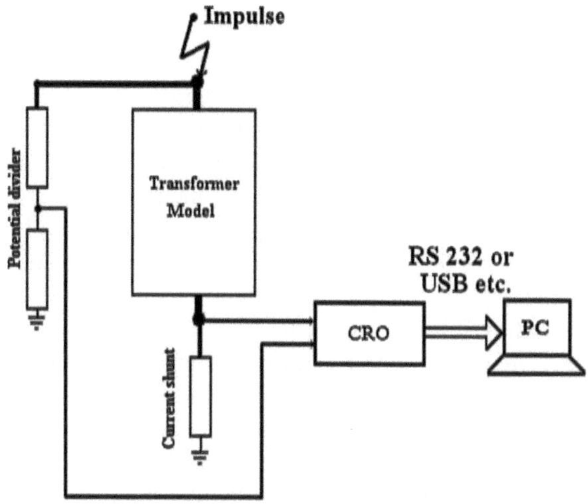

Fig. 2.3 Schematic of data acquisition setup

windings [11, 12]. However, the schemes can easily be extended or modified for other types of transformer windings.

Abetti [12] found that the response of transformer windings to transient voltages, such as lightning and switching surges, was quite different from that under power frequency voltages. Transformers have complex arrangement of coils around an iron core, and due to close spacing of these coils, they have significant stray capacitances and inductances, which exhibit complex frequency-dependent characteristics. A transformer winding when subjected to a step surge, have been seen to behave initially like a capacitive network. The capacitive voltage distribution is followed by voltage oscillations which subsequently die down to yield finally a voltage distribution determined by winding resistances.

Malewski and Poulin [13] and Vaessen and Hanique [14] have shown experimentally that for higher frequencies essentially above 1 kHz, transformers behave linearly and the iron core does not play a significant role. Similar observations are also reported by the authors in [11].

A number of high-frequency transformer models, suitable for computer-aided transient simulations using electromagnetic transient program (EMTP) have been proposed in literatures [16–21].

Roy and Biswas [11] presented a simplified lumped parameter analog model for impulse studies of transformer windings. The modeling was carried out by considering it as an isolated winding, and the iron core was replaced by air core. The authors of [7, 22–24] discussed several methods for calculation of the self-inductance, mutual inductance, series capacitance, and the shunt capacitance of transformer disc winding.

2.4.1 Conventional Methods

Earlier contributions toward fault diagnosis were made by Hagenguth [25, 26]. He introduced the concept of fault analysis by carefully studying the voltage and current waveforms recorded by cathode ray oscilloscope (CRO). According to him, failures can be detected by meticulous comparison of the recorded wave shapes of the reduced voltage wave and the full voltage wave. The neutral current method, as this method is called, still remains one of the popular methods of impulse fault detection in transformers.

Stenkvist [7] in his CIGRE report, pointed out two different connections for the measurement of impulse current—namely the line-current (tank current) connection and the neutral current connection. According to him, the tank-current method is practically more suitable for detection and location of faults. Figure 2.4 shows typical winding current wave and impulse voltage wave [4].

Oscillographic comparison method, though apparently simple and straightforward, has lost its popularity over the years due its dependency on the knowledge and experience of the experts performing the analysis.

2.4.2 Modern Analysis Tools

The problem of impulse fault current analysis using modern machine intelligence tools may be addressed in two stages:

(1) extraction of suitable features from the fault current waveforms;
(2) classification of the features for proper fault identification with suitable classification algorithm.

These two aspects are discussed in details in the following sections.

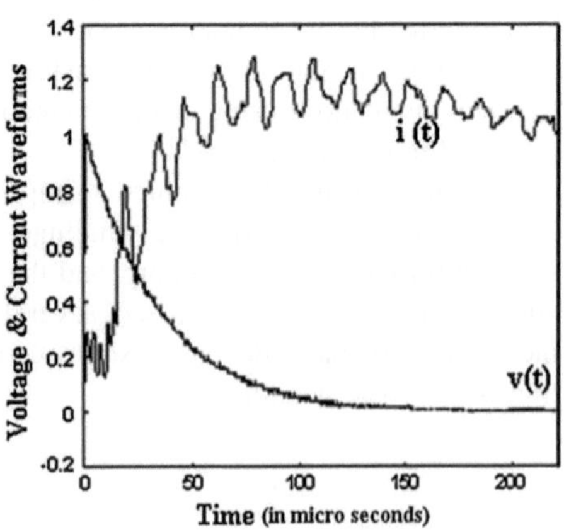

Fig. 2.4 Applied impulse voltage $v(t)$ and winding current $i(t)$ of a transformer under test

2.4.2.1 Feature Extraction Methodologies

With the advent of modern digital computers, high speed data acquisition devices, processors and analysis softwares, traditional techniques of fault current analysis have been over-shadowed by different signal/image processing tools/techniques.

Wavelet analysis has been applied successfully for transformer condition assessment and fault diagnosis in recent times by many researchers [27–30]. The inherent non-stationary pattern of transformer current waveforms during different fault conditions can be effectively classified using the frequency-selective feature of wavelet transform.

Purkait and Chakravorti [27] described the Multi-resolution analysis technique for PD detection during impulse testing of transformers. Koley [5, 30] described a technique based on wavelets for detection of impulse faults in transformers. For this purpose, they divided the HV winding of the transformer into a number of sections and they have extracted features of the faults for each of the sections from wavelet packet-based analysis. They have applied support vector machine (SVM)-based classifier to classify the extracted features for identification of faults. Their method is good enough to identify the faults within 10 % of the winding length.

Investigation results have also been reported in [6, 28, 31] on the use of wavelet analysis for the pattern recognition of winding current waveforms and classification of impulse faults. With the help of EMTP, series as well as shunt faults have been simulated at different locations along the entire length of disk-type HV winding of three different transformers. The authors studied the current waves employing continuous wavelet transform (CWT). Certain parameters were then selected for pattern classification of these fault current waves.

The authors of [32, 33] reported investigation results on the use of fractal geometry for the feature extraction and pattern recognition of winding current waveforms and classification of impulse faults therein. Fractal dimension (FD) is an intrinsic parameter of the image describing its fractal nature and lacunarity is a parameter describing the denseness (rather lack of denseness or lacuna) in the image.

Time–Frequency Domain Methods: Wavelet Transform

Any transform is basically a mathematical operation that maps a function or sequence from one domain to another. Usually mathematical transformations of different signals are performed to extract further information from that signal which are not evident from the raw (e.g. time-domain) signal. So, the advantages of using a transform are summarized below:

- Transformations, if suitably chosen, make an equation easier to solve than the original equation (e.g. Laplace transforms converts differential equations in a form that is easier to solve).

- The transform of a function gives additional or hidden information, which may not be observable in its original domain.
- The transform of a function or sequence may require less storage; hence, provide data compression or, reduction.
- An operation may be easier to apply on the transformed function, rather than the original function (like convolution)

The choice of transformation technique depends on the application.

In most of the signal processing applications, the knowledge of the frequency content of the signal plays an important role. The Fourier transform is probably the most popular transform which is used to obtain the frequency spectrum of a signal. However, Fourier transform is only suitable for stationary signals, i.e., signals whose frequency content does not change with time. The Fourier Transform, while it tells how much of each frequency exists in the signal, it does not tell at which time these frequency components occur. Signals such as image and speech have different spectral characteristics at different space and time, i.e., they are non-stationary. To analyze these signals, both frequency and time information are needed simultaneously, i.e., a time–frequency representation of the signal is needed.

To solve this problem, the short-time Fourier transform (STFT) [34, 35] was introduced. The major drawback of the STFT is that it uses a fixed window width. The Wavelet Transform, which has evolved considerably in the past few decades, provides a better time–frequency representation of the signal [35–38] than STFT.

Continuous wavelet transform. The CWT was developed as an alternative approach to the STFT to overcome the window width or, resolution problem. In the wavelet transform a signal is multiplied with a function (i.e. the *wavelet*), similar to the *window function* in the STFT, and the transform is computed separately for different segments of the time-domain signal.

The CWT is defined as follows [5, 6, 34–37],

$$\mathrm{CWT}(\tau, s) = \Psi(\tau, s) = \frac{1}{\sqrt{|s|}} \int_{-\infty}^{\infty} f(t)\Psi^{*}\left(\frac{t-\tau}{s}\right)\mathrm{d}t \tag{2.1}$$

As in the above equation, the transformed signal is a function of two variables, τ and s, the *translation* and *scale* parameters, respectively. $\psi(t)$ is the transforming function, and is called the *mother wavelet*. The term *mother wavelet* gets its name due to the important property of the wavelet analysis as explained below:

The term *wavelet* means a small wave. The smallness refers to the condition that this (window) function is of finite length. The *wave* refers to the condition that this function is oscillatory. The term *mother* implies that the functions with different region of support that are used in the transformation process are derived from one main function, or the *mother wavelet*. In other words, the *mother wavelet* is a prototype for generating the other window functions. There are various types of mother wavelets like Daubechies, Morlet, Mexican Hat, Meyer, etc., as shown in Fig. 2.5 [5, 6, 34].

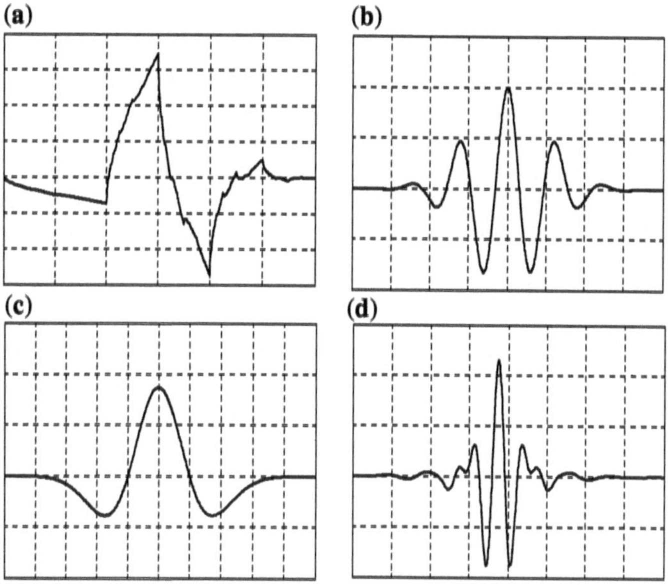

Fig. 2.5 Mother-wavelet families: **a** Daubechies2, **b** Morlet, **c** Mexican Hat, **d** Meyer

The term *translation* is used in the same sense as it is used in the STFT, i.e., it is related to the location of the window, as the window is shifted through the signal. This term, obviously, corresponds to time information in the transform domain.

The parameter *scale* in the wavelet analysis is similar to the scale used in geographical maps. As in the case of maps, high *scales* correspond to a non-detailed global view (of the signal), and low *scales* correspond to a detailed view. Similarly, in terms of frequency, low frequencies (high scales) correspond to a global information of a signal (that usually spans the entire signal), whereas high frequencies (low scales) correspond to a detailed information of a hidden pattern in the signal (that usually lasts a relatively short time). So, it can be noted that *scale* is related to the inverse of frequency. However, the inverse of scale does not give the value of frequency directly. Scale (s) and frequency (f) are related according to the following relation [34–38]:

$$f = \frac{f_c}{s \cdot \Delta} \tag{2.2}$$

Here, f_c = center frequency of the mother wavelet and Δ = sampling period of the signal.

So, *scaling* as a mathematical operation either dilates or compresses a signal. Larger *scales* correspond to dilated (or stretched out) signals and smaller scales correspond to compressed signals.

Computation of the CWT. The CWT is provided by Eq. (2.1), where $f(t)$ is the signal to be analyzed. $\psi(t)$ is the *mother wavelet* or the *basis function*. All the wavelet functions used in the transformation are derived from the mother wavelet through translation (shifting) and scaling (dilation or compression).

The mother wavelet used to generate all the basis functions is chosen depending on some desired characteristics associated with that function. The translation parameter τ relates to the location of the wavelet function as it is shifted through the signal. Thus, it corresponds to the time information in the wavelet transform. The scale parameter s is related to $\left|\frac{1}{frequency}\right|$ and corresponds to frequency information, as stated earlier.

The wavelet series is obtained by discretizing CWT. This aids in computation of CWT using computers and is obtained by sampling the time-scale plane. The sampling rate can be changed according to *scale* change without violating the *Nyquist criterion*. For example, as the scale goes higher (i.e. at lower frequencies), the sampling rate can be decreased; thus reducing the number of computations [5, 34–38]. Usually the absolute coefficient values of CWT are plotted as a 3D surface in the time–frequency domain. A typical representation of such a plot of the CWT coefficients of an impulse fault current waveform of a transformer using Daubechies4 mother wavelet is shown in Fig. 2.6.

Discrete wavelet transform (DWT). In discrete wavelet analysis, the terms that are mostly used are *approximations* and *details*. The *approximations* are the high scale, i.e., low-frequency components of the signal. The *details* are the low scale, i.e., high-frequency components. Figure 2.7 shows the basic filtering process of a signal *S* in order to obtain low frequency and high-frequency components as done in DWT [5, 6, 35].

The original signal, *S*, is passed through two complementary filters and two output signals are obtained ("*A*" and "*D*" as shown in Fig. 2.7). Here, "*A*" is the low-frequency or *approximate* and "*D*" is the high-frequency or *detailed* part of the signal *S*. Unfortunately, this operation on a real digital signal ends up with twice as much data as when started. Let, for instance, the original signal *S* consists

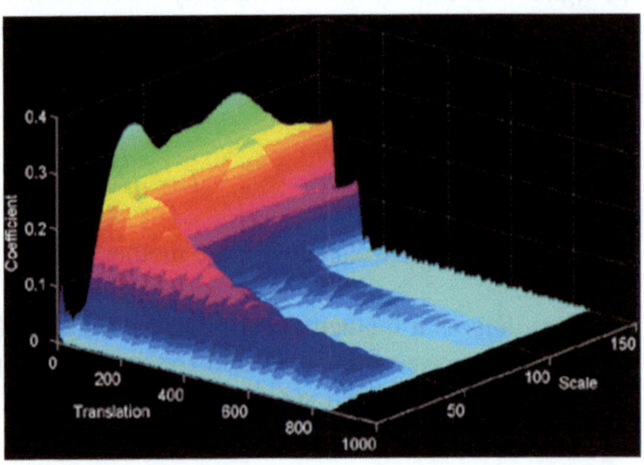

Fig. 2.6 Typical CWT surface for a PD signal

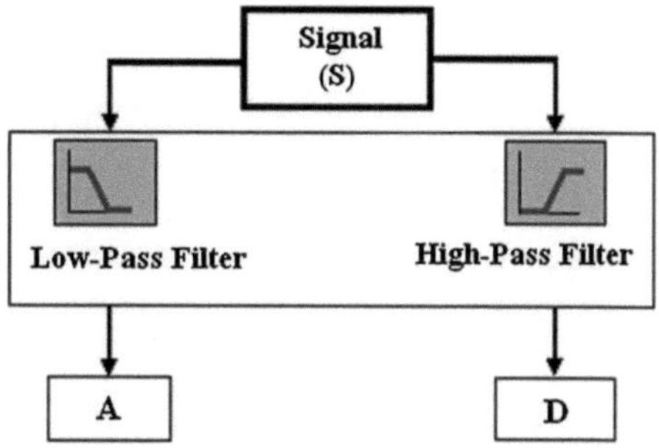

Fig. 2.7 The filtering process, at its most basic level

of 2,000 samples of data. Then the resulting signals will each have 2,000 samples, for a total of 4,000. There exists a more subtle way to perform the decomposition using wavelets, by keeping only one point out of every successive two in each of the two 4,000-length samples to get the complete information. This is the notion of *down-sampling*. It produces two sequences called *cA* and *cD* each having 1000 data points. Figure 2.8 describes the process of single step DWT. Starting from *S*, the first step produces two sets of coefficients: approximation coefficients *cA1*, and detail coefficients *cD1*. These vectors are obtained by convolving *S* with the impulse response of the low-pass filter for *approximation*, and with that of the high-pass filter for *detail*. A one-stage DWT of impulse fault current signal has been performed and shown in Fig. 2.9. In this case the detail coefficients *cD* are small and consist mainly of a high-frequency noise, while the approximation coefficients *cA* contain much less noise than the original signal.

In the case of multi-level DWT, the next step splits the approximation coefficients *cA1* in two parts using the same scheme, replacing *S* by *cA1* and producing *cA2* and *cD2*, and so on. This is called the wavelet decomposition tree and shown up to level 3 in Fig. 2.10.

Though, in the case of DWT the approximation coefficients (*cA*) are usually decomposed further so that one signal is resolved down into many lower resolution components, one can also resolve the detailed part (*cD*) depending upon the information that is required from the signal in the respective problem.

In reality, the decomposition can proceed only until the individual details consist of a single sample or pixel. It can be proved easily that given a signal *S* of

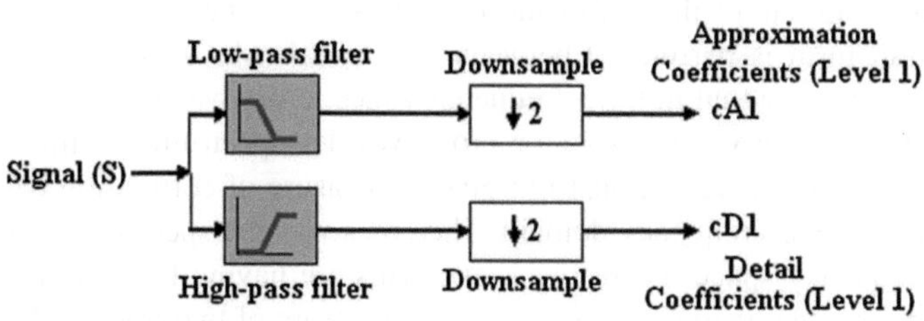

Fig. 2.8 Single step DWT

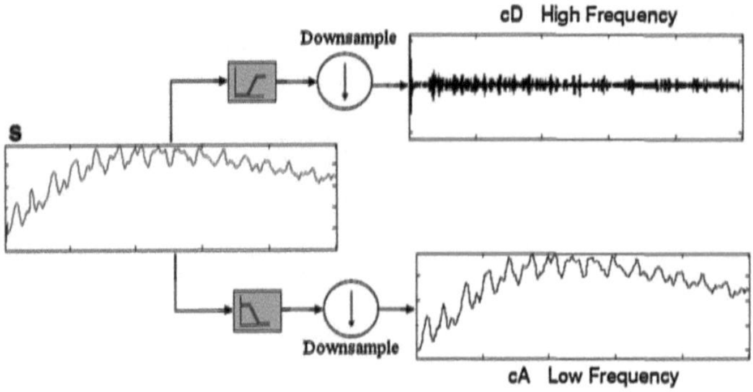

Fig. 2.9 Single step DWT of impulse fault current

Fig. 2.10 A typical Wavelet decomposition tree up to level 3

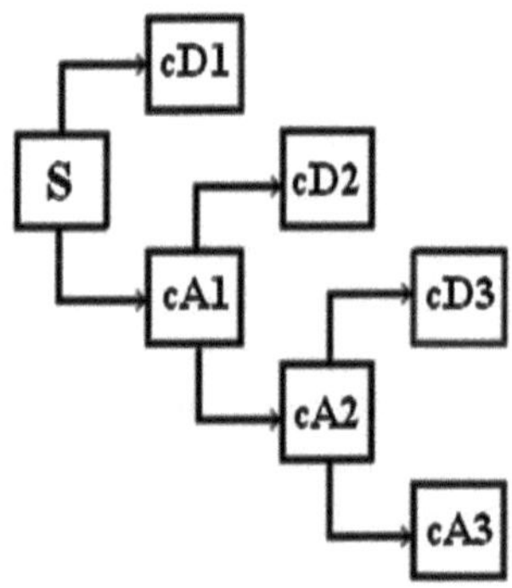

length N, the DWT consists of $\log_2 N$ levels of decomposition at most. In practice, a suitable number of levels is selected based on the nature of the signal, or on the basis of a suitable criterion.

Cross-Spectrum Analysis: Cross-Wavelet Transform

Physical significance. Cross-wavelet transform may be considered as an extension of wavelet-based analysis [39, 40]. The mathematical operation for cross-wavelet transform is similar to the computation of CWT. As the name suggests, cross-wavelet transform is performed between two signals. As it is discussed that the CWT represents a signal in time–frequency space, cross-wavelet transform of two time series gives cross-spectrum or cross-wavelet spectrum in time–frequency domain. The cross-wavelet transform gives a measure of correlation between two waveforms in time–frequency domain. The cross-wavelet spectrum shows regions in time–frequency space where two waveforms are having high common power. To demonstrate the operation, cross-wavelet spectrum of two impulse fault current waveforms of a transformer is shown in Fig. 2.11. As it is just a representative

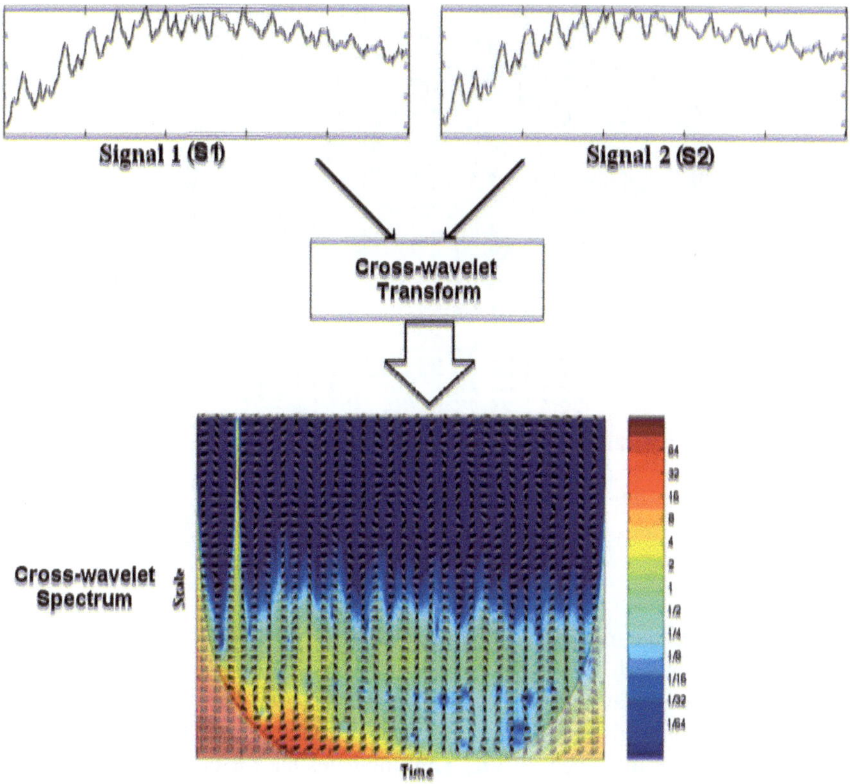

Fig. 2.11 A representation of cross-wavelet spectrum

diagram of the procedure, magnitude values are not shown in the axes and how the cross-wavelet spectrum is plotted in the colored picture, is explained in below section

Mathematical expressions. The mathematical operation for cross-wavelet transform is similar to the computation of CWT, as stated earlier. Mathematically, the cross-wavelet transform of two signals, $x(t)$ and $y(t)$ is defined as [31, 39–43]:

$$W^{xy}(s,\tau) = \frac{1}{k_{\psi}} \int_{-\infty}^{+\infty} \int_{-\infty}^{+\infty} W^{x}(a,b) W^{y*}\left(\frac{a}{s}, \frac{b-\tau}{s}\right) \frac{dadb}{a^2} \tag{2.3}$$

Here, $W^{*}(s,\tau)$ and $W^{y}(s,\tau)$ are the CWT of two signals $x(t)$ and $y(t)$ respectively with respect to a mother wavelet $\psi(t)$. The choice of mother wavelet depends on the nature of the waveform being processed. K_{ψ} is a constant, defined as, $k_{\psi} = \int_{-\infty}^{+\infty} \frac{|\Psi(\omega)|^2}{|\omega|} d\omega < \infty$. The cross-wavelet spectrum is plotted using the magnitude of W^{xy} and the phase angle, $\phi = \tan^{-1} \frac{\Im\{W^{xy}\}}{\Re\{W^{xy}\}}$. Here, $\Re\{W^{xy}\}$ and

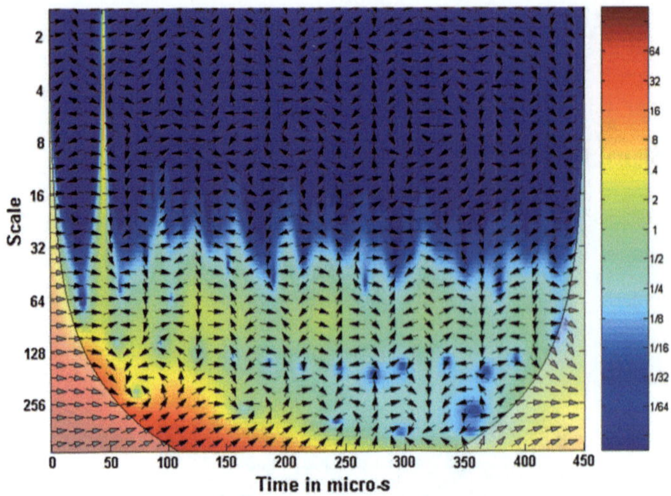

Fig. 2.12 A typical Cross-wavelet spectrum of two current waveforms obtained from impulse testing of a transformer

$\Im\{W^{xy}\}$ indicate the real and imaginary part of W^{xy} respectively. For clarity arguments are omitted.

For further explanation, the cross-wavelet spectrum shown in Fig. 2.11 has been magnified and given in Fig. 2.12. The 'U' shaped and black colored line shows the "cone of influence" (COI). COI indicates the region where edge effects due to zero padding are significant. Similar to spectral analysis, errors will occur at the edges (i.e. beginning or end) of the waveform in the case of cross wavelet because of the finite length of time series.

In Fig. 2.12, $|W^{xy}|$values at different 'time' and 'scale' are plotted. The x-axis is considered as 'time' axis and y-axis shows the 'scale', which is related to the inverse of frequency. The color of the figure at a point shows the value of $|W^{xy}|$ at that time–frequency space. The color-bar given in the right side of Fig. 2.12 indicates the value corresponding to a color. Higher the value, higher is the common power at that time–frequency point. The $|W^{xy}|$ surface for the above cross-wavelet spectrum is shown in Fig. 2.13 [31].

Black arrows in Fig. 2.12 show the phase angle at that time–frequency space. Arrows, pointing towards right indicate "in-phase" (i.e. phase difference is zero) and arrows pointing left indicate "opposite-phase" (i.e. phase difference is 180°). In general, the angles made by the arrows with the time axis indicate their respective phase angles.

Cross-wavelet transform has already been applied successfully in geosciences [40, 41] and biomedical signal processing [39]. Very recently it has also been proposed for transient analysis [31, 42, 44]. However, the cross-wavelet spectrum analysis can be used as an efficient tool in the processing of several signals obtained from different HV and power system measurements. Fault analysis in

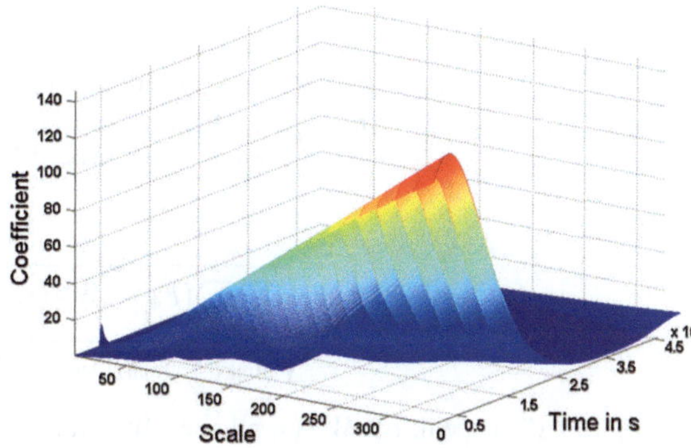

Fig. 2.13 The surface of $|W^{xy}|$ in time–frequency space

different power equipment may be considered as one of such applications. More specifically for impulse fault analysis, cross-wavelet transform is used as a method for feature extraction. It is known that location and type of fault has significant effect on the time–frequency spectrum of the impulse test waveforms. So, the cross-wavelet transform of these waveforms with one another may give some important information for localization and identification of fault in the transformer winding. Therefore, analysis of cross-wavelet spectrum is a promising technique for transformer impulse fault identification. For similar reasons, cross-spectrum analysis has been used for pattern classification of partial discharge (PD) pulses by the researchers [45]. It can be easily understood that there exists several such domains where cross-spectrum analysis can be applied.

Different types of features can be extracted from cross-wavelet spectrum. For example, in [31] many features are extracted from $|W^{xy}|$ as:

1.

$$F_1 = \frac{\sum_s \sum_\tau s\tau |W^{xy}(s,\tau)|}{\sum_s \sum_\tau |W^{xy}(s,\tau)|};$$

2.

$$F_2 = \sqrt{\frac{\sum_s \sum_\tau s^2\tau^2 |W^{xy}(s,\tau)|}{\sum_s \sum_\tau |W^{xy}(s,\tau)|}}$$

3.

$$F_3 = \frac{\sum_s \sum_\tau |W^{xy}(s,\tau)|}{|W^{xy}(s,\tau)|_{\text{peak}}};$$

4.

$$F_4 = \frac{\sum_s \sum_\tau |W^{xy}(s,\tau)|}{(s_{\max} - s_{\min})(\tau_{\max} - \tau_{\min})}$$

5.

$$F_5 = \sqrt{\frac{\sum_s \sum_\tau (F_4 - |W^{xy}(s,\tau)|)^2}{(s_{\max} - s_{\min})(\tau_{\max} - \tau_{\min})}}$$

6. $F_6 =$ "s" at peak of $|W^{xy}(s,\tau)|$, i.e.,$|W^{xy}(s,\tau)|_{\text{peak}}$
7. $F_7 =$ "τ" at peak of $|W^{xy}(s,\tau)|$, i.e.,$|W^{xy}(s,\tau)|_{\text{peak}}$

Five more features, similar to F_1 to F_5, may also be extracted from the phase angle $\varphi(s,\tau)$ data. They are:

8.

$$F_8 = \frac{\sum_s \sum_\tau s\tau|\phi(s,\tau)|}{\sum_s \sum_\tau |\phi(s,\tau)|};$$

9.

$$\text{F}_9 = \sqrt{\frac{\sum_s \sum_\tau s^2\tau^2|\phi(s,\tau)|}{\sum_s \sum_\tau |\phi(s,\tau)|}}$$

10.

$$F_{10} = \frac{\sum_s \sum_\tau |\phi(s,\tau)|}{|\phi(s,\tau)|_{\text{peak}}};$$

11.

$$F_{11} = \frac{\sum_s \sum_\tau |\phi(s,\tau)|}{(s_{\max} - s_{\min})(\tau_{\max} - \tau_{\min})}$$

12.

$$F_{12} = \sqrt{\frac{\sum_s \sum_\tau (F_{11} - |\phi(s,\tau)|)^2}{(s_{\max} - s_{\min})(\tau_{\max} - \tau_{\min})}}$$

However, one may choose some other features like, location of local peaks of $|W^{xy}|$ and $\varphi(s,\tau)$ surfaces, if any, or some more features from $|W^{xy}|$ and $\varphi(s,\tau)$ depending upon the nature of the problem.

Correlation-Based Method

Cross-correlation is a mathematical operation similar to convolution technique and measures the degree of similarity between the two signals correlated. Correlation plays an important role in the digital signal processing, especially in applications dealing with radar and sonar. In practical applications in which correlation is used, one of the sequences is contaminated by the other forms of interference. Then, the correlation sequence is a function of the statistical characteristics of that interference. In case of feature extraction for impulse fault analysis, let, the impulse fault current of healthy (no-fault) transformer winding as $A(n)$, and that in case of a particular fault condition as $B(n)$. The cross-correlation sequence of $A(n)$ and $B(n)$ is a function bearing the characteristics of that particular fault condition. The statistical characteristics of correlation sequence indicate the nature of the insulation failure like type, nature and location of faults. The cross-correlation sequence of these impulse currents is given by [46–49]

$$E_{AB}(m) = \begin{cases} \sum_{n=0}^{W-m-1} A_{n+m}B_n & m \geq 0 \\ E_{BA}(-m) & m < 0 \end{cases} \quad (2.4)$$

where $m = -W, \ldots, -1, 0, 1, \ldots, +W$, represents time shift parameter and the subscript AB represents the signals correlated. Let, W be the finite number of samples of each of the signals then the resultant cross-correlation sequence will have $2W - 1$ number of samples. For simplicity E_n has taken as the cross-correlation sequence of the two impulse currents, then sample features which may be extracted from the cross-correlation sequence are given from Eqs. (2.5a) to (2.5c).

$F1$ = Maximum value of the sequence (E_{max})

$$F2 = \frac{\sum_{n=-W}^{W} nE_n}{E_{max}} \quad (2.5a)$$

$$F3 = \frac{\sum_{n=-W}^{W} nE_n}{\sum_{n=-W}^{W} E_n} \quad (2.5b)$$

$$F4 = \frac{\sum_{n=-W}^{W} |n|E_n}{\sum_{n=-W}^{W} E_n} \quad (2.5c)$$

However, one may choose other features like index of the maximum value, kurtosis, skewness and root mean abscissa depending upon the nature of the pattern identification problem in hand.

Fractals Based Method

Fractals have been found to be successful to provide a description of naturally occurring phenomena and shapes, where conventional and existing mathematical models were found to be inadequate. In recent years, this technique has attracted increased attention for classification of textures and objects present in images and natural scenes [50–53].

The applications of fractals in the field of electrical insulation commenced in early 1980s. Niemeyer et al. [54] introduced a new stochastic model to describe the discharge pattern of dielectric breakdown in terms of fractal properties.

Fuji et al. [55] presented results to show that the 'electrical tree' shapes of the discharge process within the insulation material have fractal properties.

Satish and Zaengl [56] discussed the results on the use of fractal features for recognition of 3D PD patterns. They computed two fractal features, like the FD and lacunarity from 3D discharge patterns and analyzed their ability to discriminate among various discharge patterns.

Candela et al. [57] described a novel PD defect identification method by means of statistical and fractal parameters and a neural network.

Several researchers [58–61] have also reported suitable and effective use of fractal geometry for pattern classification of waveforms and profile-like curves. A variety of algorithms are available for the computation of FD.

Presently there is an increasing trend on the application of fractal techniques in the field of signal processing and pattern recognition, as it has solved many pattern recognition problems efficiently [62, 63]. That is why, fractal techniques have been applied for PD pattern recognition [57] and in the field of classifying impulse faults of transformers [32, 64]. Researchers in [64] have successfully classified type and location of fault within 33 % of transformer winding with the help of two fractal features—FD using box count algorithm, and Lacunarity.

In this section, the description of different Fractal features effective for impulse fault analysis of transformer winding is given. These fractal features include calculation of FD using box count and Higuchi algorithm, Lacunarity and Approximate Entropy (ApEn), as they were found to be efficient for this purpose by the researchers [5, 64].

Basic concept of fractals. The word fractal was first coined by the polish scientist Benoit Mandelbrot [50]. A *fractal* is a rough fragmented geometric shape that can be subdivided in parts, each of which (at least approximately) is a reduced size or copy of whole (alternatively called *self sufficient* functions or patterns). Fractal is any pattern that reveals greater complexity as it is enlarged. Figure 2.14 shows one such fractal shape.

This quality is usually called 'self-similarity' and can be also looked at as the object being composed of many copies of itself, each of which is scaled down with possible translations and rotations. However, many fractal objects found in nature are not self-similar, but similar only in a statistical sense, and hence referred to as statistically self-similar, i.e., these are only statistically scale invariant when magnified [50].

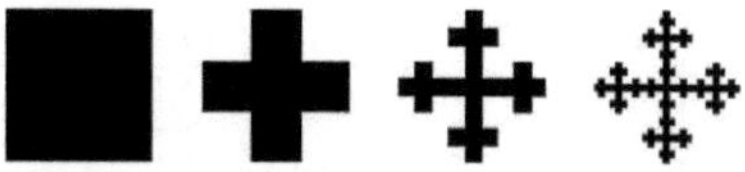

Fig. 2.14 Typical fractal shape

Box count method. FD is an important characteristic of fractals because it contains information about their geometric structure. The term FD was introduced by Mandelbrot [50]. According to him, a set A in a Euclidean n *dimensional*-space, is said to be self-similar when A is the union of N distinct (non-overlapping) copies of itself, each of which has been scaled down by a ratio r in all coordinates. The fractal dimension D of the set A is given by the relation [5, 50, 64]:

$$Nr^D = 1$$

or,

$$D = \frac{\log N}{\log\left(1/r\right)} \tag{2.6}$$

Let, the entire set A is covered by a single n-dimensional box of size $L_{\max}$ which can be a multiple of the side length L in the x–y plane. If the set A is now scaled down by the ratio r, then the number of boxes of size $L = rL_{\max}$ needed to cover the whole can be derived from (2.6) as:

$$N(L) = \frac{1}{r^D} = \left[\frac{L_{\max}}{L}\right]^D \tag{2.7}$$

The simplest way to estimate D from (2.7) is to divide the *n-dimensional* space into a grid of boxes with side length L (e.g. square of side L for the case of two-dimensional space) and to count the number of non-empty boxes. If such a box contains at least one image point, it is called a non-empty box. For a fixed L, let the total number of non-empty boxes be $N(L)$. Several values of L are chosen and the corresponding $N(L)$ are determined. Fractal dimension D is then obtained from the slope of a least square linear fit on the plot $\log(L)$ versus $\log(N(L))$ [5, 64].

Higuchi method. Higuchi's algorithm for the calculation of FD regarded as one of the most effective and stable estimator of FD [5, 60, 64].

Considering $x(1), x(2), x(3), \ldots, x(N)$ as the time sequence to be analyzed, let us construct k new time series x_m^k as

$$x_m^k = \left\{x(m), x(m+k), x(m+2k), \ldots, x\left(m + \left\lfloor\frac{N-m}{k}\right\rfloor k\right)\right\}, \quad \text{for}\, m = 1, 2, \ldots, k$$

where m indicates the initial time value, $k$indicates the discrete time interval between the points and $\lfloor a \rfloor$ means integer part of a. For each of the time series x_m^k, the average length $L_m(k)$ is computed as

$$L_m(k) = \frac{\sum_{i=1}^{\lfloor \frac{N-m}{k} \rfloor} |x(m+ik) - x(m+(i-1)k)|(n-1)}{\lfloor \frac{N-m}{k} \rfloor k} \tag{2.8}$$

where N is the total length of the data sequence x and $(N-1)/\lfloor (N-m)/k \rfloor k$ is a normalization factor. An average length is computed for all time series having the same delay (or scale) k, as the mean of the k lengths $L_m(k)$ for $m = 1, \ldots, k$. This procedure is repeated for each k ranging from 1 to$k_{\max}$, yielding a sum of average lengths $L(k)$ for each k as indicated in (2.8)

$$L(k) = \sum_{m=1}^{k} L_m(k) \tag{2.9}$$

The total average length for scale k, $L(k)$, is proportional to k^{-D}, where D is the FD by Higuchi's method. In the curve of $\ln(L(k))$ versus $\ln(1/k)$, the slope of the least squares linear best fit is the estimate of the FD [65].

Lacunarity. It has been observed by the researchers [5, 50, 64] that FD alone is insufficient for the purpose of discrimination, since two differently appearing surfaces can have the same value of D. To overcome this, Mandelbrot [50] introduced the term called *lacunarity* χ, which quantifies the *denseness* or *compactness* of an image surface. Many definitions of this term have been proposed and the basic idea is to quantify the gaps or lacunae present in the given pattern.

A procedure for calculating Lacunarity is given in [52]. Let $P(m, L)$ be the possibility that there are m points within a box of side L (i.e. square of side L), which is centered about an arbitrary point on the waveform. Then $P(m, L)$ is normalized, as below for all L

$$\sum_{m=1}^{n} P(m, L) = 1 \tag{2.10}$$

where n is the number of possible points within the box of side length L. Let, the image is of size $L_{\max} \times L_{\max}$. If the entire waveform is overlayed with boxes of size L, then the number of boxes with m points inside the box is $(L_{\max}/m)P(m, L)$.

Defining $G(L)$ and $G^2(L)$, Voss [52] defined lacunarity as:

$$G(L) = \sum_{m=1}^{n} mP(m, L) \tag{2.11}$$

$$G^2(L) = \sum_{m=1}^{n} m^2 P(m, L) \tag{2.12}$$

$$\chi(L) = \frac{G^2(L) - [G(L)]^2}{[G(L)]^2} \tag{2.13}$$

Approximate entropy. ApEn is a "*regularity statistic*" that quantifies the unpredictability of fluctuations in a time series [66]. ApEn reflects the likelihood that "similar" patterns of observations will not be followed by additional "similar" observations. *A* time series containing many repetitive patterns has a relatively small ApEn; a less predictable (i.e. more complex) process has a higher ApEn [5, 67].

The algorithm for computing ApEn has been published [66]. Here a brief summary of the calculations, as applied to a time series of impulse current responses $i(t)$of transformer under test has been provided. Given a sequence S_N, consisting of N samples of impulse current response $i(1), i(2), \ldots., i(N)$ the values of two input parameters, m and r has been chosen, to compute the approximate entropy, $\mathrm{ApEn}(S_N, m, r)$, of the sequence. The second of these parameters, m, specifies the pattern length, and the third, r, defines the criterion of similarity. Denoting a subsequence or *pattern* of m samples, beginning at measurement t_0 within S_N, by the vector $p_m(t_0)$, two patterns, $P_m(t_0)$ and $P(j)$, are *similar* if the difference between any pair of corresponding measurements in the patterns is less than r, i.e., $|i(t_0+k) - i(j+k)| < r$ for $0 \le k < m$

Now considering the set P_m of all patterns of length m, i.e., $p_m(1), p_m(2), \ldots, p_m(N = m+1)$,

within S_N, let us define

$$C_{tm}(r) = \frac{n_{tm}(r)}{N - m + 1} \tag{2.14}$$

where $n_{tm}(r)$ is the number of patterns in P_m that are similar to $P_m(t_0)$ (given the similarity criterion r). The quantity $C_{tm}(r)$ is the fraction of patterns of length m that resemble the pattern of the same length that begins at interval t_0. Now after calculating $C_{tm}(r)$ for each pattern in P_m, and defining $C_{tm}(r)$ as the mean of these $C_{tm}(r)$ values. The quantity $C_{tm}(r)$ expresses the prevalence of repetitive patterns of length m in S_N. Finally, the ApEn of S_N, for patterns of length m and similarity criterion r, is defined as:

$$ApEn(S_N, m, r) = \ln\left[\frac{C_m(r)}{C_{m+1}(r)}\right] \tag{2.15}$$

2.4.2.2 Classification Methodologies

Faults in different parts of a transformer winding during impulse tests generate distinct signatures in the time–frequency spectrum. Although discrimination of a large set of such signatures is difficult and some time impossible for the human eye, this complex task can be successfully handled by modern classification algorithm like ANN, fuzzy logic-based classifiers, etc., provided proper features are extracted from the response signals.

De and Chatterjee [33, 68] proposed an ANN approach for detection and diagnosis of fault nature and fault location in oil-filled power transformers during impulse testing. In their work, the "self-organizing feature map" (SOFM) algorithm with Kohonen's learning and learning vector quantization-based classification have been successfully applied to the problem of fault current waveform pattern classification. The studies on impulse fault diagnosis using the above mentioned methods showed that the localization of the faults could be done within 25 % of the winding length.

Purkait et al. [2] described development of an adaptive fuzzy classifier, which can effectively determine various classes or categories of series and shunt faults in a wide range of power transformers. They performed the classification of faults successfully for 33 % of the winding length.

Authors in [69] have applied Fuzzy ARTMAP on frequency response data to localize impulse faults in power transformers. They found that localization of faults can be successfully done within 20 % of the winding length.

In recent years SVM and rough set theory (RST)-based classifiers are also used successfully by the researchers [5, 6, 31].

Overview of Different Classification Methods

Construction of a machine capable of learning from experience has been the object of both philosophical and technical debate for a long time. The technical aspect of the debate has received enormous impetus from the advent of electronic computers. It has been demonstrated that machines can display a significant level of learning ability, though the boundaries of this ability are not clearly defined yet [5, 70]. However, a machine, capable of learning from experience through systematic training, is called 'learning machine'.

It is worth mentioning here that, as ANN and fuzzy logic-based classifiers are being used by the researchers successfully for quite a long time, they are not discussed in this chapter. However, there should not be any doubt that they are indeed very powerful tools for such fault classifications.

In *supervised* learning, the machine is given a *training set* or, *examples* (called *inputs*) with associated labels (called *output/target* values). The inputs are in the form of feature vectors, so that the input space is a subset of $\Re^n$. Once the attribute vectors are available, a number of sets of hypotheses could be chosen for the problem which relates the input and output vectors. Among these, linear functions are the best understood and simplest to apply. Traditional statistics and the classical neural network literature have developed many methods for discriminating between two classes of instances using linear functions, as well as methods for interpolation using linear functions [71].

Linear classification. Linear classification is frequently performed by using a real-valued function $f : X \subseteq R^n \rightarrow R$ in the following way: the input $x = (x_1, \ldots, x_n)$is assigned to the positive class, if $f(x) \geq 0$, and otherwise to the negative class. Consider the case where $f(x)$is a linear function of $x \subseteq X$, so that it can be written as

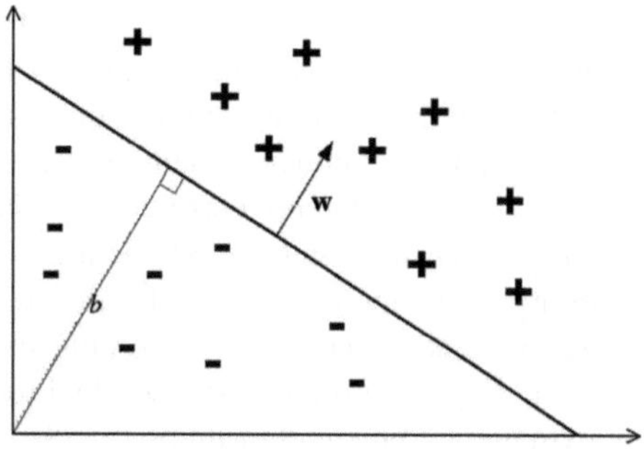

Fig. 2.15 A separating hyperplane (*w*, *b*) for a two-dimensional training set

$$f(x) = \langle w \cdot x \rangle + b = \sum_{i=1}^{n} w_i x_i + b$$

where $(w, b) \in R^n \times R$are the parameters that control the function.

A geometric interpretation of this kind of hypothesis is that the input space X is split into two parts by the hyperplane defined by the equation $\langle w \cdot x \rangle + b = 0$, as shown in Fig. 2.15. The vector w defines a direction perpendicular to the hyperplane, while varying the value of b moves the hyperplane parallel to itself.

The theory of linear discriminants was developed by Fisher in 1936, while neural network researchers studied perceptrons in the early 1960s, mainly due to the work of Rosenblatt [72]. The quantities w and b are the weight vector and bias, terms borrowed from the neural network literature.

Multi-class discrimination. The problem of two-class discrimination studied so far can also be solved by defining a weight vector w_i and a bias b_i for each class. Each time a new instance has to be classified, both functions are evaluated, and the point x is assigned to class 1 if $\langle w_1 \cdot x \rangle + b_1 \geq \langle w_{-1} \cdot x \rangle + b_{-1}$, or to class -1 otherwise. This approach is equivalent to discrimination using the single hyperplane (w, b), with the substitutions $w = w_1 - w_{-1}$, $b = b_1 - b_{-1}$.

For a multi-class classification problem the output domain is $Y = \{1, 2 \ldots m\}$. The generalization of linear learning machines to the m-class case is straightforward: to each of the m-classes are associated a weight vector and a bias, (w_i, b_i), $i \in \{1, \ldots, m\}$, and the decision function is given by

$$c(x) = \arg \max_{1 \leq i \leq m} (\langle w_i \cdot x \rangle + b_i)$$

Geometrically this is equivalent to associating a hyperplane to each class, and to assigning a new point x to the class whose hyperplane is furthest from it. The input space is split into m simply connected and convex regions. Algorithms for learning the m hyperplanes simultaneously from data exist, and are extensions of the basic procedures outlined above [5].

Linear regression. The problem of linear regression consists of finding a linear function $f(x) = \langle w \cdot x \rangle + b$ that best interpolates a given set S of training points labeled from $Y \subseteq R$. Geometrically this corresponds to a hyperplane fitting the given points.

This problem has been studied since the eighteenth century, and the best-known solution is that proposed independently by Gauss and Legendre of choosing the line that minimizes the sum of the squares of the distances from the training points. This technique is known as method of least squares, and is known to be optimal in the case of linear targets corrupted by Gaussian noise.

Support vector machine. SVMs are a system for efficiently training the linear learning machines in the kernel-induced feature spaces. An important feature of these systems is that, while enforcing the learning biases necessary for generalization, they also produce 'sparse' dual representations of the hypothesis, resulting in extremely efficient algorithms [5, 73]. Another important feature of the support vector approach is that due to Mercer's [74] conditions on the kernels the corresponding optimization problems are convex and hence have no local minima.

The aim of support vector classification is to devise a computationally efficient way of learning 'optimum' separating hyperplanes in a high dimensional feature space, where 'optimum' hyperplanes are the ones optimizing the generalization bounds and by 'computationally efficient' means algorithms able to deal with sample sizes of the order of say, 100,000 instances.

Maximal margin classifier: The simplest model of SVM, which was also the first to be introduced, is the so-called maximal margin classifier. It works only for linearly separable data in the n-dimensional feature space, and hence cannot be used in many real-world situations. Nonetheless it is the easiest algorithm to understand, and it forms the main building block for the more complex SVMs.

The maximal margin classifier optimizes the bound of error by separating the data with the maximal margin hyperplane, given that the bound does not depend on the dimension of the space and this separation can be sought in any kernel-induced feature space. The maximal margin classifier finds the maximal margin hyperplane in an appropriately chosen kernel-induced feature space.

This strategy can be implemented by reducing it to a convex optimization problem: minimizing a quadratic function under linear inequality constraints. In case of linear classifiers there is an inherent degree of freedom, due to the fact that the function associated with the hyperplane (w, b) does not change in rescaling the hyperplane to $(\lambda w, \lambda b)$, for $\lambda \in R^+$. However, there will be a change in the margin as measured by the function output as opposed to the geometric margin. The margin of the function output is referred as the functional margin. It is possible to optimize the geometric margin by fixing the functional margin to be equal to 1 and minimizing the norm of the weight vector [73]. It is worth mentioning here that, hyperplanes with functional margin 1 are sometimes known as *canonical hyperplanes.* If w is the weight vector realizing a functional margin of 1 on the positive point x^+ and the negative point x^-, its geometric margin can be computed as follows. Now, a functional margin of 1 implies

$$\langle w \cdot x \rangle + b = +1$$

$$\langle w \cdot x \rangle + b = +1$$

and to compute the geometric margin w must be normalized. The geometric margin γ is then the functional margin of the resulting classifier

$$\begin{aligned}\gamma &= \frac{1}{2}\left(\left\langle \frac{w}{\|w\|_2} \cdot x^+ \right\rangle - \left\langle \frac{w}{\|w\|_2} \cdot x^- \right\rangle\right) \\ &= \frac{1}{2\|w\|_2}(\langle w \cdot x^+ \rangle - \langle w \cdot x^- \rangle) \\ &= \frac{1}{\|w\|_2}\end{aligned}$$

Hence, the resulting geometric margin will be equal to $1/\|w\|_2$ [5, 73, 74].

Support vector regression: The support vector method can also be applied to the case of regression, maintaining all the main features that characterize the maximal margin algorithm: a nonlinear function is learned by a linear learning machine in a kernel-induced feature space while the capacity of the system is controlled by a parameter that does not depend on the dimensionality of the space.

The classification approach is motivated by seeking to optimize the generalization bounds given for regression. Figure 2.16 shows an example of a one dimensional linear regression function with an ε-insensitive band. The variable ζ measure the cost of the errors on the training points. These are zero for all points inside the band. Figure 2.17 shows a nonlinear regression function.

RST-based classification. RST theory was developed by Zdzislaw Pawlak [42, 43, 75–79] in the early 1980s. It deals with the analysis of data tables. The data can be acquired from measurements or from human experts. The main goal of the rough set analysis is to synthesize approximation of concepts from the acquired data. The term *classification* concerns any context in which some decision is taken or a forecast is made on the basis of currently available knowledge or information, as stated earlier.

RST can be applied for classification of data by simplifying a decision table. RST-based rule generation shows that RST is an effective methodology for classification of data patterns where the information system contains imprecise,

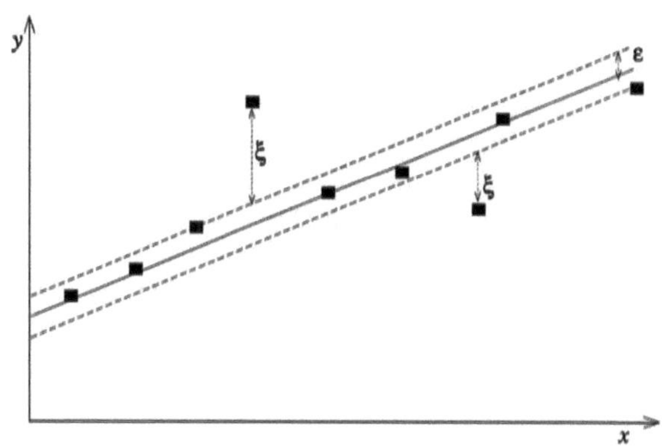

Fig. 2.16 Example of a one dimensional linear regression function with an ε-insensitive band

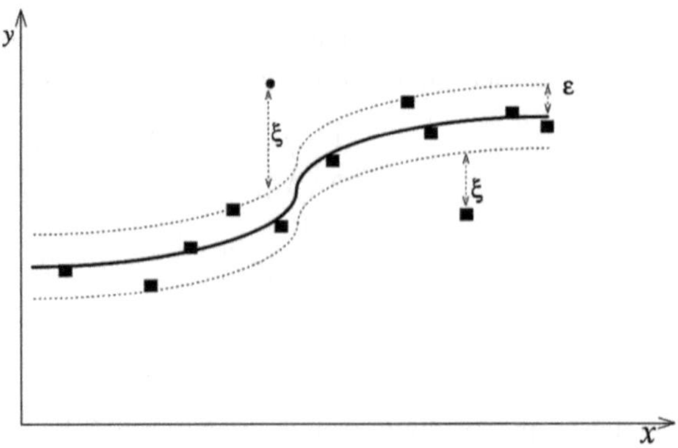

Fig. 2.17 A nonlinear regression function

superfluous and inconsistent data. It is known that prior to classification of data, some features are extracted from the data so that a classification algorithm can classify the data patterns efficiently. Selection of proper features for efficient classification is not an easy task. For example, in the present problem of impulse fault classification significant features may be extracted by using any method explained in the previous sections; but the difficulty of choosing appropriate features is an issue. Improper selection of feature vector may make the data table inconsistent or, it may contain superfluous data requiring unnecessarily higher time of processing. In this context RST-based classification is very effective. That is why RST has successfully been used for these types of problems [42, 43, 75–79].

Overview of RST: Even though details of the RST can be found in [42, 43, 75–79], a brief overview of RST is given in the following paragraphs to understand the classification procedure. In RST, data is presented in a decision table in which each row represents an object (e.g. impulse fault information at different disc position) and each column represents an attribute. For example, in cross-wavelet transform-based feature extraction for impulse fault analysis as explained in [31] the decision table contains extracted features (F_1–F_{12}) as the 12 condition attributes and the fault location and type as the decision attribute.

Mathematically, information system $T = \langle U, Q, V, f \rangle$. Here, U is the finite set of objects and Q is the set of attributes. $V = \bigcup_{q \in Q} V_q$, where V_q is the domain of the values of q and f denotes decision function as, $f : U \times Q \to V$. If the table is having a large number of attribute values, i.e., card(V_q) is very high for some $q \in Q$, then there is a very low chance of classifying a new object by the rules generated directly from the table. Here, card() means cardinality operator, which means "number of elements of a set". Therefore, discretization of the decision table is required for large real-valued decision table. Discretization of a data table indicates some partitioning of the attribute values. In many cases maximal discernible (MD) heuristic may be followed which is discussed in details in [79].

In short, to obtain the decision rules for classification of faults using RST, the following steps are followed:

Step 1: *The data table is discretized.* This means that each condition attribute values are divided into ranges. MD heuristic may be followed which is discussed in details in [79], as stated in the previous section

Step 2: Identical Attributes and cases are eliminated to simplify the table by removing superfluous data

Step 3: Dispensable attributes are removed

Step 4: *Reducts and CORE are obtained.* The sample computation of *Reduct* and *CORE* is available in [42]

Step 5: Decision rules are generated in "IF_THEN" form using the final data table of *CORE* and *Reducts*

An example of RST-based Rule generation for classification [6, 31]: Let us consider Table 2.2 as a data table with three features (condition attributes) and one decision attribute (fault type). Here, 'SE1' indicates the series fault in the winding location L1. Similarly 'SH2' implies shunt fault in location L2 and so on.

In RST, for different attributes, objects are called indiscernible, i.e., similar, if they are characterized by the same information. If $P \subseteq Q$ and $x_i, x_j \in U$, then x_i and x_j are indiscernible with respect to the set of attributes P, if $f(x_i, q) = f(x_j, q), \forall q \in P$. For example, let $P = \{F_1\}$. The part of the discretized decision table given in Table 2.2 shows that, at objects $1, 3, 4$ are indiscernible with attribute F_1, as all these objects are having same value (i.e. value = 7) for feature F_1. Similarly objects 5 and 6 are also indiscernible using F_1 as they possesses same values (i.e. value = 5) for these objects. An *elementary set* is the set of all indiscernible objects. So, for $P \subseteq Q$, an equivalence relation on U, called P-indiscernibility relation is given by I_P. Considering the previous example of $P = \{F_1\}$, it can be found that the P-elementary set includes elementary sets like $\{1, 3, 4\}$, $\{2\}$, and $\{5, 6\}$, because all the elements of each of these sets are having same attribute values for attribute F_1. The equivalence classes of the partition induced by the P-indiscernibility relation are called *information granules*. This is explained later.

For any rough set Y, $\underline{P}Y$ and $\overline{P}Y$ are called P-lower and P-upper approximation of Y and defined as, $\underline{P}Y = \{x \in Y | I_P(x) \subseteq Y\}$ and $\overline{P}Y = \{x \in Y | I_P(x) \cap Y \neq \varphi\}$ respectively. This indiscernibility relation can reduce a decision table. This can be done by keeping only one element of an equivalence class and also keeping those

Table 2.2 Truncated decision table for illustration

Object	Condition attributes			Decision attribute
	F_1	F_2	F_3	
1	7	6	1	SE1
2	3	6	0	SE1
3	7	6	0	SH1
4	7	2	0	SH1
5	5	0	5	SE2
6	5	6	5	SE2

attributes which preserve the indiscernibility relation. In other words, keeping all the information intact and removing the superfluous attributes. Thus obtained *minimal* sets of attributes are called *Reduct*. The CORE is the set of relations occurring in every *Reduct*, i.e., $\mathrm{CORE}(P) = \cap\mathrm{RED}(P)$. Therefore, CORE represents the *most important* part of the knowledge. From the CORE and Reducts one can generate the decision rules. Usually these rules are considered in "IF…THEN" formats. This is illustrated in the following paragraphs.

For a given subset $P \subseteq Q$, an attribute $q \in P$ is dispensable in P if and only if, $I_P = I_{(P-\{q\})}$; otherwise q is indispensable. If every element in P is indispensable then P is called *independent* otherwise *dependent*. Let $P \subseteq Q$ and $D \subseteq Q$ have equivalence relations in U. The *P-positive* region of D is indicated as, $\mathrm{POS}_P(D) = \bigcup_{Y \in I_D} \underline{P}Y$. In other words, it denotes the set of elements that can correctly be classified into D-elementary sets obtained from I_D using the knowledge described by I_P. If $q \in P$ and $\mathrm{POS}_P(D) = \mathrm{POS}_{(P-\{q\})}(D)$ then q is D-dispensable in P, otherwise q is D-indispensable in P. If the set of attributes G $(G \subseteq P)$ is a D-independent in P and$\mathrm{POS}_G(D) = \mathrm{POS}_P(D)$, then G is called D-reduct of P or in general *Reduct* of P.

All these definitions can be explained using Table 2.2. For example, if P is taken as, $P = \{F_1, F_2, F_3\}$, and D = {*'Location and type of fault'*} (i.e. decision attribute), then $I_P = \{1\}, \{2\},\{3\}, \{4\}, \{5\}$ and $\{6\}$; $I_D = \{1, 2\}, \{3, 4\}$, and $\{5, 6\}$. Also, $\mathrm{POS}_P(D) = \{1,2,3,4,5,6\}$. If the attribute F_1 is removed from P then, $\mathrm{POS}_{(P-\{F_1\})}(D) = \{1,4,5,6\}$. Clearly, $\mathrm{POS}_{(P-\{F_1\})}(D) \neq \mathrm{POS}_P(D)$. Therefore the attribute F_1 is D-indispensable in P. Similarly, removing attribute F_2 gives, $\mathrm{POS}_{(P-\{F_2\})}(D) = \{1,2,3,4,5,6\} = \mathrm{POS}_P(D)$. Therefore attribute F_2 is D-dispensable in P. Again, for attribute F_3 it is easy to observe that, $POS_{(P-\{F_3\})}(D) = \{2,4,5,6\} \neq POS_P(D)$. Similarly, F_3 is also D-indispensable in P. Thus, the set $\{F_1, F_3\}$ is the D-reduct of P. Therefore, the simplified or reduced form of Table 2.2 is given in Table 2.3. '–' indicates "don't care" (i.e. dispensable) condition. It can be said that, attribute values $(F_1 = 7 \wedge F_3 = 1) \vee (F_1 = 3 \wedge F_3 = 0)$ are the characteristic for decision class 'SE1'. '$\wedge$' and '$\vee$' are logical "AND" and "OR" operators respectively. Similarly $(F_1 = 7 \wedge F_3 = 0)$ is the characteristic of decision class 'SH1' and $(F_1 = 5 \wedge F_3 = 5)$ is the characteristic of 'SE2'. These are called *information granules*. Intersections of these

Table 2.3 Reduced form of Table 2.2

Object	Condition attributes			Decision attribute
	F_1	F_2	F_3	
1	7	–	1	SE1
2	3	–	0	SE1
3	7	–	0	SH1
4	7	–	0	SH1
5	5	–	5	SE2
6	5	–	5	SE2

Table 2.4 Decision rules obtained from CORE and Reducts

Decision rule no.	Statement of the rule	
	If	Then
1	$(F_1 = 7 \wedge F_3 = 1) \vee (F_1 = 3 \wedge F_3 = 0)$	Location and type of fault is 'SE1'
2	$(F_1 = 7 \wedge F_3 = 0)$	Location and type of fault is 'SH1'
3	$(F_1 = 5 \wedge F_3 = 5)$	Location and type of fault is 'SE2'

reduct values for each of the decision class (i.e. SE1, SH1, and SE2) will give the CORE for the respective class. For the decision class 'SE1' no such CORE value is obtained from Table 2.3, as the intersection of $(F_1 = 7 \wedge F_3 = 1)$ and $(F_1 = 3 \wedge F_3 = 0)$ is null. Similarly, for decision class 'SH1' the intersection of $(F_1 = 7 \wedge F_3 = 0)$ and $(F_1 = 7 \wedge F_3 = 0)$ gives CORE values $F_1 = 7$ and $F_3 = 0$. Again, for 'SE2' the CORE values are $F_1 = 5$ and $F_3 = 5$. Furthermore, this reduced Table 2.3 can be used to generate decision rules. The decision rules obtained from the Reduct and CORE values are given in Table 2.4 [31].

2.4.3 *How to Build an Expert System for Impulse Fault Analysis*

The structure of the designed expert system (ES) has the following functional modules: (a) user interface (UI), (b) knowledge base: consisting of the data-base and the rule-bases, (c) inference engine, (d) explanation module, and (e) updating module.

The different functional modules of the ES are interrelated with each other [80].

The necessary input data required for the data-base may be the voltage and current wave records at reduced impulse voltage level and at full impulse voltage level (BIL), respectively. The ES stores and retrieves data from the data-base before the actual fault detection analysis begins. After the analysis, the ES comes out with a decision regarding possible fault and its location in the winding.

With this UI, the user can ask the ES to identify and locate the fault following the rules of fault diagnosis using the combined time and frequency domain knowledge bases. The ES may also have an explanation module, which displays the decision tree to the user to provide transparency in decision making to the user.

For fault diagnosis, the waves to be analyzed are added to the database of the ES via the UI. Then the UI interacts with the user to read the transformer rating and invokes the inference engine which accordingly determines the necessary diagnostic indices. The fault diagnosis algorithm follows an IF_THEN_ELSE structure.

It combines the heuristic rules framed on the basis of the acquired knowledge and conventional mathematical pattern recognition parameters to identify the fault current waveform pattern to arrive at the conclusion regarding the type and

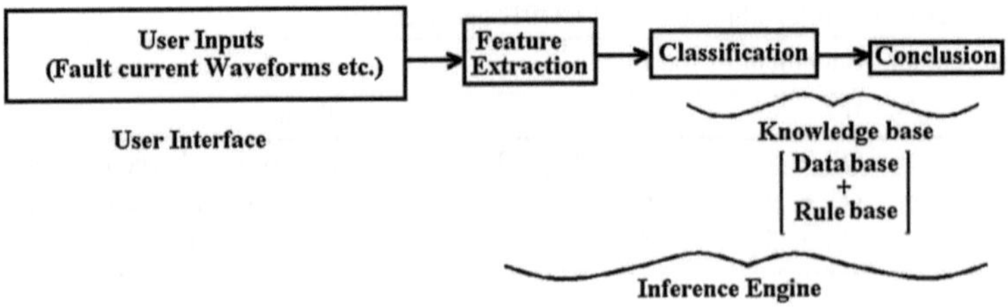

Fig. 2.18 Schematic of ES

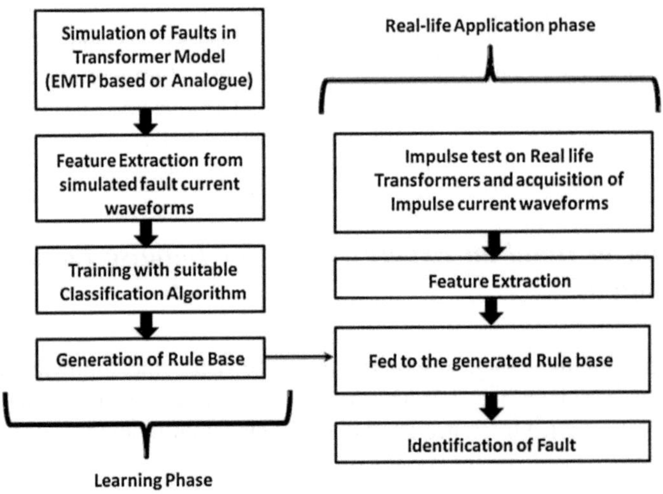

Fig. 2.19 Logical Steps for ES operation

location of the fault. A typical schematic for ES operation is given in Fig. 2.18. The basic logical steps of operation of the ES are given in Fig. 2.19. The ES can be developed based on these figures in any high level languages [80, 81].

2.5 Recent Challenges and Research Directions

In most of the cases referred previously, the fault within the transformer winding is considered as a *static* fault. The faults created due to the defects that are present in insulation prior to the application of impulse sequences and do not vary with the propagation of the impulse wave through the winding, are named as static faults. They can be assumed as the permanent faults. However, in many practical cases there exist faults which are not static. The fault that is developed and varying during the propagation of impulse voltage wave along the winding is termed as *dynamic* fault. Moreover, there may be more than one fault in the winding. Identification of such faults are difficult than that for the static faults, i.e., simultaneous occurrence of faults of different topologies (static and dynamic). The dynamics of such fault conditions are complex. Therefore, there are growing interests among the researchers for identification of such faults. Authors in [82] have reported a scheme using cross-wavelet transform-based feature extraction and wavelet network-based classification for such problems. However, there is ample scope for the researchers to contribute in this direction.

Acknowledgments The authors acknowledge with thanks the immense contribution of Dr. Saibal Chatterjee, Prof. Prithwiraj Purkait, Dr. Chiranjib Koley and Mr. P. Rajamani towards the completion of this chapter. Most of the results discussed in this chapter are the outcome of their experimentation and research.

References

1. Guide to the lightning and switching impulse testing of power transformers and reactors. IEC Standard, Publication 60076-4, 2002
2. Purkait P, Chatterjee A, Chakravorti S, Bhattacharya K (2003) Translationally adaptive fuzzy classifier for transformer impulse fault identification. IEE Proc Gen Transm Distrib 150(1):33–40
3. Hagenguth JH (1944) Progress in impulse testing of transformers. AIEE Trans 63:999–1005
4. Mamishev AV (1996) Analysis of high impedance faults using fractal technique. IEEE Trans Power Syst 11(1):435–440
5. Koley C (2007) Advanced techniques for identification of faults in transformers due to impulse stresses. Ph. D Thesis, Faculty of Engineering and Technology, Jadavpur University
6. Dey D (2009) Studies on intelligent digital signal processing tools for measurements in high voltage systems. Ph. D Thesis, Faculty of Engineering and Technology,Jadavpur University
7. Stenkvist E (1952) Study of fault detection and failure location during surge testing of transformers. CIGRE 1952, Supplement to paper no. 12-125
8. Satish L (1998) Short-time Fourier and wavelet transforms for fault detection in power transformers during impulse tests. IEE Proc Sci Meas Technol 145(2):77–84
9. Purkait P, Chakravorti S (2002) Time and frequency domain analyses based expert system for impulse fault diagnosis in transformers. IEEE Trans Electr Insul 9(3):433–445
10. Zhang Y, Ding X (1996) An artificial neural network approach to transformer fault diagnosis. IEEE Trans Power Deliv 11(5):1836–1841
11. Roy CK, Biswas JR (1994) Studies on impulse behavior of a transformer winding with simulated faults by analogue modeling. IEE Proc C 141:401–412
12. Abetti PA (1958) Bibliography on the surge performance of transformers and rotating machines. Trans AIEE 77:1150–1168
13. Malewski R, Poulin B (1988) Impulse testing of power transformers using the transfer function method. IEEE Trans Power Deliv 3:476–490
14. Vaessen PTM, Hanique E (1992) A new frequency response analysis method for power transformers. IEEE Trans Power Deliv 7:384–390
15. Abetti PA (1953) Transformer models for the determination of transient voltages. Trans. AIEE, pp. 468–480
16. Vaessen PTM (1988) Transformer model for high frequencies. IEEE Trans Power Deliv 3:1761–1768

17. de Leon F, Semlyen A (1994) Complete transformer model for electromagnetic transients. IEEE Trans Power Deliv 9:231–239
18. Gharehpetian GB, Mohseni H, Moller K (1998) Hybrid modeling of inhomogeneous transformer winding for very fast transient over-voltage studies. IEEE Trans Power Deliv 13:157–163
19. Degeneff RC (1978) A method for constructing terminal models for single-phase n-winding transformers. In: Proceedings of IEEE PES Summer Meeting, paper no. A 78 539-9
20. Morched A, Marti L, Ottevangers J (1993) High frequency transformer model for the EMTP. IEEE Trans Power Deliv 8:1615–1626
21. Fergestad PI, Henriksen T (1974) Inductance for the calculation of transient oscillation in transformers. IEEE Trans PAS, vol. PAS-93, pp. 510–517
22. Chatterjee S (1994) Analog studies on impulse performance of a H.V. transformer winding with tapping coils. M.E.E Thesis, Jadavpur University, Kolkata
23. Grover FW (1973) Inductance calculations. ISA Publication, New York
24. Maxwell JC (1904) A treatise on electricity and magnetism, 3rd edn, vol. II. Clarendon Press, Oxford
25. Hagenguth JH, Meador JR (1952) Impulse testing of power transformers. AIEE Trans 71:697–704
26. Wang M, Vandermaar AJ, Srivastava KD (1999) Condition monitoring of transformers in service by the low voltage impulse test method. In: Eleventh international symposium on high voltage engineering (Conf. Publ. No. 467), vol 1, pp 45–48
27. Purkait P, Chakravorti S (2001) Pattern classification of impulse faults in transformers by wavelet analysis. IEEE Trans Electr Insul 9(4):555–561
28. Purkait P, Chakravorti S (2003) Wavelet transform based impulse fault pattern recognition in distribution transformers. IEEE Trans Power Deliv 18(4):1588–1589
29. Koley C, Purkait P, Chakravorti S (2006) Wavelet aided SVM tool for impulse fault identification in transformers. IEEE Trans Power Deliv 21(3):1283–1290
30. Purkait P, Chakravorti S (2003) Impulse fault classification in transformers by fractal analysis. IEEE Trans Dielectr Electr Insul 10(12):109–116
31. Dey D, Chatterjee B, Chakravorti S, Munshi S (2008) Rough-granular approach for impulse fault classification of transformers using cross-wavelet transform. IEEE Trans Dielectr Electr Insul 15(5):1297–1304
32. Purkait P, Chakravorti S (2004) Can fractal techniques be used for analysis of impulse fault patterns in distribution transformers. Int J Electr Power Syst Res (Elsevier) 68:258–267
33. De A, Chatterjee N (2002) Recognition of impulse fault patterns in transformers using Kohonen's self-organizing feature map. IEEE Trans Power Deliv 17(2):489–494
34. Mathworks, Inc. Matlab Toolbox, http://www.mathworks.com/access/helpdesk/help/toolbox/wavelet/wavelet.html
35. Stark HG (2005) Wavelets and signal processing, 1st edn. Springer, Berlin
36. Daubechies I (1990) The wavelet transform, time-frequency localization and signal analysis. IEEE Trans Inf Theory 36:961–1005
37. Ruessink BG, Coco G, Ranasinghe R, Turner IL (2006) A cross-wavelet study of alongshore non-uniform near shore sandbar behavior. In: Proceedings of joint conference on neural networks, pp. 4310–4317, Vancouver, Canada
38. Mallat S (1999) A wavelet tour of signal processing, 2nd edn. Elsevier, Amsterdam
39. Plett MI (2007) Transient detection with cross wavelet transforms and wavelet coherence. IEEE Trans. Signal Processing 55(5):1605–1611
40. Torrence C, Compo GP (1998) A practical guide to wavelet analysis. J Am Meteorol Soc 79(1):61–79
41. Li X, Yao X, Jefferys JRG, Fox J (2005) Computational neuronal oscillations using Morlet wavelet transform. In: Proceedings of IEEE Annual Conference on Engineering in Medicine and Biology, pp. 2009–2012, Shanghai, China
42. Pawlak Z (1991) Rough sets: theoretical aspects of reasoning about data. Kluwer, Boston

43. Peng JT, Chien CF, Tseng TLB (2004) Rough set theory for data mining for fault diagnosis on distribution feeder. IEE Proc Gen Transm Distrib 151:689–697
44. Rajamani P, Chakravorti S (2012) Identification of simultaneously occurring dynamic disc-to-disc insulation failures in transformer winding under impulse excitation. IEEE Trans Dielectr Electr Insul 19(2):443–453
45. Dey D, Chatterjee B, Chakravorti S, Munshi S (2010) Cross-wavelet transform as a new paradigm for feature extraction from noisy partial discharge pulses. IEEE Trans Dielectr Electr Insul 17(1):157–166
46. Chandaka S, Chatterjee A, Munshi S (2009) Cross-correlation aided support vector machine classifier for classification of EEG signals. Expert Syste Appl 36:1329–1336
47. Chandaka S, Chatterjee A, Munshi S (2009) Support vector machine employing cross-correlation for emotional speech recognition. Measurements 42:611–618
48. Rajamani P, Dey D, Chakravorti S (2011) Cross-correlation aided wavelet network for classification of dynamic insulation failures in transformer winding during impulse testing. IEEE Trans Dielectr Electr Insul 18:521–532
49. Rajamani P, Dey D, Chakravorti S (2010) Cross-correlation aided fuzzy c-means for classification of dynamic fault in transformer winding during impulse testing. Electric Power Compon Syst 38(13):1513–1530
50. Mandelbrot B (1983) The fractal geometry of nature. Freeman, New York
51. Lovejoy S, Schertzer D (1986) Scale invariance, symmetries, fractals and stochastic simulations of atmospheric phenomena. Bull Am Meterol Soc 67(1):21–32
52. Voss RF (1985) Random fractals: characterisation and measurement. In: Pynn R, Skjeltrop A (eds) Scaling phenomena in disordered systems. Plenum press, New York, pp. 1–11
53. Keller JM, Chen S, Crownover RM (1989) Texture description and segmentation through fractal geometry. Comput Vis Graph Image Process 45:150–166
54. Niemeyer L, Pietronero L, Wisemann HJ (1984) Fractal dimension of dielectric breakdown. Phys Rev Lett 6(6):661–674
55. Fuji M, Watanabe M, Kitani I, Arii K, Yoshino K (1991) Fractal character of dc trees in polymethylmethacrylate. IEEE Trans Electr Insul 26:159–162
56. Satish L, Zaengl WS (1995) Can fractal features be used for recognizing 3-D partial discharge patterns? IEEE Trans Dielectr Electr Insul 2:352–359
57. Candela R, Mirelli G, Schifani R (2000) PD recognition by means of statistical and fractal parameters and a neural network. IEEE Trans. Dielectr Electr Insul 7(1):87–92
58. Sevcik C (1998) A procedure to estimate the fractal dimension of waveforms. Complex Int J 5. http://www.complexity.org.au/ci_louise/vol05/sevcik/sevcik.html
59. Esteller R, Vachtsevanos G, Echauz J, Litt B (2001) A comparison of waveform fractal dimension algorithms. IEEE Trans Circuits Syst I 48(2):177–183
60. Katz MJ (1988) Fractals and the analysis of waveforms. Comput Biol Med 18:145
61. Vanouplines P. Rescaled range analysis and the fractal dimension of pi. [Online]. Available: http://homepages.vub.ac.be/-pvouplin/pi/compfdim.htm
62. Krivda A, Gulski E, Satish L, Zaengl WS (1995) The Use of Fractal Features for Recognition of 3-D Discharge Patterns. IEEE Trans Electr Insul 2(5):889–892
63. Ganapathiraju A, Hamaker JH, Picone J (2004) Applications of support vector machines to speech recognition. IEEE Trans Signal Process 52(8):2348–2355
64. Koley C, Purkait P, Chakravorti S (2007) SVM classifier for impulse fault identification in transformers using fractal features. IEEE Trans Dielectr Electr Insul 14(6):1538–1547
65. Wong KL (2004) Application of very-high-frequency (VHF) method to ceramic insulators. IEEE Trans Electr Insul 11(6):1057–1064
66. Ho KKL, Moody GB, Peng CK, Mietus JE, Larson MG, Levy D, Goldberger AL (1997) Predicting survival in heart failure case and control subjects by use of fully automated methods for deriving nonlinear and conventional indices of heart rate dynamics. Circulation 96(3):842–848
67. Pincus SM (1991) Approximate entropy as a measure of system complexity. Proc Natl Acad Sci USA 88:2297–2301

68. De A, Chatterjee N (2001) Impulse fault diagnosis in power transformers using self-organising map and learning vector quantisation. IEE Proc Gen Transm Distrib 148(5):397–405
69. De A, Chatterjee N (2004) A fuzzy ARTMAP fault classifier for impulse testing of power transformer. IEEE Trans Electr Insul 11(6):1026–1036
70. Penrose R (2002) The emperor's new mind. Oxford University Press, Oxford
71. Anthony M, Biggs N (1992) Computational Learning Theory, vol 30. Cambridge tracts in theoretical computer science. Cambridge University Press, Cambridge
72. Rosenblatt F (1959) The perceptron: a probabilistic model for information storage and organization in the brain. Psychol Rev 65:386–408
73. Cristianini N, Shawe-Taylor J (2000) Support vector machines. Cambridge University Press, Cambridge
74. Mercer J (1909) Functions of positive and negative type and their connection with the theory of integral equations. Phil Trans R Soc London, A 209:415–446
75. Cabral JE, Pinto JP, Gontijo EM, Filho JR (2004) Fraud detection in electrical energy consumers using rough sets. In: Proceedings of IEEE Conference on Systems, Man and Cybernetics, pp 3625–3629
76. Kusiak A (2001) Rough set theory: a data mining tool for semiconductor manufacturing. IEEE Trans Electron Packag Manufact 24:44–50
77. Pal SK, Mitra P (2004) Case generation using rough sets with fuzzy representation. IEEE Trans Knowl Data Eng 16:292–300
78. Yao YY (1999) Rough sets, neighborhood systems and granular computing. In: Proceedings of IEEE conference on electrical and computer engineering, pp. 1553–1558, Canada
79. Nguyen HS (1997) Discretization of real value attributes: a boolean reasoning approach. PhD thesis, Dept. of Mathematics, Warsaw Univ., Poland
80. Purkait P (2002) Artificial intelligence technique based impulse fault diagnosis in transformers. Ph.D thesis, Faculty of Engineering and Technology, Jadavpur University
81. Purkait P, Chakravorti S (2003) Investigations on the usefulness of an expert system for impulse fault analysis in distribution transformers. Electric Power Syst Res, Elsevier 65(2):149–157
82. Rajamani P, Chakravorti S (2012) Identification of simultaneously occurring dynamic disc-to-disc insulation failures in transformer winding under impulse excitation. IEEE Trans Dielectr Electr Insul 19(2):443–453

Chapter 3
Partial Discharge Measurement and Analysis

3.1 Introduction

With the increased requirement of reliability of high voltage equipment in the deregulated power market, diagnosing health of insulation has assumed very high importance. Partial Discharge (PD) in the insulation of electrical equipment is a sign of dielectric defects as well as a cause of further degradation of its insulation system, which may ultimately lead to failure of the apparatus. Therefore, early detection of PD sources may prevent failures and hence save revenue loss due to damage and/or interruption in service. PD detection has gained increased significance with more and more use of dry type equipment in power sector, for which PD is now believed to a major cause of failure. Thus PD detection, measurement, and classification constitute an important tool for quality assessment of insulation systems utilized in HV power system.

PD phenomenon is manifested in various measurable quantities such as chemical, electrical, mechanical, optical, electromagnetic energy, etc. Modern PD detection systems use either one or a combination of these measurable quantities, to identify as well as to locate the PD with the help of signal processing and machine learning algorithm.

This chapter discusses different diagnostic techniques by which partial discharge detection and localization could be performed not only in laboratory condition but also on-site.

3.2 What is Partial Discharge?

According to IEC 60270:2000 [1], partial discharge is a localized electrical discharge that only partially bridges the insulation between conductors and which can or cannot occur adjacent to a conductor.

In general, partial discharges occur when the electric field intensity at a certain location in the insulation or on the surface of the insulation exceeds the dielectric strength of the insulation as a consequence of local electric field concentration.

S. Chakravorti et al., *Recent Trends in the Condition Monitoring of Transformers*, Power Systems, DOI: 10.1007/978-1-4471-5550-8_3,

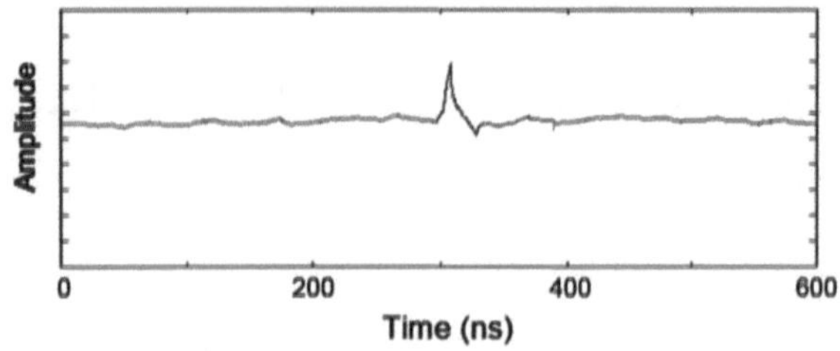

Fig. 3.1 Typical PD pulse

Generally, such discharges appear as pulses having duration of much less than 1 μs. With the availability of GHz bandwidth oscilloscopes, it is now generally agreed that PD pulses may have rise-times as short as 1–2 ns, which should, however, not be taken to infer that all partial discharges do have such rapid rise-times. In contrast to the short rise-times characterizing the pulses generated by spark discharges, glow discharges exhibit typically longer rise-times. A typical PD pulse is shown in Fig. 3.1

In this context it would be rather inappropriate to omit 'corona' from the discussion. The term 'corona' has been ingrained over many years in the terminology of discharge detection and its replacement by the term 'partial discharge' is only relatively recent. According to IEC definition, corona is a form of partial discharge that occurs in gaseous media around conductors which are remote from solid or liquid insulation. Thus, it may be stated that all coronas are partial discharges, but all partial discharges are not coronas.

PD sources within an insulation could be of various types. Typical examples are (i) voids in insulation, (ii) delamination at insulation interfaces, (iii) cracking/fissures in insulation, (iv) metallic protrusions, (v) metal particles to name a few. The technological reasons behind PD sources could be contamination, manufacturing errors, incorrect design, etc. Another major reason for appearance of PD in insulation is aging due to electrical, thermal, and mechanical stresses in operation.

In a broader sense, there are six types of PD, as follows: (a) Corona discharges occurring at sharp points on the electrodes; (b) surface discharges occurring in gases or in liquid if there is a strong tangential component of electric field on the dielectric surface; (c) internal discharges occurring in gas-filled cavities, but oil-filled cavities can also break down and cause gaseous discharges afterwards; (d) electric trees starting from sharp conducting particles or from a cavity in solid insulation; (e) floating part discharges occurring in the cases of poorly grounded components in or near a HV circuit and (f) contact noise occurring in the cases of bad contacts or poor grounding.

3.2.1 *Stress Mechanisms Activated by PD*

Several types of stresses act on insulation due to partial discharge which causes slow degradation of insulation which often go unnoticed. The stresses on insulation that are caused by partial discharges could be classified into following categories:

Electrical Stresses: (a) charges generated and deposited due to partial discharge may cause a space charge field that could act in addition to the applied electric field resulting into further aggravation of local field concentration; (b) PD-enhanced sites may experience local electronic breakdown.

Thermal Stresses: (a) partial discharge injects energy into the surrounding environment of its occurrence causing local increase in temperature; (b) possibility of restructuring of chemical bonds due to increased temperature; (c) gas molecules may migrate along the thermal gradient.

Particle Impact Stresses: (a) particles may strike surfaces at both the ends of a discharge channel with considerable velocity causing degradation of the surface; (b) particles may contain enough energy for bond scission, which is particularly important for cellulose insulation; (c) chemical changes may take place due to particle impact, e.g., cross-linking in polymers making materials stiffer.

Chemical Stresses: (a) changes in chemical structure are caused by particle impact, thermal, and mechanical stresses; (b) possibility of exothermic reactions that may cause damage to solid insulation.

Mechanical Stresses: (a) vibration forces may be generated due to interaction of charges caused by PD trapped in the solid structure with the applied electric field; (b) attraction and repulsion between trapped charges may cause local mechanical stresses.

Degradation of insulating material is caused by synergetic interaction of all the above-mentioned stresses caused by PD.

3.2.2 *PD Inception and Extinction in Closed Cavity*

In an effort to have a better understanding of PD phenomenon, this subsection presents a simplistic explanation of the processes that are possibly involved in PD inception and extinction in closed cavities.

3.2.2.1 PD Inception in Void

For simplicity let there be a rectangular void in between two parallel plate electrodes in a 2D system as shown in Fig. 3.2. Then the potential distribution will be capacitive in nature and the potential that will appear across the void will depend on its dimension w.r.t. the whole arrangement as shown in Fig. 3.3. The field intensity within the void will be given by $E_{void} = \frac{V_2}{d_1}$. If the field intensity exceeds the breakdown strength of the gas in the void, then partial discharge will take place within the void. The voltage V at which PD starts within the void is known as PD

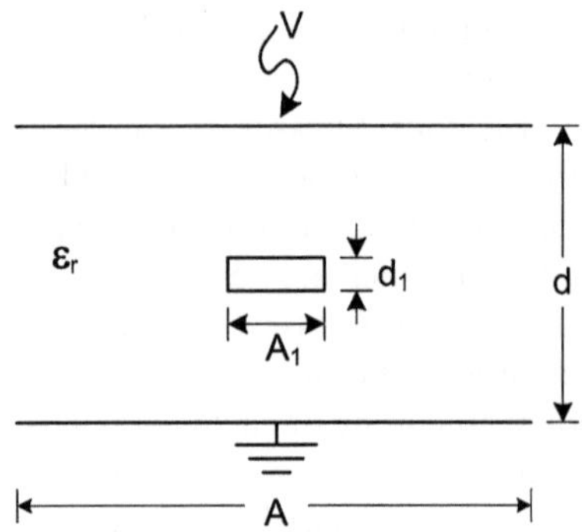

Fig. 3.2 Void in parallel plate arrangement

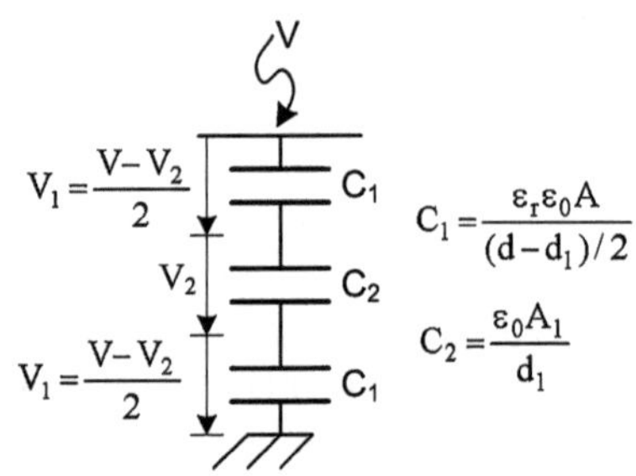

Fig. 3.3 Equivalent capacitive network

inception voltage. The voids are generally filled with gas, mostly air. Breakdown strength of air at STP is 3 kV_p/mm. However, this breakdown strength is a function of the gas pressure as shown in Fig. 3.4.

At a pressure below 1 atm up to Pm, the breakdown strength of air is less than 3 kVp/mm. Since, the pressure within the void is in general slightly less than 1 atm, the breakdown strength of gas in void can be assumed to be anything between 2.5 and 3 kV_p/mm for the analysis of partial discharges.

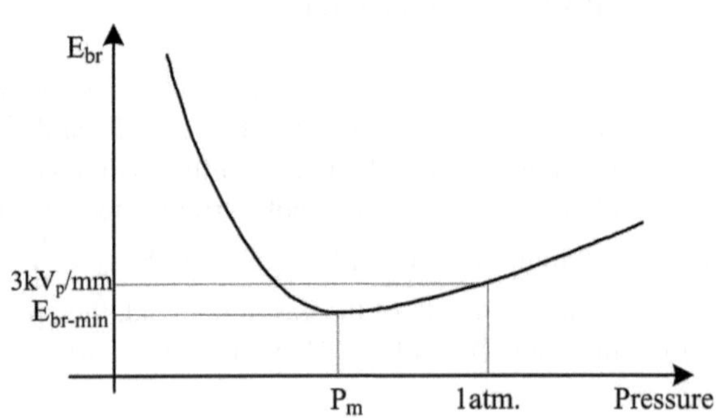

Fig. 3.4 Breakdown strength of air as a function of pressure

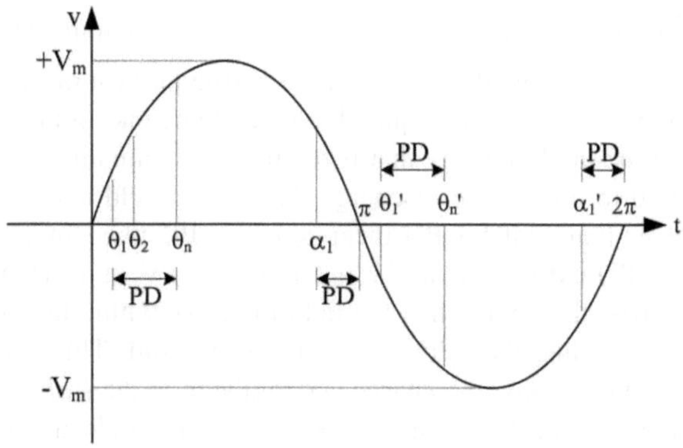

Fig. 3.5 Sinusoidal applied voltage

3.2.2.2 Generation of PD Pulse Under Sinusoidal Applied Voltage

In the case of sinusoidal applied voltage, the instantaneous voltage between the plates increases with time from the zero crossing of the positive half-cycle. Say, the instantaneous voltage corresponding to the phase angle θ_1 as shown in Fig. 3.5 is such that the field intensity within the void exceeds the breakdown strength of gas in void and PD starts within the void. The positive and negative charges generated by PD will move toward the void surfaces according to the direction of the applied field. If the void surfaces are considered to be perfectly insulating, then the charges will not be able to move after reaching the void surface and will get accumulated there as shown in Fig. 3.6. Say, this charge is q and the capacitance of the void is C_2. So, a voltage due to the accumulated charges will appear across the void given by $V_3 = \frac{q}{C_2}$ and it acts in opposition to V_2 of Fig. 3.3. Hence, the voltage across the void becomes $(V_2 - V_3)$ and the field intensity within the void

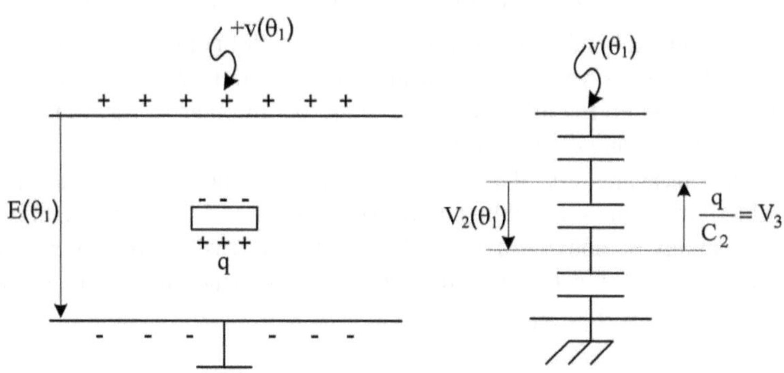

Fig. 3.6 Charge accumulation across the void due to PD at the phase angle θ_1

becomes $\frac{V_2 - V_3}{d_1}$. If the rate of accumulation of charge within the void is higher than the rate of rise of applied voltage, which is true in most of the cases, the field intensity within the void decreases quickly due to charge accumulation. Again, it is well-known that the field intensity that is required to maintain a discharge is lower than that required to initiate the discharge. Hence the field intensity below which PD stops is lower than that for PD inception. For PD phenomenon in voids, the field intensity at PD extinction can be taken as 1–2 kV_p/mm. Due to fast charge accumulation across the void, when the field intensity within the void goes below this PD extinction value, then PD stops within the void. This generates one PD pulse at phase angle θ_1. Say, the charge accumulated at the surfaces of the void is q_1 when the first PD pulse is extinguished. Since, the dielectric surrounding the void is assumed to be a perfect insulator; hence this charge will remain at the surface of the void. So, the next PD pulse will occur corresponding to the phase angle θ_2, for which the difference between the voltage across the void due to applied sinusoidal voltage and the opposing voltage across the void due to accumulated charge at the extinction of the first PD pulse causes a field intensity in the void in excess of the PD inception strength. Thus, the condition for the inception of second PD pulse is as follows:

$$E_{void}(\theta_2) = \frac{v(\theta_2) - \frac{q_1}{C_2}}{d_1} > \text{PD inception strength.} \qquad (3.1)$$

In effect, the second PD pulse occurs at a voltage higher than that required for the first PD pulse. At the extinction of the second PD pulse, the total charge across the void will be q_2, where $q_2 = q_1 + q_2'$ and q_2' = charge accumulated due to the second PD pulse only. In this way, PD pulses will occur at increasingly higher voltages and the accumulated charge across the void will go on increasing. After a certain phase angle θ_n before the positive peak of the sinusoidal applied voltage, the rate of rise of applied voltage becomes small. Say, the total charge accumulated across the void at the extinction of the PD pulse corresponding to the phase angle θ_n is q_n. Then, the voltage across the void acting in opposition to the applied voltage will be $V_n = \frac{q_n}{C_2}$. After the phase angle θ_n, due to lower rate of rise of voltage, the applied voltage between the two plates will not be sufficient enough to cause a field intensity within the void in excess of PD inception strength. Therefore, there will be no PD after the phase angle θ_n. In other words, PD will occur between the phase angles θ_1 and θ_n.

After the phase angle θ_n, the total charge q_n across the void remains trapped as the conductivity of the surrounding dielectric is assumed to be zero. Then, between the phase angles θ_n and α_1 of Fig. 3.5, the field intensity within the void will always be less than the PD inception strength and hence there will be no PD. Therefore, for $\theta_n < \theta < \alpha_1$,

$$E_{void} = \frac{v(\theta) - \frac{q_n}{C_2}}{d_1} < \text{PD inception strength} \qquad (3.2)$$

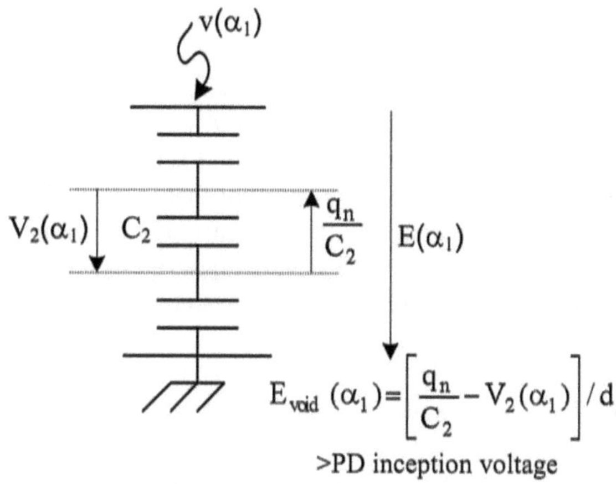

Fig. 3.7 PD inception at $\theta = \alpha_1$ due to accumulated charge

But, at α_1, the applied voltage between the plates becomes too small. Hence, the difference between the voltage across the void due to q_n and the voltage across the void due to the instantaneous applied voltage at α_1 will be large enough to cause field intensity within the void in excess of PD inception strength as shown in Fig. 3.7. However, in this case the voltage across the void due to accumulated charge q_n will be higher than that due to applied voltage. Hence, the field intensity within the void will act in opposite direction to that due to the applied field. Thus, PD will again occur in the void but in such a way that positive and negative charges will start accumulating at the upper and lower surfaces of the void, respectively, as shown in Fig. 3.8. These charges will cause reduction of the total accumulated charge across the void, which in turn will cause extinction of this PD pulse. At the extinction of the PD pulse at the phase angle α_1, the total charge across the void will be $q_1'' = q_n - q_1'$, where q_1' is the charge due to the PD pulse

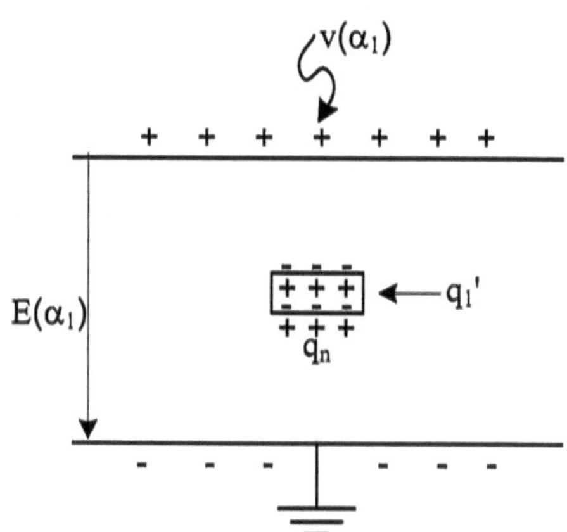

Fig. 3.8 Charges at phase angle α_1 causing reduction of accumulated charge

at α_1. In this way there will be PD pulses between α_1 and π which will gradually reduce the charge accumulated across the void to zero for all practical purposes.

Then in the negative half-cycle, again PD will take place between the phase angle θ'_1 and θ'_n and then between α'_1 and 2π. Such PD pulses will then reappear in the subsequent cycles of the applied sinusoidal voltage.

However, it is to be noted here that in real-life PD pulses do not follow such deterministic pattern. On the contrary, PD phenomenon is stochastic in nature due to the following: (i) the electric field at which PD inception occurs (known as critical field for PD inception) is probabilistic and (ii) the electric field at which PD extinguishes (known as residual field) is also probabilistic in nature. Moreover, the void surfaces are not perfectly insulating, particularly under the action of the discharges. Depending upon the byproducts of discharge that are generated within the cavity for different dielectric media surrounding the void, the PD pattern changes from one solid dielectric to another.

3.3 Partial Discharge Measurement

The occurrence of partial discharges in electrical equipment was recognized by researchers even at the beginning of the twentieth century [2]. At that time, the presence of PD was usually detected either visually or audibly for discharges of sufficiently high intensities that were either directly visible or located relatively near to the observer. Detection of discharges in terms of their influence on the dissipation factor ($\tan\delta$) value became possible following the development of the Schering Bridge [3] in the 1920s. But it was not until the development of the cathode ray tube and the oscilloscope that the uncertainties associated with the partial discharges were understood properly. In 1927, Lloyd and Starr [4] reported their observations of the voltage waveforms in the case of wire-ground plane geometry using the oscilloscope for the first time and in 1932 Gemant and von Philippoff [5] reported similar observations for discharging spark gaps. Major advancement in the detection of discharges came with the development of circuits suitable for the detection of internal discharges in high voltage equipment, e.g., in transformers by Tykociner et al. [6] in 1939, in cables by Quinn [7] in 1940, in bushings by Paine and Brown [8] in 1940, and also in other insulation systems in early 1940s [9]. In 1953, Mason [10] demonstrated that discharges may have pronounced detrimental effects upon solid insulation.

It is now well-known that PD signal manifest itself in many physical forms. The early PD detection systems were based on measurement of electrical signals and employed analog instrumentation. The target then was to measure discharge inception and extinction voltages with the PD pulse patterns displayed on the CRO with a power frequency time base and calibrated ordinate scale. In those cases the charge transfers associated with the discrete PD pulses could be estimated visually and the phase relationship between the pulses and the applied voltage could only be noted approximately. Development of crystal controlled pulse counters in 1950s made it possible to count the PD pulse per unit time [11] and to determine the pulse density of discharge patterns [12]. Differential pulse height analyzers [13] and single channel pulse height analyzers [14] were developed soon thereafter. PD pulse height distribution analysis [15], the area of discharge pulse interval and discharge pulse phase distribution measurements were developed rapidly thereafter in the 1970s [16].

Considering the short rise-time of PD pulses, detection of such pulses at bandwidths up to 1 GHz is suitable. However, it should also be kept in mind that the energy content of a PD pulse is inversely related to frequency. Commercially available conventional PD detectors for conducting routine tests on cables, capacitors, and transformers have narrow bandwidth of 30–400 kHz. These detectors are charge integrating devices and may be calibrated directly to provide the charge transfers associated with detected PD pulses. Research-related work, where faithful reproduction of the PD pulse shapes is of utmost importance, warrants use of higher bandwidths for PD detection. In order to get improved pulse resolution, wider bandwidths are employed for PD site location in cables ($\sim$20 MHz), rotating machines (800 kHz–1 GHz) and bus ducts as well as compressed gas cables and GIS ($\sim$1 GHz). Partial discharge measurements on transformers are carried out using narrow band (40–250 kHz) detectors, in accordance with the IEC Specifications [1]. Narrowband detection is to be preferred due to internal resonances between the windings of transformers, which tend to predominate below 200 kHz.

In 1990s, fast response digital circuits were introduced for PD measurement [17]. Since then the use of digital techniques in PD pulse measurement and related data acquisition has been growing markedly. In this respect it should be mentioned that the peak PD pulse magnitude determined by the digital system will not generally be the same as the true magnitude determined by the analog circuit in real time because of dependence of the digital system upon its sampling rate, bandwidth, and storage capacity. The variety of digital circuits, available and in-use for PD measurements resulted in the publication of a paper by the IEEE Committee on Digital Measurement Techniques [18] and a subsequent paper with invited discussions by experts on PD measurement [19].

Over the past decade, there is a growing trend for non-destructive PD measurement for predictive diagnostic tests under on-site condition. In such cases electrical method of PD measurement has been found to be not so useful. Based on the physical manifestation of PD and availability of acoustic and ultra-high frequency sensors, it is now possible to perform such noninvasive predictive diagnostics tests on-site. Different methods of PD measurement are shown schematically in Fig. 3.9.

In the light of present day requirement, the goals of PD measurement are as follows: (a) Pass/fail criterion during acceptance test, (b) Detection of defects, (c) Identification of defects, (d) Localization of defects, and (e) Risk Assessment.

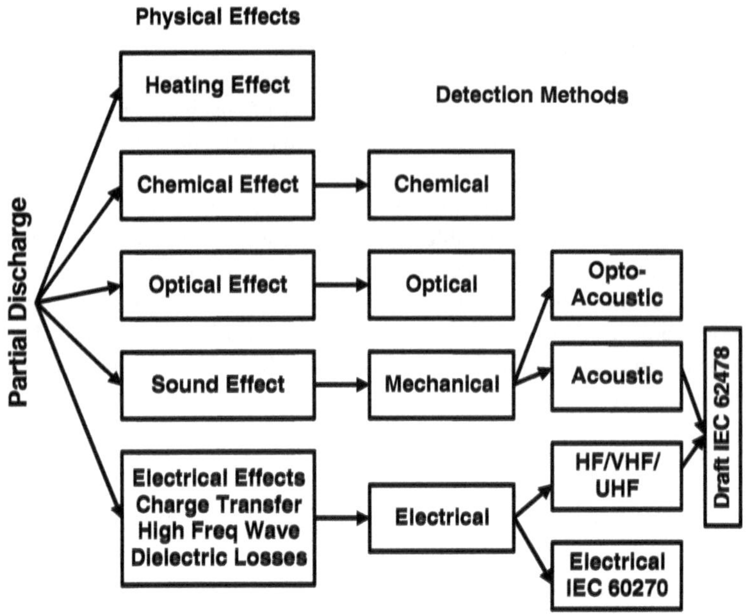

Fig. 3.9 Physical effects of PD and consequent detection methods

3.3.1 Electrical Method of PD Measurement

The basic test circuit for PD measurement by electrical method as per IEC standard [1] is shown in Fig. 3.10. Corresponding to Fig. 3.10 a real-life partial discharge test setup suitable for PD measurement up to 100 kV in the High Voltage Laboratory of Jadavpur University is shown in Fig. 3.11.

The basic operating principle is as follows. PD pulses, as generated within the test sample, being of high frequency will pass through the capacitive potential divider (as the impedance of the capacitive divider will be low at high frequency) and will not appear at the measuring terminal on the left-hand side of Fig. 3.10. Again the PD pulses will pass through the coupling capacitor which will offer low impedance to the high frequency PD pulse. But the inductor will offer high impedance to the high frequency PD signal and hence the PD signal will appear at the measuring terminal on the right-hand side of Fig. 3.10. The purpose of the damping resistor is to eliminate any unwanted oscillation in the measurement. Similarly, the power frequency signal will see high impedance offered by the coupling capacitor and hence the high voltage will appear across the coupling capacitor and not across the measuring impedances. The capacitance C_1 and C_2 will act as a capacitive potential divider and the power frequency applied voltage can be measured at the measuring terminal on the left-hand side of Fig. 3.10 to get the voltage reference in relation to PD pulses.

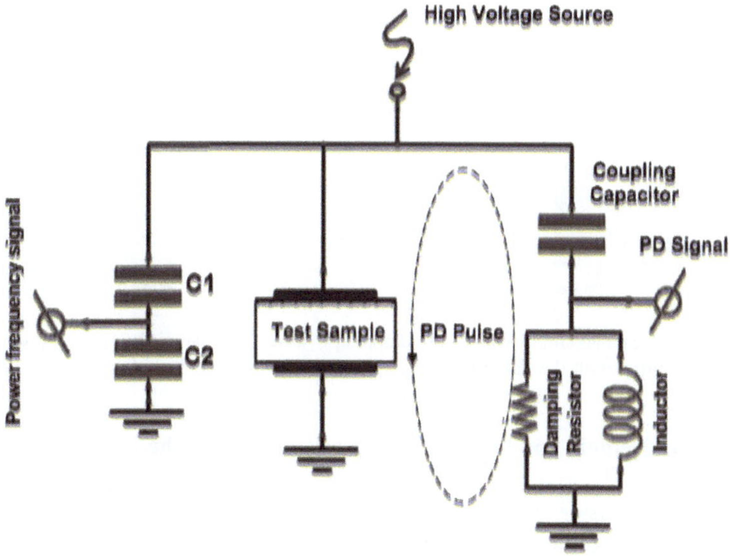

Fig. 3.10 Basic test circuit following IEC 60270

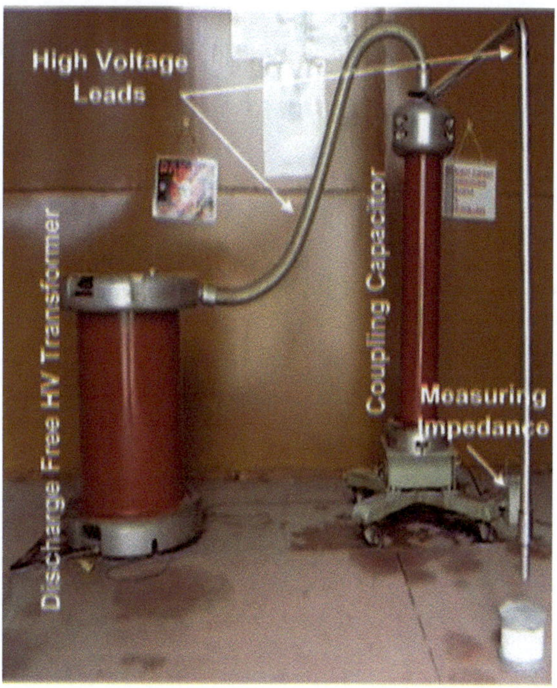

Fig. 3.11 Partial discharge test setup in the High Voltage Laboratory of Jadavpur University

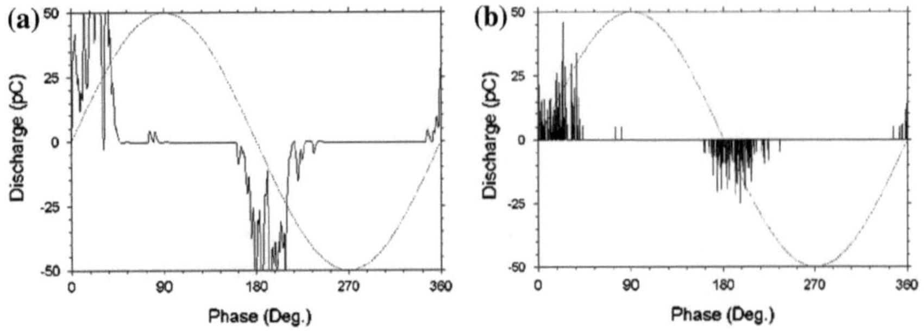

Fig. 3.12 **a** PD data acquired without the damping resistor for a sample at a given test voltage. **b** PD data acquired with the damping resistor for the same sample at the same test voltage

3.3.1.1 Requirement for the Damping Resistor Across the Inductor

The damping resistor that is used in parallel with the inductor, as shown in Fig. 3.10, has a significant role to play in damping out the unwanted oscillations. Figure 3.12a, b shows the data acquired from the same sample at the same test voltage without and with the damping resistor across the inductor.

3.3.1.2 Requirement of Appropriately Designed High Voltage Conductors

The high voltage leads used in the PD test setup must have smooth surfaces and rounded edges. Otherwise, the HV leads will introduce PD due to electric field concentration at the thin leads. This phenomenon is shown in Fig. 3.13a, b, where the data are acquired in the absence of any test sample. From Fig. 3.13 it can be

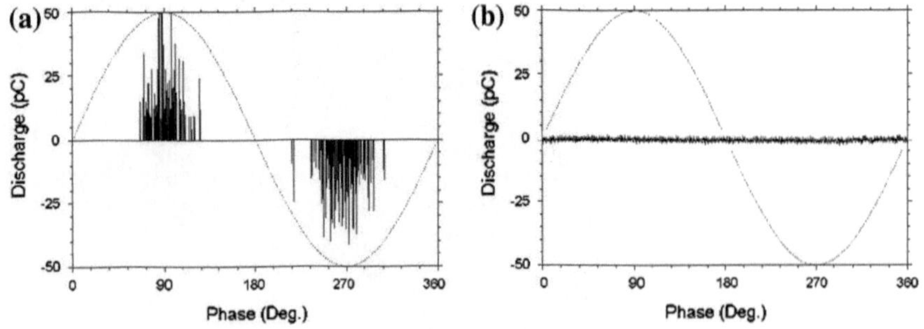

Fig. 3.13 **a** Data acquired with thin high voltage leads at a given test voltage without any test sample. **b** Data acquired with large diameter conductors as high voltage leads at the same test voltage without any test sample

seen that the thin wires introduce significant amount of unwanted PD in the system, which are not present when the thin wires are replaced by conductors having larger diameter.

3.3.1.3 Significance of Shielded Laboratory

The increasing use of commercial, military, and scientific electrical products and electronic devices has raised the problem of electromagnetic interference (EMI), also known as electromagnetic environmental pollution [20]. As a consequence the investigation of low-level sensitive signals, such as PD signal, suffers from this EMI producing erratic results. So there is an increased need for EMI shielding. Different metal-plates and foils, several other structures like bounded-junction wire-mesh screens [21] are adopted for shielding.

A double-shielded partial discharge testing laboratory has been constructed at Jadavpur University. The laboratory having dimensions of $7.5 \times 4 \times 6$ m^3 is shielded throughout using welded copper plates. Inside the first layer of shielding, another structure of dimension $3.9 \times 3.75 \times 3$ m^3 has been constructed using welded Galvanized Iron (GI) sheets that form the second layer of shielding to eliminate environmental noises as much as possible in the acquired signal [22]. Figure 3.14a shows the layout and Fig. 3.14b shows the photograph of the double-shielded laboratory at Jadavpur University.

Utility of the double-shielded laboratory can be seen from the data acquired in three different locations: (a) in the open space outside the laboratory, (b) inside the single shielded section of the laboratory and (c) inside the double-shielded section of the laboratory. The corresponding data are shown in Fig. 3.15a, b, and c, respectively. These data were recorded for the same test sample at the same test voltage.

(a) (b)

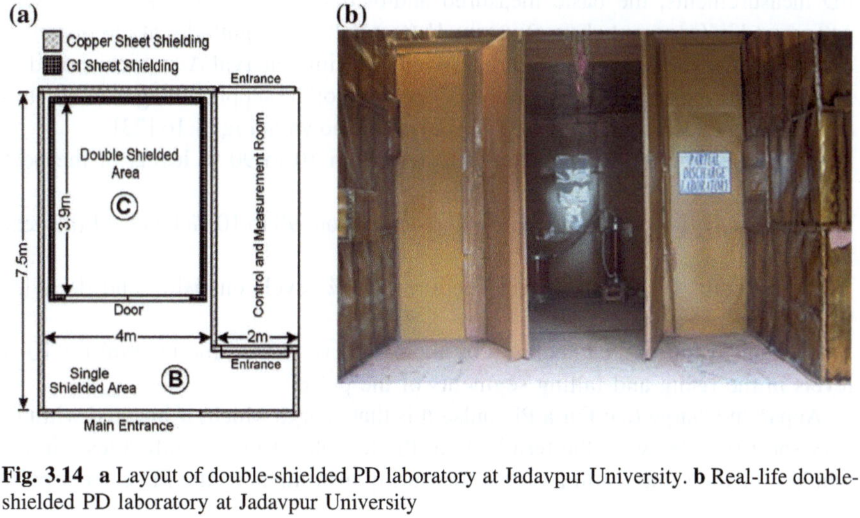

Fig. 3.14 **a** Layout of double-shielded PD laboratory at Jadavpur University. **b** Real-life double-shielded PD laboratory at Jadavpur University

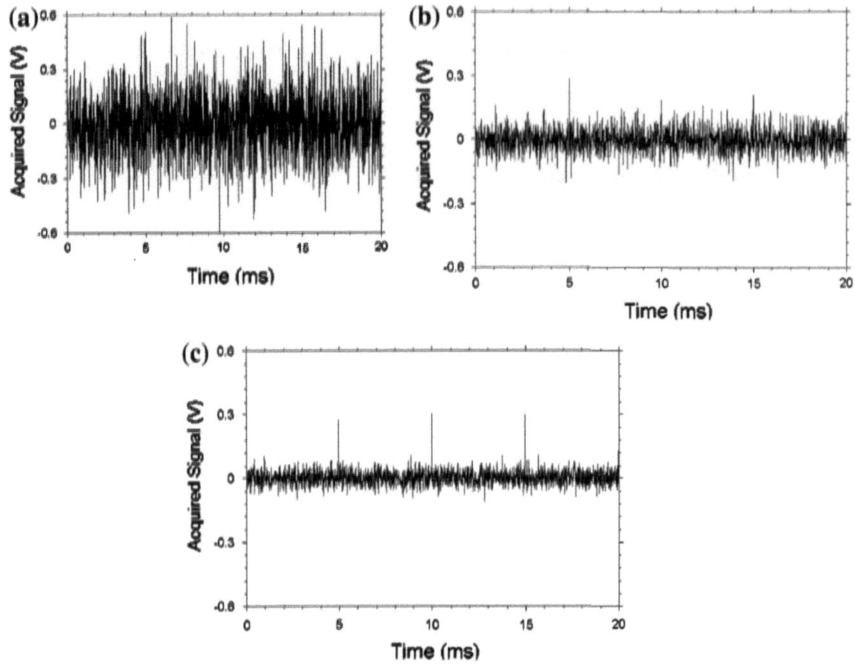

Fig. 3.15 **a** Data acquired outside the shielded laboratory. **b** Data acquired inside the single shielded section of the laboratory. **c** Data acquired inside the double-shielded section of the laboratory

3.3.1.4 PD Pulse Characterization Parameters

The PD pulse parameters that characterize its shape in the time-domain are defined in such a way that they reveal the various inherent attributes of the discharge pulse. In PD measurements, the basic measured and/or displayed quantities are: q–t data pattern and q_i (apparent charge) versus V_i (test voltage magnitude) at the instant of discharge. The measurement is carried over for a time interval Δt, which should be long enough compared to the duration of one cycle of the supply voltage. Parameters describing the idealized shape of a PD pulse are shown in Fig. 3.16 [23].

Pulse rise-time (t_r): time required to rise from 10 to 90 % levels of the peak value.

Pulse decay time (t_d): time required to decay from 90 to 10 % levels of the peak value.

Pulse width (t_w): time interval between 50 % levels on rising and decaying sides of the peak value.

Area under pulse: area enclosed by the q–t curve in the time interval for 10 % levels in the rising and falling segments of the pulse.

Apparent Charge (q): For a PD pulse it is that charge which, if injected within a very short time between the terminals of the test object in a specified test circuit, would give the same reading on the measuring instrument as the PD current pulse

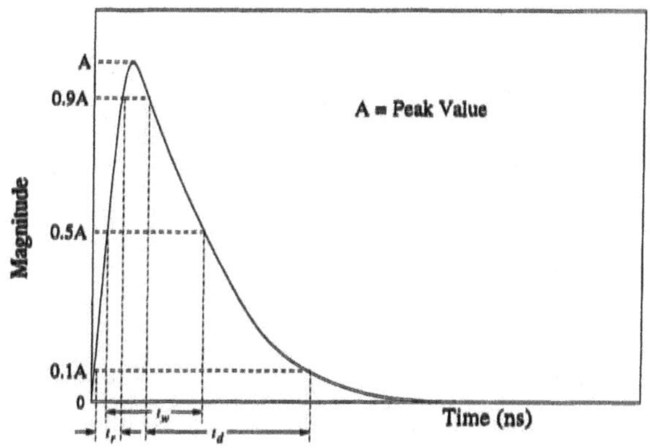

Fig. 3.16 Shape of an idealized PD pulse

itself. The apparent charge is usually expressed in pico-coulombs (pC). The apparent charge is not equal to the amount of charge locally involved at the site of the discharge, which cannot be measured directly.

There are certain derived parameters which are commonly related to PD pulses. The definitions of these are given in IEC 60270 [1]. In the following definitions N is the total number of discharges within the observation time interval T_{ref}, which is normally taken as one second.

Pulse Repetition Rate (n): It is the ratio between the total number of PD pulses recorded in a selected time interval and the duration of this time interval. In practice, only pulses above a specified magnitude or within a specified range of magnitudes are considered.

Average Discharge Current (I): It is the sum of the absolute values of individual apparent charge magnitudes q_i during a chosen reference time interval T_{ref} divided by this time interval, i.e.,

$$I = (|q_1| + |q_2| + |q_3| + \cdots + |q_i|)/T_{\mathrm{ref}}. \tag{3.3}$$

Average Discharge Power (P): It is the average pulse power fed into the terminals of the test object due to apparent charge magnitudes q_i during a chosen reference time interval T_{ref}:

$$P = (q_1 u_1 + q_2 u_2 + \cdots + q_i u_i)/T_{\mathrm{ref}} \tag{3.4}$$

where u_1, u_2,… u_i are instantaneous values of the test voltage at the instants of occurrence t_i of the individual apparent charge magnitudes q_i. The sign of the individual values must be observed.

Quadratic Rate (D): It is the sum of the squares of the individual apparent charge magnitudes q_i during a chosen reference time interval T_{ref} divided by this time interval:

$$D = \left(q_1^2 + q_2^2 + q_3^2 + \cdots + q_i^2\right)/T_{\mathrm{ref}}. \tag{3.5}$$

3.3.1.5 PD Data Patterns

There are three different types of data patterns commonly used for the analysis and evaluation of PD. They are: (a) phase-resolved data, (b) time-resolved data, and (c) data without phase/time information.

Phase-Resolved Data

For phase-resolved PD data, the test voltage is assumed to be held constant and the voltage phase angle is divided into selected number of small windows. The PD detector acquires all the individual quasi-integrated pulses within a window and quantify each of these PD pulses by their discharge magnitude (q), the corresponding phase angle (ϕ) at which they occur, average discharge current (I) and the discharge rate (n).

The most commonly plotted univariate distributions are: (a) q–ϕ, (b) n–ϕ, and (c) I–ϕ.

The bivariate distribution data is ϕ–q–n plot. This 3-D plot is regarded as one of the more complete forms of graphical representation of a PD pulse distribution [24]. The PD pulse height distribution (PDHD) or pulse amplitude histogram (PAH) is the discharge rate (n) versus pulse amplitude (q) [25]. Figure 3.17a shows two types of ϕ–q plots and b shows a typical ϕ–q–n plot.

Time-Resolved Data

True shapes of the partial discharge pulses cannot be ascertained from the phase-resolved data. On the other hand, the true shapes of the individual PD pulses are displayed in the time-resolved data pattern, i.e., q–t plot. Typical q–t plot is shown in Fig. 3.18. A major advantage of this q–t pattern is that there exists some direct relationship between the type of the defect and the shape of the PD signal. Moreover, less expensive instrumentation is required for time-resolved data acquisition than phase-resolved measurements. Since aging of insulation markedly changes the shape of the PD pulse, hence aging can be more readily observed from these data patterns.

PD Data Without Phase or Time Information

Often measurement of partial discharge magnitude is carried out by varying amplitudes of test voltage, which gives q–V plot. This is done to demonstrate the PD hysteresis effect in a test cycle, which occurs due to the fact that a lower applied voltage is necessary to sustain a discharge once it has been initiated. In such q–V plots as shown in Fig. 3.19, neither phase nor time information is available.

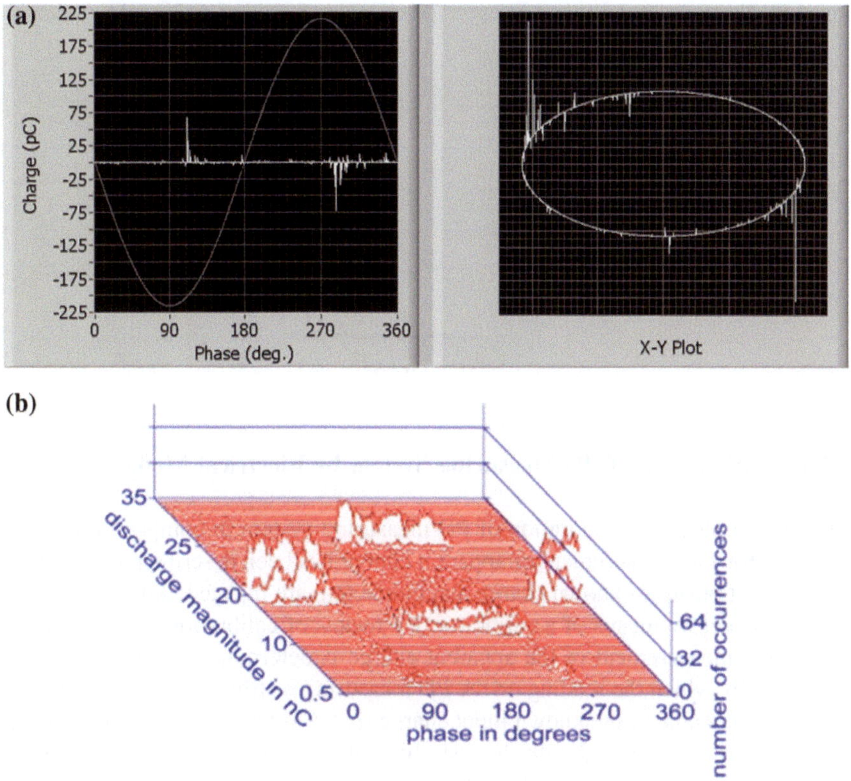

Fig. 3.17 **a** Two different types of ϕ–q plots. **b** Typical ϕ–q–n plot

Fig. 3.18 Typical q–t plots

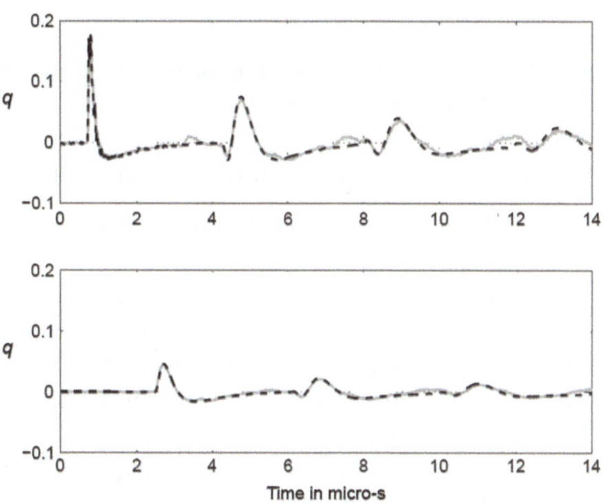

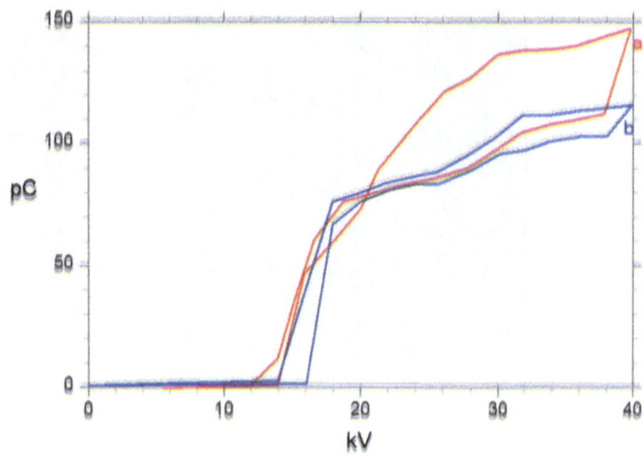

Fig. 3.19 Typical q–V plot

3.3.1.6 Calibration of PD Measuring System by Electrical Method

The methodology for calibration of PD measuring system in complete test circuit intended for the measurement of apparent charge has been described in IEC 60270 [1]. Current pulses of known charge magnitude (q_0) are injected at the terminals of the test object by means of a calibrator. By means of calibration, the scale factor (k) for the measurement of the apparent charge is determined. Scale factor (k) is the factor by which the reading on the PD measuring instrument is to be multiplied to obtain the value of the known input charge (q_0). Calibration of the complete test arrangement is performed with the test object de-energized and at one magnitude within the relevant range, which is often 50–200 % of the specified PD magnitude. As the capacitance of the test object affects the circuit characteristics, calibration is done with each new test object.

3.3.1.7 Advantages and Disadvantages of Conventional Electrical Method

The main advantages of electrical method of PD measurement are as follows:

(i) Calibration with respect to the PD charge is possible in terms of apparent charge.
(ii) Due to standardized calibration procedure the PD values that are measured in pC can be mutually compared between different circuits and test objects.
(iii) Identification of PD source type is possible.
(iv) Can be adopted for testing the insulation integrity for most of the high voltage equipment.
(v) Can be used for acceptance testing as accuracy and sensitivity are high under controlled environmental conditions.

On the other hand, the main disadvantages of electrical method are:

(i) Localization of PD source is very difficult.
(ii) Sensitivity decreases with increased capacitance of test object.
(iii) Prone to electromagnetic interference, particularly for on-site measurement.
(iv) Not suitable for on-line and continuous PD monitoring.

3.3.2 Acoustic Method of PD Measurement

3.3.2.1 Introduction

Instead of recording an electrical signal, acoustic PD detection aims to record the acoustic signal that is generated during a partial discharge. When a current streamer is formed within a void during PD, the material around the hot steamer get vaporized, which causes an explosion of mechanical energy, which propagates through the insulation in the form a pressure wave. Suitable acoustic sensor along with appropriate data acquisition system could be used to detect and analyze this wave. In order to comprehend the possibilities and limitations of acoustic PD detection, it is important to note the complexities involved in acoustic measurement technique as applied to electrical equipment [26, 27].

The ranges of frequency that is used in the acoustic method of PD detection are approximately 20–250 kHz, i.e., the ultrasonic range, as well as the audible range from approximately 100 Hz–3 kHz. In electrical equipment like transformer or switchgear, material dimensions are of the order of several cm, while the wavelength varies from mm to tens of cm over the relevant frequency range. Thus, the acoustic coupling between different materials in an equipment becomes important for understanding acoustic wave propagation for localization of PD source. During propagation of an acoustic wave through a structure comprising various materials, the intensity of the wave decreases as the distance from the source increases. This is because of several mechanisms such as geometrical spreading of acoustic waves, acoustic absorption, i.e., conversion of acoustic energy to heat and scattering of the wave [28]. When an acoustic wave propagates from one medium to another having a different acoustic impedance, reflection, and refraction take place similar to that for electromagnetic wave propagation. For large differences in acoustic impedance only a small fraction of incident wave is transmitted to the second medium, e.g., between oil and steel the transmission coefficient is 0.11 and that between air and steel is 0.0016 [28]. Again, acoustic waves bend around corners and obstacles as a result of diffraction. When several obstacles are present, diffraction is often called scattering. Gas bubbles present in gas or liquid media cause scattering of acoustic waves. Moreover, velocity of sound cannot be taken as constant quantity because in real-life media it depends on the frequency, which is known as dispersion. In gaseous media, within the range of frequency relevant to acoustic PD detection, i.e., less than 1 MHz, dispersion is small. But in a complex structure of power equipment, where several boundaries involving different media are present, dispersion could be significant. In liquid media, velocity of sound varies from 900 to 2000 m/s and decreases monotonically with temperature. Acoustic properties of common gases and liquids used in power equipment are available in [28].

Normally only pressure waves can propagate through gaseous and liquid media. But both pressure and shear waves can propagate through a solid medium, which are often termed as longitudinal and transverse waves that are created simultaneously in a PD event. Acoustic properties of solids depend on both the composition and production process of material such as metals, glasses, ceramics, polymers, etc. Absorption coefficient of solids varies over several decades. Moreover, losses occur at boundaries having dissimilar materials such as porous material like paper or pressboard.

With proper understanding of the above-mentioned complexities, appropriate acoustic PD detection system having good sensitivity can be implemented, which can localize PD sources accurately within the complex structure of electrical equipment.

3.3.2.2 Conventional Electrical Method Vis-à-Vis Acoustic Method

While making PD measurement as per IEC 60270, electromagnetic interference can be controlled in a laboratory or may even in factory settings, but is very difficult in the case of field testing. Moreover, it has been noted that the sensitivity of electrical method decreases with increased capacitance of test object [29]. From practical considerations, in the case of large power transformer or switchgear, simple detection of PD sources is of not much help unless information on the location of the PD sources is provided. Electrical PD detection methodology is not suitable for such localization of PD sources. On all these accounts acoustic PD detection is advantageous, as acoustic method is noninvasive, can be used even when the equipment is in operation and is immune to electromagnetic interference. The sensitivity of acoustic method does not vary with the capacitance of the test object and it can provide valuable information on the location of PD sources in electrical equipment [28]. Consequently, acoustic PD measurement is widely used for on-site as well as on-line PD detection for GIS, transformers, switchgears, and other electrical equipment.

3.3.2.3 Acoustic PD Monitoring System

Acoustic PD detection system is simple comprising a sensor, preamplifier, filter, and a data acquisition system. The block diagram of such a system is shown in Fig. 3.20.

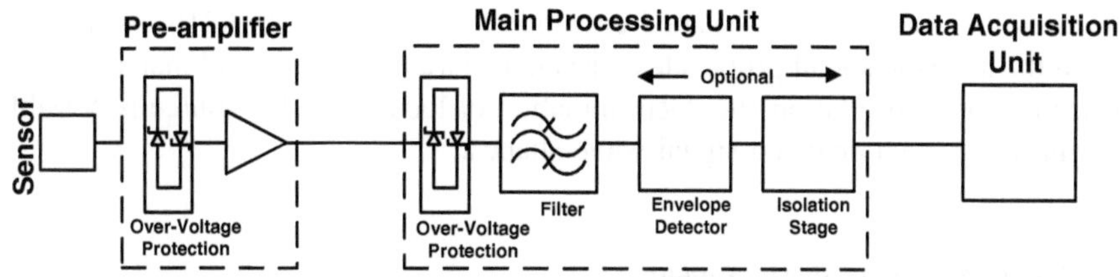

Fig. 3.20 Block diagram of an acoustic PD detection system

The sensor signal is amplified by a low noise preamplifier, which is normally connected to the sensor by a short length of cable. Now-a-days, acoustic sensors are also having internal preamplifiers. A cable of suitable length connects the preamplifier to the filter unit, which consists of a high pass filter with switchable cut-off frequency. Sometimes, the filter unit also has an envelope detector and an isolation stage [30]. The filter unit along with the data acquisition system is normally placed on a movable cart so that it can be moved around for possible mounting of the sensor at different locations. For on-site testing the total cable length from the filter unit to the sensor could vary from 10 to 30 m. Although the acoustic sensor itself is immune to electromagnetic interference, the electrical circuitry associated to acoustic PD detection system is prone to EM interference and thus needs to be shielded adequately against EM interference. All the connections must be done using double-shielded cables and the characteristic impedance of all the electrical components must be properly matched.

The shape of the detected signal depends on the source, the acoustic wave propagation path and the sensor. In the case of transformers, depending on the propagation path of the signal, very often two types of acoustic signals are obtained, viz. arrowhead-shaped and egg-shaped signals, as shown in Fig. 3.21 [31]. The arrowhead-shaped signal, shown in Fig. 3.21a, is obtained when there is a minimal attenuation, i.e., the propagation path is mainly through insulating oil.

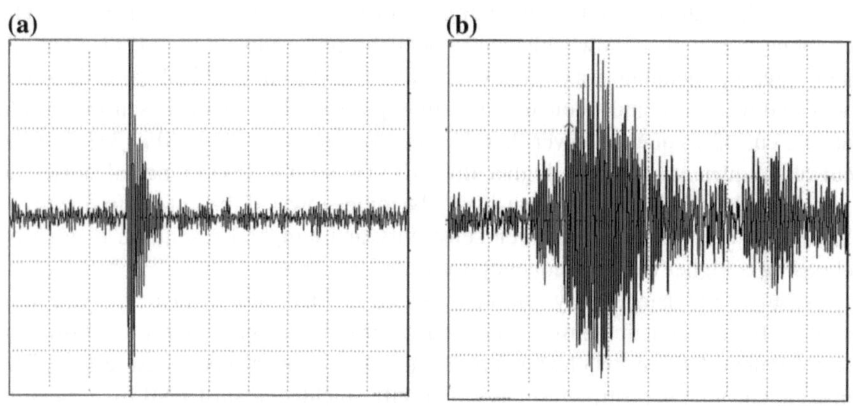

Fig. 3.21 **a** *Arrow head-shaped* acoustic PD signal [32]. **b** *Egg-shaped* acoustic PD signal [32]. Reproduced with kind permission © HV Technologies, Inc)

On the other hand, the egg-shaped envelope as shown in Fig. 3.21b, is obtained when the acoustic signal propagates through different types of solid materials such as iron, copper, insulation, etc. Determination of location of PD source is possible by analyzing such detected signal wave shapes.

3.3.2.4 Types of Acoustic Sensors

The type of acoustic sensor depends on the medium in which they are to sense the acoustic wave. Microphones are normally used in air and other gases. Most commonly used sensor is condenser microphone which has wide bandwidth, high accuracy and wide dynamic range. Hydrophones made from piezoelectric materials are used in liquid media, which have bandwidth greater than 150 kHz. For electrical power equipment, acoustic PD detection systems are divided into two categories: external and internal systems. In the external system, which is most commonly used, the sensors are mounted externally on the enclosure of the equipment. On the other hand, in the internal system, the sensors are mounted within the enclosure. For sensing acoustic PD signal by means of externally mounted sensors, piezoelectric sensors based on ferroelectric material such as Lead Zirconium Titanate (PZT) are presently wide in use. Normally either accelerometers or acoustic emission sensors are used. The output of an accelerometer is a charge that is proportional to the acceleration of the surface (in the direction normal to the base of the sensor) to which they are fixed ($pC{\cdot}s^2/m$). Accelerometers have a useful frequency range up to 50 kHz and are fixed to the surface using magnet, glue, pin screw, etc. Acoustic Emission (AE) sensors give a voltage signal as output that is proportional to the velocity of the surface (in the direction normal to the base of the sensor) on which they are mounted ($V{\cdot}s/m$). AE sensors operate in resonant condition and have a useful frequency range from 30 kHz to 1 MHz and are fixed to the surface using tape, elastic bands, magnetic hold-downs, etc. [33]. A thin layer of acoustic coupling agent is used between the base of the sensor and the mounting surface to ensure good sensitivity. The coupling agents that show satisfactory acoustic properties up to 100 kHz are silicon grease, meltable wax, adhesive wax, etc. From practical considerations adhesive wax is the most convenient coupling agent [30]. Acoustic sensors should have a cut-off frequency lower than 50 kHz for detecting PD in SF_6 [34]. But the cut-off frequency should be higher than 10 kHz so that environmental noises do not affect the sensitivity. The sensitivity becomes lower if the cut-off frequency is higher. So a compromise between sensitivity and temporal resolution is done by choosing a higher cut-off frequency [30].

Acoustic PD detection system can also be divided into contact type and non-contact type. Acoustic sensors mounted externally on the metallic enclosure of transformers, GIS or switchgears are examples of contact type measurement. The outer shell of an acoustic sensor is commonly shielded against EMI by a metallic enclosure. If the surface of the equipment under test in ungrounded type, e.g., cast-resin type transformers, use of contact type measurement may result in significant amount of electric field disturbances. Consequently in such cases, non-contact type measurement has to be performed. In non-contact type measurement it has to be kept in mind that acoustic signals get attenuated rapidly in air as distance of the sensor from the test object is increased. Moreover, there is a recommended minimum safe distance of operation for every system voltage level. Thus, non-contact type measurement for equipment without any grounded metallic enclosure must be carried out at a distance greater than the minimum safe distance.

Properly shielded acoustic PD monitoring system is insensitive to electromagnetic interferences. Comparison of background acoustic noise level in laboratories and on-site does not show significant differences [30]. But two forms of mechanical noise do affect on-site acoustic measurements. Continuous mechanical noise of frequency around 100 Hz appear in substations from magnetostriction in the core material of power and instrument transformers. Low frequency external noises also arise from normal speech, walking, etc. Such noises are practically non-existent above 10 kHz and can be effectively eliminated using a high pass filter. On the other hand, high frequency noises such as clattering of metal tools can also be eliminated without any difficulty. Problems arise due to transient noise caused by rain, snow, hail, sand, etc. Such noises have a broader frequency spectrum and can have features similar to PD signals. If the PD signals have good correlation with power frequency voltage signal, then the transient noises can be eliminated using suitable discrimination techniques based on such correlations [33].

Acoustic PD signals could be measured as either a peak or RMS value. The amplitude of an acoustic wave generated by a single PD event is proportional to the square-root of the mechanical energy released during that PD event. But the travel time of an acoustic wave from a PD source to the sensor is usually appreciable and hence the detected acoustic PD signal for a single PD event has duration in milliseconds range. But the electrical pulses due to PD events are of much shorter duration and the pulse rate is also high. As a result, the peak value of the signal as detected by acoustic sensor will not be related to the energy of a single PD event. The RMS value of the detected acoustic signal will depend on both the energy of a single PD event as well as the PD pulse rate. Hence, the RMS value is more closely related to the total energy released during a power frequency cycle. As the acoustic signal level is strongly dependent on propagation path, it is not possible to make absolute measurements in the case of electrical equipment for which the structure is normally complex. Acoustic sensors are calibrated according to procedures laid down in international standards, viz. ISO 5347-3 for accelerometers and ASTM E976 and ASTM E1106 for AE sensors.

3.3.2.5 Localization of PD Source by Acoustic Method

The changes in the magnitude and shape of the acoustic PD signal as it propagates away from the PD source to the sensor should be understood for proper localization of PD sources by acoustic method. Attenuation of acoustic signal is caused by (i) spreading of geometric wave-front, (ii) multiple propagation paths causing division of the wave, (iii) transmission losses as the wave propagates from one medium to another and also at discontinuities within the structure, and (iv) absorption in different materials. The last three mechanisms are frequency dependent and in some cases can reduce the sensitivity to very low level. On the other hand, changes in the shape of the acoustic PD signal are caused by (i) frequency-dependent velocity which causes different frequency components of the wave to arrive at the sensor location at different times, (ii) frequency-dependent propagation paths which results in different wave components to arrive at the sensor location with different time lags and (iii) removal of higher frequency components due to absorption in materials [33, 35]. Rise-time and the duration of the signal increase as the acoustic wave propagates further away from the source.

There are several techniques of PD source localization using acoustic sensors:

(i) Searching the highest amplitude of the signal: It is based on the fact that the amplitude of the detected signal will be highest when the sensor is closest to the PD source. In this case only one sensor is moved around the equipment until the position for maximum signal amplitude is found.
(ii) Observing the rise-time of the signal: The rise-time of the detected signal will be small when the sensor is close to the source. In this case also one sensor is used for detection by placing it at different positions around the equipment.
(iii) Time of flight measurement: If the signal is repetitive in nature, then two sensors are used and the times of arrival of the signals at the two sensor locations are noted. If the PD source is located away from the both the sensors, then the source is closer to the sensor for which the arrival time is less. If the source is in between the two sensors, then the location of the source can be found from the difference in the times of arrivals of the signals at the two sensor locations. In fact the source location can be found to be on a hyperboloid, which can be determined from the analyses of the signal arrival times at the two sensor locations [33].

However, an accurate all acoustic PD source localization technique uses four acoustic sensors. The sensor which detects the acoustic wave first, triggers the other three sensors simultaneously and the arrival time at the other three sensors are recorded. This results in three timing measurements starting from the reference sensor which is the sensor detecting wave first as shown in Fig. 3.22. Considering the velocity of acoustic wave to be v_s, the following equations could be written involving the time of flights as shown in Fig. 3.22 [36].

$$(x_s - x_1)^2 + (y_s - y_1)^2 + (z_s - z_1)^2 = (v_s t_0)^2 \tag{3.6}$$

$$(x_s - x_2)^2 + (y_s - y_2)^2 + (z_s - z_2)^2 = (v_s (t_0 + t_{12}))^2 \tag{3.7}$$

$$(x_s - x_3)^2 + (y_s - y_3)^2 + (z_s - z_3)^2 = (v_s (t_0 + t_{13}))^2 \tag{3.8}$$

$$(x_s - x_4)^2 + (y_s - y_4)^2 + (z_s - z_4)^2 = (v_s (t_0 + t_{14}))^2 \tag{3.9}$$

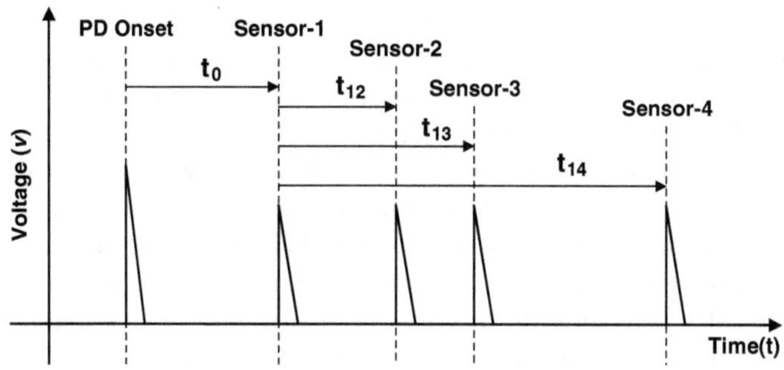

Fig. 3.22 PD source localization by time of flight measurements using four acoustic sensors

where, the coordinates of PD source is (x_s,y_s,z_s) and the coordinates of the four sensor locations are (x_i,y_i,z_i) for $i = 1,\ldots,4$.

The PD site is therefore located at the point of intersection of the hyperboloids given by Eq. (3.6) through (3.9).

3.3.2.6 Typical Applications of Acoustic PD Detection

One area where acoustic method is very effective is detection of corona discharge in air. It is because of the fact that propagation of acoustic wave through air is like that through infinite space. Moreover, absorption of sound in air is negligible over the distances where the measurements are taken in practice. Equipments for commercial use are now available for this purpose.

Acoustic PD detection is in use for GIS for more than two decades [34], in which case the acoustic sensors are mounted externally on the outer metallic enclosure of GIS. A good indication of PD is the periodicity of the acoustic signals at 50/60 or 100/120 Hz, where the periodicity depends on the occurrence of PD in one or both the positive and negative half cycles of the voltage wave [33].

Currently, acoustic PD detectors are available commercially in which acoustic sensors are mounted on hot-sick. In this way the sensors could be brought close to switchgear or other equipment under live operating condition. These detectors are suitable for PD detection in air. In the case of cable termination of high voltage equipment, the sensors are mounted in such a way that they are in mechanical contact with the cable termination with proper acoustic impedance matching [33].

Due to significant reduction of acoustic signals caused by absorption in cable insulation, acoustic sensors must be in contact with the cable to get reasonable sensitivity. In practice it has been found that a narrowband detection circuit tuned to about 2 kHz improve sensitivity [36]. The sensor is moved along the cable to determine the location where the signal level is highest. A major problem in the case of many high voltage cables is the presence of armor. Due to poor acoustic transmission through air-filled spaces in the armor, acoustic coupling between cable insulation and the sensor becomes weak.

Long back in 1978, it was identified that acoustic sensors to be used for PD detection in transformers must be resonant at frequencies lower or higher than the frequencies of magnetostriction-related noise, which are around 50–60 kHz [37]. When the PD occurs outside the winding then the propagation path of acoustic signals from the source to the sensor is simple, as the wave propagates through transformer oil. As mostly externally mounted sensors are used for PD detection in transformers, a low pass filter is needed to eliminate all the frequencies that travel faster in metallic tank than in oil. Then the arrival times can be computed based on acoustic signal propagation in oil only. Typically, for commissioning purposes, multiple sensors are employed to achieve greater sensitivity. But for in-service testing, one sensor is usually moved around the transformer for PD detection. When the PD occurs within transformer winding, then the propagation path is very complex and the sensitivity could be severely reduced for those PD events [33].

3.3.2.7 Advantages and Disadvantages of Acoustic Methodology

The main advantages of Acoustic method of PD measurement are as follows:

(i) It is noninvasive, relatively inexpensive and easy to be used.
(ii) Greater immunity to electromagnetic interference in the case of on-site testing.
(iii) Efficient location of PD sources in GIS and transformers in service.
(iv) In the case of GIS, sensitive to common PD defects and hence able to determine PD defect type.

On the other hand, the main disadvantages of Acoustic method are:

(i) Calibration of the acoustic signal with respect to the PD charge is not possible.
(ii) Not so convenient for continuous monitoring of PD.
(iii) Not very useful for risk estimation of the defects in transformers, because of high and complex attenuation of acoustic signal.
(iv) Time consuming.

3.3.3 Ultra-High Frequency Method of PD Measurement

3.3.3.1 Introduction

The principle of the Ultra-High Frequency (UHF) technique was developed more than 25 years ago, and is now well-known worldwide after being adopted by many leading manufacturers and utilities. To detect partial discharge activity under conditions of on-site on-line testing, UHF PD diagnosis principle is increasingly being used. This method is based on sensing the electromagnetic emissions from discharge sites in the insulation. Discharges in insulation could originate from insulation defects such as protrusions, particles, and voids in insulation, but could also be failures induced from other imperfections such as poor contacts. A typical PD pulse has a rise-time of less than 1 ns (which recent measurements have indicated could be less than 70 ps) and a width of a few ns, implying in frequency-domain a bandwidth of several hundreds of MHz. The UHF PD diagnosis is based on capturing those pulses. This method yields higher signal-to-noise ratio under on-site conditions compared to the conventional PD measurement based on IEC 60270. Consequently, the sensitivity of online PD detection is higher in UHF method compared to conventional electrical methodology. UHF PD monitoring system has been found to be able to detect insulation defects in the case of on-site monitoring owing to its high sensitivity and high signal-to-noise ratio [38]. It is known that the eletromagnetic waves are attenuated more at higher frequencies as they propagate. Since EM noise of higher frequency is originated further from the UHF sensor than the PD source within the equipment, the noise will be attenuated more than the PD, even if the frequency content of the PD and the noise is similar, resulting in higher signal-to-noise ratio in UHF measurement. Moreover, data obtained from continuous on-site monitoring provide a good basis for evaluating the strengths and weaknesses of the UHF technique and for developing useful signal interpretation strategies.

The UHF measuring technique was first applied on gas-insulated switchgear [39] and subsequently has also been applied to power transformers [40]. In order to apply the UHF technique, the ultra-high frequency signals have to be captured by a sensor or coupler. Due to the fact that the transformer or GIS is a closed metallic box, the signals will remain inside and special measures have to be installed in order to couple these signals out of the transformer or GIS.

3.3.3.2 Conventional Electrical Method Vis-à-Vis UHF Method

IEC 60076-3 describes the electrical method used for PD testing in factory acceptance testing, which is based on IEC 60270, a generic standard for PD testing that aims to measure apparent discharge magnitude and is adopted for testing the insulation integrity for most of the high voltage equipment. But it is to be mentioned here that the measurement according to IEC 60270 only reflects the magnitude of the charge related to the voltage drop and current flow within the dielectric caused by the discharge. It is true that this is related to events within the defect and the true charge flowing due to the discharge, but the actual relationship between these and the apparent charge magnitude remains unclear as long as the proportions of void and series dielectric remain hidden. Another problem has also been identified with the methodology of IEC 60270 in the case of transformers. Within the windings of a transformer a single PD event propagates in a variety of modes, each with their own frequency content, propagation speed, and attenuation factor. It has been recognized that winding geometry produces a range of resonances for electromagnetic wave and its propagation influences the measurement significantly. Although IEC 60270 approach toward measurement of apparent charge is used for factory acceptance test, the reality is that more useful information may be obtained from alternative methods [41, 42]. Alternative methods also become more relevant when measurement of PD is to be done in the presence of high levels of interference. The sources of such interferences could be communications and broadcast transmissions, corona as well as surface discharges elsewhere on-site.

A good alternative is the use of UHF probe inserted in an oil valve or access port allowing transformer tank to screen interference. In this way the electromagnetic radiation from the PD will be detected but it will not be affected by winding resonances. It will also help in achieving a more uniform measurement for transformers having a wide range of ratings and designs. UHF method is a technique that can be used on-site, installed even when the transformer remains energized and can provide useful results in a high interference environment. Modern signal processing tools can be used to monitor and process UHF signals, to enhance signal-to-noise ratio and to give phase-resolved patterns as shown in Fig. 3.23, which provide an effective means of identification of the type of defects causing PD.

3.3.3.3 UHF PD Monitoring System

The features of the UHF discharge pulses most useful for interpretation purposes are their amplitude, instant occurrence on the voltage wave and time interval between the pulses. All UHF PD monitoring system focuses on these parameters for identification of insulation defects. Typically, UHF PD monitoring system uses UHF sensors, which are basically antenna for UHF signals. These sensors are connected via preamplifier and multiplexer to spectrum analyzer and digital oscilloscope. Commonly, the frequency range of the amplifier is 2 MHz–2.6 GHz

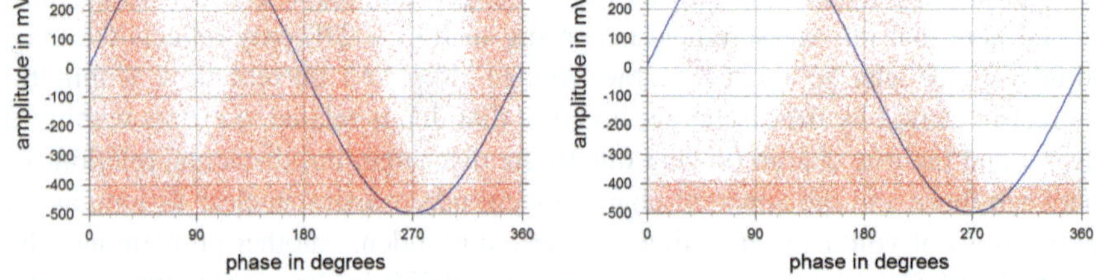

Fig. 3.23 Typical Phase-Resolved PD (PRPD) patterns as obtained from UHF sensors. **a** PRPD pattern for major PD activity. **b** PRPD pattern for no major PD activity

with a gain of +40 dB, while the sampling rate and bandwidth of the oscilloscope are 5 GS/s and 1 GHz, respectively. The main components of a high performance PD monitoring system are the following: (a) UHF sensors, (b) PD converter box which converts UHF signal to low frequency signal and performs noise reduction, (c) Main Processing unit which monitors the data stream, and (d) Postprocessing unit which displays and archives the data. Very often there is an analysis tool or an expert system for identification of PD defects. Optical converter units are also used which collects the data from a UHF sensor and then transmits via an optical fiber cable to the PD converter box that is located at a far away place on-site. The sensitivity limit of UHF signals is commonly set at −80 dBm in the frequency range. To enhance PD sensitivity of a UHF sensor, it is necessary to match the impedances of the sensor, transmission cable, and receiving instrument. As impedance of the transmission cable and receiving instrument is usually 50–75Ω, radiation impedance of the sensor should have similar value.

The dynamic range of UHF/LF converter input covers all signals between −80 to 0 dBm. Currently, portable UHF PD monitoring systems are in use, which can monitor signals continuously in the time-domain and can capture several events (often six events) on each channel in fixed time interval (often 15 min). A PD event is the one that is triggered by a discharge pulse that exceeds a certain predefined threshold. The recorded pulses could be visualized later as 3D display.

Frequency spectrum of on-site UHF noise shows that such noises are mostly due to modulated carriers in the range below 1 GHz. Thus for UHF measurements below 1 GHz, suitable signal processing is necessary to eliminate these noises. Since these noises are not dependent on the equipment, it is possible to select narrowband detection to eliminate these interferences.

3.3.3.4 Types of UHF Sensor

In general, the field emissions caused by PD pulses can be coupled using capacitive or inductive sensors. Capacitive sensors use the electric component of the transient field. The capacitive coupling generally depends on the dimensions of the sensor, distance to the test object, and the electric field intensity of the emitted waves.

Inductive sensors, also called high frequency current transformers (HFCT), couple the magnetic component of the field and are usually made in form of coil or toroid. Thus, the voltage induced in the coil is proportional to the rate of change of the current passing through the coil. The magnetic coupling depends on the number of turns, distance to the test object, the magnetic field strength outside the cable, and the frequency. Sensor coils can be made of ferrite core to make them more sensitive. But in that case they are no longer linear, which is undesirable from the calibration point of view.

While the capacitive sensor captures broadband frequency components of an internal PD in the range of 500–1000 MHz, the inductive sensor provides a different FFT spectrum with the highest signal-to-noise ratio in the range of 200–350 MHz [43].

One important aspect of on-site PD monitoring by UHF method is its susceptibility to corona disturbances. In time-domain often the corona pulses look very similar to the PD pulses due to internal defect in the insulation. Practical experience shows that the spectral range of corona activity is mostly in the range up to 300 MHz. The FFT spectrum of the pulses picked up by the capacitive sensor facilitates a clear distinction between the corona and the PD discharges that are in the range of 700–1000 MHz. The inductive sensor, on the contrary, sees the corona in the air with a very similar Fourier spectrum as a real PD in the range of 200–350 MHz [43].

Comparison of capacitive and inductive sensors has revealed that that capacitive sensor generally has higher sensitivity than the inductive one [43]. This is due to the narrower band of the inductive sensor, limited by the self resonance frequency in the upward direction and the high pass filter in the downward direction. The former is determined by the relative values of the inductance and the parasitic capacitance of the coil. Considering all the above the main focus hereafter is on capacitive sensors.

An additional aspect that needs to be studied in detail is the ability of the sensor to give a good response to the wave-front of a UHF signal. For example, a small, simple sensor may have a low sensitivity but a good initial response, because the energy buildup is not limited by propagation delays within the coupler structure. On the other hand, a more complex broadband sensor may have good sensitivity but requires more time for the energy to build up in its output signal. This aspect is particularly important in power transformers, where accurate measurement of arrival times of signals at a set of sensors is needed for locating PD sources in three dimensions.

Four types of UHF sensors are mostly in use as shown in Fig. 3.24: (a) Window sensor, the structure of which is a spiral antenna. These are designed for external mounting on dielectric windows added to power transformer or GIS, (b) Internal disk sensor, commonly referred to as a capacitive coupler, are used extensively on GIS, (c) Barrier sensor uses a dipole element and are designed for external mounting having exposed edges of insulating epoxy gas barriers where UHF signals can escape in the case of GIS and (d) Transformer probe for insertion through an oil drain valve of a transformer even when the transformer is in service. Due to the restriction imposed on the probe diameter by the oil drain valve, this device is a simple monopole antenna [44]. The disk sensor and the transformer probe have the largest and least complex responses due to their simple structure. The window sensor based on a spiral antenna has the most complex structure. In addition to the above-mentioned sensors, other types of UHF sensors are also in use, viz. hatch sensor, wave guide sensor, directional electromagnetic sensor, etc.

The energy content U of the sampled signals can be approximated by a discrete summation of the form

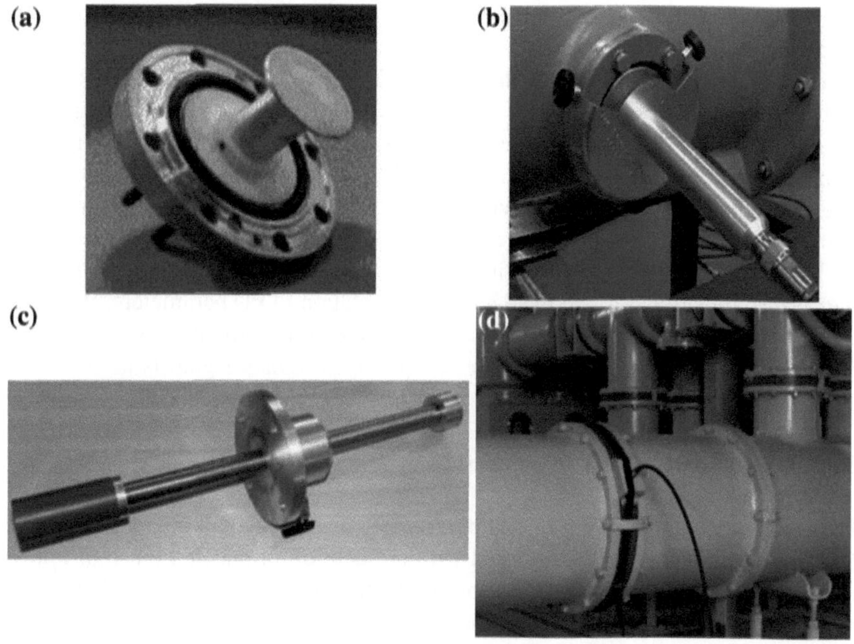

Fig. 3.24 Different types of UHF sensors. **a** UHF internal disk sensor. **b** UHF Window sensor for GIS. **c** UHF oil valve sensor for transformer. **d** UHF barrier sensor for GIS

$$U = \frac{\Delta t}{R}\sum_{i=1}^{N} V_i^2 \tag{3.10}$$

where, Δt is the time (often 100 ps) between the samples of sensor output voltage V_i, R is the input impedance of the measurement system (50 Ω), and N is the number of samples in the data. For UHF sensors the total accumulated energy is in the pJ range. The window sensor energy is about 0.25 pJ and the internal sensor has the largest response around 2.5 pJ, for N = 1000. Moreover, the buildup of energy at the output of the disk and probe sensors is faster and more well-defined than that of the window or barrier sensors.

From the above discussion, it may be seen that the UHF sensors could be broadly grouped into two categories, viz. internal and external sensors. Their usage also depends on practical constraints. For example, in a service-aged power transformer in operation, dielectric windows are not present and the only possibility is to install an internal UHF sensor through the oil drain valve. On the other hand, the external window sensor is an effective tool to monitor PD in existing GIS equipped with no internal sensor. External mounting allows the sensors to be removed or replaced without affecting the equipment.

Practical experience shows that both external and internal sensors are capable of detecting partial discharge levels of several nC without any problem [45].

It means that critical insulation defects can be detected using both the sensors. Depending on the application, the internal sensor is slightly more sensitive compared to the external sensor. In the case of GIS monitoring, external dipole type UHF sensor has been developed, which has sensitivity of 2 pC at the signal-to-noise ratio of 3 where the distance from PD sources to the UHF sensor is about 1 m. Experimental results show that in the range of 300 MHz–1.3 GHz, the internal sensor has higher sensitivity up to 700 MHz and the external sensor has higher sensitivity above that range [45].

The output of UHF sensors is in most of the cases a voltage signal. However, the derived quantities should have a direct correlation to PD parameters. When the direct output voltage of the UHF sensor is used then this correlation is linear, while it is quadratic when the power or the energy of the sensor signal is used [46].

3.3.3.5 Sensitivity Check of UHF Sensor

For quality assurance testing of equipment, the conventional method based on IEC 60270 is used as it can be calibrated with the help of an injected pulse of known magnitude. The sensitivity verification of the apparent charge of PD, which is the most important factor according to the conventional IEC 60270 method focuses on the kHz range. The conventional method is difficult to adopt for sensitivity verification of on-site GIS, because most of the PD-detecting sensors usually form a small coupling capacitance that is insufficient to detect the kHz components of the PD pulse. On the other hand, PD-detecting sensors used in the UHF method can detect the MHz-GHz component of the PD pulses without large coupling capacitance.

In the UHF method, the magnitude of detected signals depends strongly on setup, location of defect and sensor, so that an overall transfer function cannot be determined. It may be noted here that energy is more objective criterion to compare the UHF and IEC60270 methods. The latter reacts on a charge, which eventually represents the stored energy. Peak voltage in the case of UHF method, on the contrary, is more a parameter of the discharge current growth rate and can differ greatly for different types of PD defects and strongly depends on damping characteristics of materials on the way from defect's origin to the sensors. Thus, the calibration of UHF method in terms of apparent charge is impossible. In other words, calibration of the UHF method cannot be done in a way similar to that of IEC 60270 method. For this reason, dual port sensitivity verification has been proposed by CIGRE WG 15/33.03.05 [47]. Here, the purpose is not only to confirm the detection sensitivity of the sensor itself to a UHF pulse, but also to provide a functional check of the whole measuring system including sensors and PD acquisition system. This will ensure that typical defects can be detected by UHF measuring system. In the case of GIS, dual port sensitivity check involves pulse injection into one sensor while detecting signals at the other sensor. In the case of transformers, pulse injection is useful not only for sensitivity check, but also to test the capability of each sensor in use [48].

Wherever possible, sensitivity check may also be performed under laboratory conditions. For this purpose a conventional PD measurement according to the IEC 60270 standard is carried out simultaneously with the UHF diagnostics. To establish functionality between the key parameters of both methods, namely the peak voltage of the UHF pulses and the corresponding values of the apparent charge, a series of simultaneous measurements need to be performed.

3.3.3.6 Locating PD Sources Using UHF Methodology

In recent years extensive investigations in laboratories have confirmed that PD detection using UHF technology results in higher or at least the same sensitivity as detection by PD measurement as set out in IEC 60270 for several equipment like transformers, GIS, switchgears, etc.

In the case of GIS or GIL, at locations where the enclosure is interrupted by a dielectric material such as epoxy resin, glass inspection window, etc., the UHF signals will couple out of the system. These 'dielectric windows' offer the possibility to pick-up these UHF signals by means of external couplers. The amplitude of the signals reduces not only due to damping but also due to reflections at discontinuities such as circuit breakers and disconnectors. As consequence of UHF signal attenuation, several couplers have to be installed in an extended GIS or GIL for proper localization of PD sources in such metal enclosed gas-insulated systems.

Power transformers being important key components in the electrical power system, the following PD magnitudes have been categorized by CIGRE [49]: (a) Defect free 10–50 pC, (b) Normal deterioration <500 pC, (c) Questionable 500–1,000 pC, (d) Defective condition 1,000–2,500 pC, (e) Faulty (Irreversible) >2,500 pC, (f) Critical >100,000–1,000,000 pC. Thus, it can be concluded that partial discharge levels of 500 pC or more are of interest in the case of power transformers. Studies on sensitivity of UHF technique have shown that the UHF technique is sensitive enough to detect PD levels as low as a few tens of pC.

In the case of new power transformers, a flexible approach may be taken during the design stage to facilitate PD localization using UHF sensors. In this case, dielectric windows through which the UHF signals can propagate can be installed at desired locations to achieve maximum sensitivity. Externally, installed UHF sensors could then be used to capture these signals. A second method is to incorporate additional oil valves in order to slide in UHF sensor inside the transformer tank. If three or more UHF sensor installations are possible for one particular transformer, then localization of the PD source might be possible by using time-of-flight measurements [48, 50]. When locating a PD source using time-of-flight measurement, it is usually necessary to determine the arrival times of the signals at two or more sensors and use the time differences to localize a PD source, as shown in Fig. 3.25. For this purpose, it is necessary to know the relative positions of the sensors in a coordinate system mapped onto the HV equipment

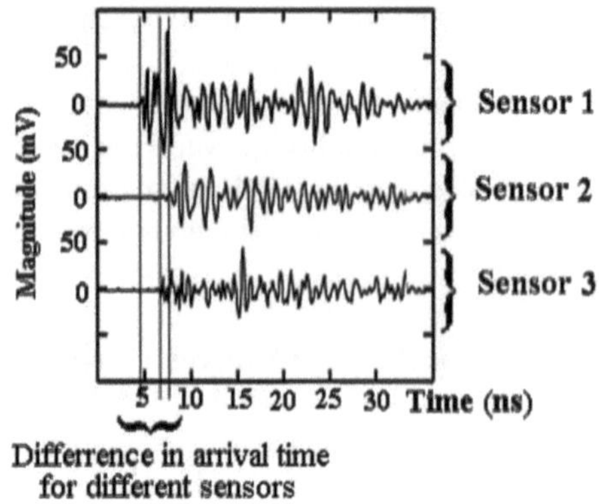

Fig. 3.25 Typical flight of time measurement using UHF sensors

under test and the velocity of UHF signal propagation. In the bulk oil within a power transformer, the velocity of UHF signal has been measured as 2×10^8 m/s. The convenient figure, which is being used when discussing PD location in transformers, is 20 cm/ns [44].

In the case of GIS or GIL, the velocity of propagation is 30 cm/ns. If two sensors are placed at a separation distance of 1 m, then the difference in arrival time at the two sensors will be 3.3 ns. By using proper bandwidth of UHF measuring system and by placing the sensors appropriately, the PD source can be determined by triangulation.

Factors that influence the arrival time of UHF signal from the PD source to the UHF sensor are [44]:

(a) Noise: It obscures the initial part of the UHF waveform until it attains a level sufficient to be detected.
(b) Measurement bandwidth: An approximate relationship for the rise-time t_r of a measurement system with a first-order bandwidth of B Hz is $t_r = 0.35/B$. Therefore, if the signals are captured with a bandwidth of 1 GHz, the minimum observable rise-time of the UHF sensor output signal would be 0.35 ns, which corresponds to 7 cm of UHF signal propagation in oil.
(c) Sensor response: Measurement of the step response of typical UHF sensors has shown that there are small but potentially significant differences in their speed of response.
(d) PD current pulse rise-time: If the current pulse is 10 ns wide, then it would not be reasonable to expect localization of a PD source to within 10 cm of its true location. The pulse itself would occupy around 2 m of distance in terms of its radiated electromagnetic transient. In such a case it is difficult to identify the point on the long current pulse that could be considered as being responsible for the "start" of the radiated UHF signal. A current pulse will radiate an electric field in proportion to the rate of change of current. The time-domain waveforms of both will stretch as the current pulse lengthens, the consequence being to make the definition of a time of arrival at a sensor more difficult [44].

Another aspect that is to be ensured while making any timing measurement to locate discharge sources is the use of equal length of identical cables for making connection to the sensors. In practice, attenuation of these cables is chosen around 4 dB at 500 MHz and around 8 dB at 1500 MHz.

3.3.3.7 Advantages and Disadvantages of UHF Methodology

The main advantages of UHF method of PD measurement are as follows:

(i) Greater immunity to noise, which means that UHF method can be employed for PD measurement in noisy environment, e.g., when the equipment is operational on-site.
(ii) Accuracy of UHF method is good enough to locate defects within the equipment.
(iii) Pulse shapes and associated frequencies are preserved by UHF sensors. This aspect is important as different types of defects may generate PD signals of different shapes and frequencies, which in many cases helps in PD identification and localization.

On the other hand, the main disadvantages of UHF method are:

(i) In UHF method the PD cannot be calibrated in terms of apparent charge.
(ii) The UHF method is less flexible. This is because the UHF sensors and associated accessories are usually specific to one type of equipment, e.g., the UHF measuring arrangement for GIS cannot be used for PD detection in transformers.
(iii) As the UHF measuring components work at very high frequencies, these are more expensive.

3.4 Combined Use of Different PD Detection Methods

While electrical or UHF PD measurements can confirm PD activity, a three-dimensional localization of PD source is necessary for risk evaluation. Normally, three space coordinates and one time information related to a single PD event are used for such localization, which demand use of four sensors. It is not possible by electrical method, but is possible by UHF method. However, in practice in most of the equipment access is limited to three sensors or less. But, with externally mounted acoustic sensors, there is no such limit on the number of sensors to be used. Hence, acoustic method is popular for localization of PD source in electrical equipment. But, acoustic sensors are more sensitive to external mechanical noises than to the acoustic signals generated by internal PD sources. Thus to increase the sensitivity of acoustic measurements, acoustic method is combined with the more sensitive electrical or UHF method. In such combined application, electrical or UHF signals are used as trigger signals which activate the acoustic sensors during PD measurements. Electrical method can be used in laboratory or factory testing, while UHF method is preferred for on-site testing. For example, in the case of transformers, one UHF sensor can be mounted through an oil drain valve and three acoustic sensors are mounted on the transformer tank externally. More plausible decision for or against PD activity can be taken by such combined measurement because of the following reasons: (a) mechanical noise is typically not associated with an electrical/electromagnetic signal and (b) electrical or electromagnetic noise could not generate mechanical signals that have coherent phase relationship. Thus, combined application of electrical/UHF and acoustic methods provide increased immunity to false alarms in on-site testing due to acoustic signals generated by rain, snow, etc. The electrical/UHF triggered averaging of acoustic PD signals has also been found to increase acoustic sensitivity for small discharges significantly due to removal of unwanted acoustic noise in the signals.

3.4.1 Absolute Time Approach in Combined Application of Electrical/UHF and Acoustic Methods

By triggering acoustic measurement with electrical/UHF signal, a proper starting point could be obtained for acoustic signals for computation of arrival times at different acoustic sensor locations. In other words, the instant of occurrence of PD is taken as known, which is the triggering instant of acoustic sensors. This is because of the fact that the travel time of electrical or UHF signal from PD source within an electrical equipment to the measuring location is negligible for all practical purposes. Typically, the propagation delay of electrical or UHF signal is in nanosecond range, while that for acoustic signal is in microsecond or sometimes in millisecond range. That is why this technique is also known as absolute time measurement and the schematic of the absolute arrival times of signals at three sensor locations are shown in Fig. 3.26a [46]. In Fig. 3.26a, the instant of triggering by electrical/UHF signal is shown as PD onset.

The following three equations can then be written in terms of the coordinates of unknown PD source (x_s, y_s, z_s), the known coordinates of three sensor locations (x_i,y_i,z_i) for $i = 1,\ldots, 3$, the assumed velocity of acoustic signal (v_s) and the measured absolute arrival times t_i for $i = 1,\ldots, 3$:

$$(x_s - x_1)^2 + (y_s - y_1)^2 + (z_s - z_1)^2 = (v_s t_1)^2 \tag{3.11}$$

$$(x_s - x_2)^2 + (y_s - y_2)^2 + (z_s - z_2)^2 = (v_s t_2)^2 \tag{3.12}$$

$$(x_s - x_3)^2 + (y_s - y_3)^2 + (z_s - z_3)^2 = (v_s t_3)^2 \tag{3.13}$$

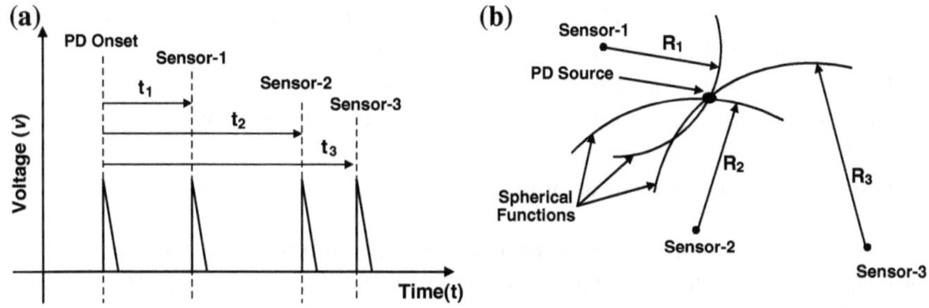

Fig. 3.26 PD source localization using three acoustic sensors combined with Electrical/UHF signal. **a** Schematic representation of absolute arrival times. **b** Intersection of spherical functions

Equation (3.11) through (3.13) represent spherical functions having radii R_i as shown in Fig. 3.26b, where R_i is defined as

$$R_i^2 = \left(v_s t_i^2\right) \tag{3.14}$$

The three spherical functions intersect at the location of PD site.

3.5 Analysis of Partial Discharge Signal

3.5.1 Importance of Denoising

Noise contamination is one of the significant problems of PD detection. PD signals have high frequency components making it difficult to distinguish them from noises. This noise often subsumes completely the low Level PD signals picked up by the detectors. It makes PD detection an intricate problem. To restore PD signals that are mixed with various types of noises is a major challenge in PD measurement. Thus, postprocessing using sophisticated signal processing techniques is needed.

The ability to discriminate PDs from noise improves with an increased knowledge of both the PD and noise characteristics. Such knowledge might include PD bandwidth, dominant frequency, waveform, phase information relative to power frequency voltage, PD attenuation characteristics, the type of noise, the noise phase position, the noise bandwidth, and attenuation characteristics [54, 58].

PDs are normally low-level pulses with fast rise-time and short duration, typically no more than a few hundred nanoseconds. Its waveform varies depending on the fault type, location of fault, insulation characteristics and physical connection. PDs cover a wide signal bandwidth. So, any or all of the following effects can cause interference that may corrupt the PD signals [51–54, 57]:

1. Discrete spectral interference (DSI), e.g., communication and amplitude modulation (AM)/frequency modulation (FM) radio emissions.
2. Repetitive pulses, e.g., from power electronics.
3. Random pulses, e.g., from switching operations or lightning or RF corona emitted from HV equipment.
4. Other noise sources such as ambient and amplifier noise.

PD pulses get contaminated even in a shielded laboratory environment. In this case, noise level is low compared to that in the outdoor or on-field PD measurements. However, this low noise too, deforms the signal up to such an extent that it becomes difficult to detect the defect type and location from the recorded signals, even for a human expert. These noises are mostly random in nature. So, denoising schemes are often used to overcome this problem.

3.5.2 Various Denoising Methodologies

In recent years Wavelet-based denoising techniques are being extensively used for this purpose. Several works related to PDs have been reported in literatures in the past few decades, too many to mention here. As the present study regarding PD detection is related to wavelet-based analysis, only related prior works have been referred here. Ma et al. [51] applied continuous wavelet transform for PD pattern classification. They also used wavelet transform to separate PDs from electrical noise and proposed a wavelet transform-based threshold value selection method [52].

Hang et al. [53] described application of the wavelet transform to extract PDs from narrowband interference. Zhang et al. [54] employed wavelet transform to remove narrowband interference and white noise in on-line PD measurement. Ming et al. [55] reported wavelet applications for characterization of PDs based on laboratory studies.

Satish et al. [56] employed wavelet transform, 128 tap FIR filtering and IIR filtering for PD denoising and compared their performances with respect to various noise sources. In this method the wavelet based decomposition up to level 10 is performed. But this method requires significant human expertise to judge and detect the PDs from the noise. A typical figure of decomposed PD signal is shown in Fig. 3.27.

Shim et al. [57, 58] proposed the use of wavelet transforms to improve their capability for detection and location of PDs in shielded distribution cables. Tian et al. [59] applied wavelet transform denoising techniques to improve signal–to-noise ratio (SNR) for detecting PDs within a HV cable joint.

The method proposed recently by Zhang et al. [54, 60] is an efficient method for PD denoising for outdoor cable inspection. Their method also employs wavelet based decomposition. This method is effective but is semi-automatic from the operational point of view. Selection of noise reference and determination of PD inception voltage require human intervention and expertise.

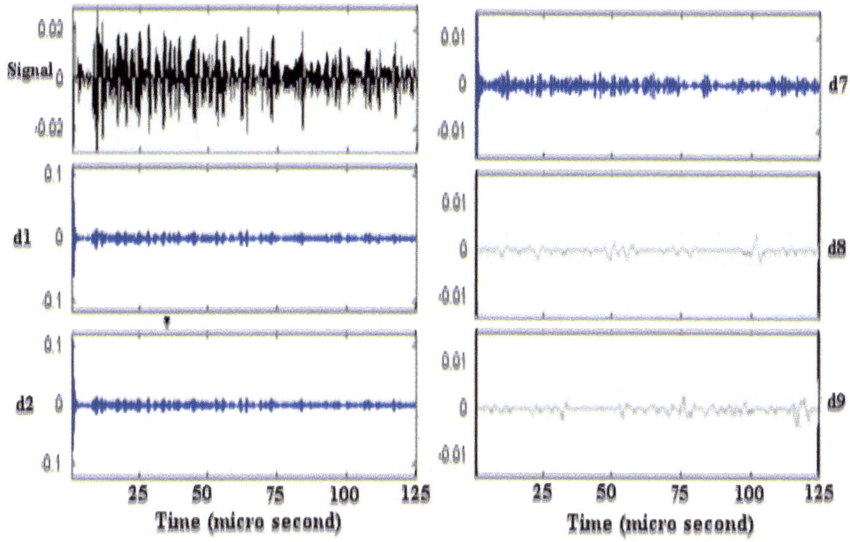

Fig. 3.27 Wavelet decomposition of noisy PD signal at various levels

The significant problem with these denoising methods is that they require an expert operator for the denoising scheme to work. The expertise is required for selection of threshold values of denoising, observation of PD inception voltage and noise level present in that case, etc. In recent times researchers have not only developed [92] a scheme to eliminate the effect of noises from PD pulses but also have done it in a way that requires minimal human expertise, so that the scheme can be used in an intelligent, integrated, and automated PD detection unit. Cross-wavelet transform method is used in this type of analysis.

3.5.3 Cross-Wavelet Transform-Based Feature Extraction from PD Pulses

In this method, noisy PD signals are recorded from the sample under test. These waveforms are then superimposed on the reference power frequency signal. Cross-wavelet transform between the superposed waveform and the corresponding original noisy waveform is obtained. Various features similar to those explained for impulse fault classification may be extracted from the cross-wavelet spectra to classify the defect type [81–84, 92, 103].

The reason behind the choice of these two waveforms (i.e., PD pulses and the superposed waveform) to obtain the cross-wavelet spectra is explained below:

It is known that the nature of PD patterns depend on the phase locations and frequency contents of the pulses. So, the defect may be classified on the basis of these informations if they can be extracted from these PD pulses [54]. As cross-wavelet transform gives a measure of correlation between two waveforms in time–frequency domain, it is a possible choice for extracting the hidden features from the PD pulse patterns to classify the type of defects. In the present case two types of signals are recorded: (a) the noisy PD pulses; (b) the power frequency reference signal. So, the possible cross-wavelet spectrum which can be used for feature extraction to classify PD patterns may be one of the followings:

(i) Cross-wavelet spectrum obtained from the cross-wavelet transform of noisy PD pulse and the power frequency reference signal.
(ii) Cross-wavelet spectrum obtained from the cross-wavelet transform of noisy PD pulse with itself.
(iii) Cross-wavelet spectrum obtained from the cross-wavelet transform of PD pulse and power frequency reference signal superposed on noisy PD pulse.

In the first case, as the frequency content of the noisy PD signal is very high with respect to the power frequency signals, the coefficient values for cross-wavelet spectrum are low and close to each other, thus giving poor performance in classification.

In the second case, the phase information is lost because two waveforms used in cross-wavelet transform are same. So, there occurs a loss of information, causing degraded performance of the classification algorithm.

It has been found during experimentation by the authors that the third one gives acceptable performance for classification of the PD pulses using the scheme studied in this work. A pictorial representation of the scheme is shown in Fig. 3.28. However, there exist other possibilities of selecting different combinations of waveforms for some cases. So, the reader may use or may modify the method suitably to get acceptable performance.

Using the data acquisition setup detailed in IEC 60270, PD signals are usually recorded over one full cycle of power frequency wave (i.e. 20 ms assuming 50 Hz frequency). One of such recorded data is shown in Fig. 3.28 and shows the waveform when this noisy data is superposed on the power frequency reference signal. Cross-wavelet transform between these two waveforms are performed for all the data sets. For identification of PD patterns seven features may be extracted from the $|W^{xy}|$, in a way similar to that followed in the case of impulse fault identification, explained in Chap. 2. The features (from F_1 to F_7) are described below [92, 102, 103]:

$$F_1 = \frac{\sum_s \sum_\tau s\tau |W^{xy}(s,\tau)|}{\sum_s \sum_\tau |W^{xy}(s,\tau)|} \tag{3.15}$$

$$F_2 = \sqrt{\frac{\sum_s \sum_\tau s^2\tau^2 |W^{xy}(s,\tau)|}{\sum_s \sum_\tau |W^{xy}(s,\tau)|}} \tag{3.16}$$

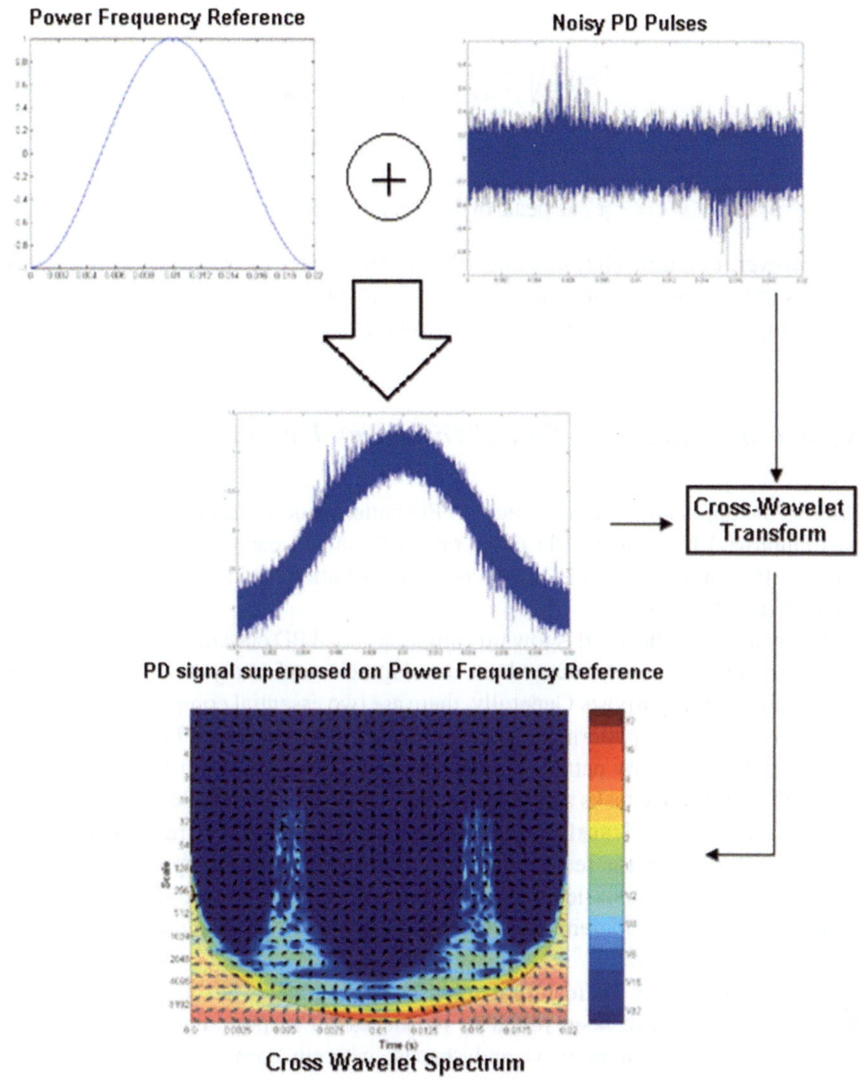

Fig. 3.28 Scheme to obtain cross-wavelet spectrum for feature extraction

$$F_3 = \frac{\sum_{s}\sum_{\tau} |W^{xy}(s,\tau)|}{|W^{xy}(s,\tau)|_{\text{peak}}} \tag{3.17}$$

$$F_4 = \frac{\sum_{s}\sum_{\tau} |W^{xy}(s,\tau)|}{(s_{\max} - s_{\min})(\tau_{\max} - \tau_{\min})} \tag{3.18}$$

$$F_5 = \sqrt{\frac{\sum_s \sum_\tau (A_4 - |W^{xy}(s,\tau)|)^2}{(s_{max} - s_{min})(\tau_{max} - \tau_{min})}} \tag{3.19}$$

$$F_6 = \text{``s''at peak of} |W^{xy}(s,\tau)| \text{i.e.} |W^{xy}(s,\tau)|_{peak} \tag{3.20}$$

$$F_7 = \text{``}\tau\text{''at peak of} |W^{xy}(s,\tau)| \text{i.e.} |W^{xy}(s,\tau)|_{peak}. \tag{3.21}$$

However, one may choose some other features like, location of local peaks of $|W^{xy}|$ surfaces, if any, or some features from phase angle information $\varphi(s,\tau)$ depending upon the number of patterns to be classified.

3.5.4 Classification of Partial Discharge Patterns

Prior to the application of computers to PD pattern studies, digital instrumentation was employed to measure PD pulse rate [61] and to carry out the pulse height analysis [62] and to record PD pulse phase distributions [63] as well as pulse time interval distributions [64].

In recent times, the trend is toward computer-aided PD classification. Thus, much of the research is, now, focused on development and analysis of sophisticated algorithms for this purpose. Generally, there are two essential components of almost all the algorithms for pattern classification [65], i.e., formation of Feature Vector or Fingerprint followed by pattern recognition (classification). It must be emphasized that both of the above tasks should be viewed together for an efficient classification.

Discrete Wavelet Transform (DWT) is used for feature extraction of PD by Jang et al. [66] where the feature vector is constructed using the approximation and detail coefficients in a six-level DWT. Carminati et al. have used different mother wavelets in [67] for efficient classification of the aging phenomena due to partial discharge degradation.

The use of fractals for feature extraction of PD images has also been reported [68–70]. Fractals have been found to be successful in providing descriptions of naturally occurring phenomena and shapes, wherein conventional and existing mathematical models were found to be inadequate. In recent years, this technique has attracted increased attention for classification of textures and objects present in images and scenes [71, 72].

Researchers like Niemeyer et al. [73] and Fuji et al. [74] presented results to show that the 'electrical tree' shapes of the discharge process within the insulation material has fractal properties and also pointed out that fractal dimension is correlated to the shape of the tree.

ANN-based classifiers are also in use, such as backpropagation network [75], modular neural network [76], cascaded neural network [77], etc. Kohonen Self Organizing Feature map [78] and learning vector quantization method [79] are other methods which have successfully been used for PD detection and classification.

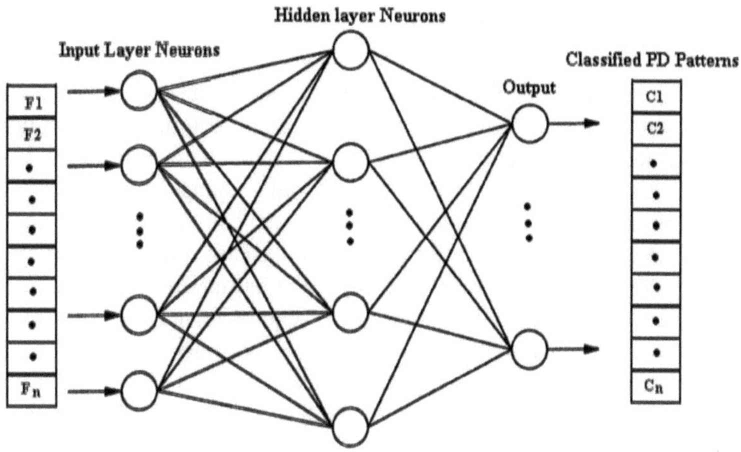

Fig. 3.29 Backpropagation Neural Network

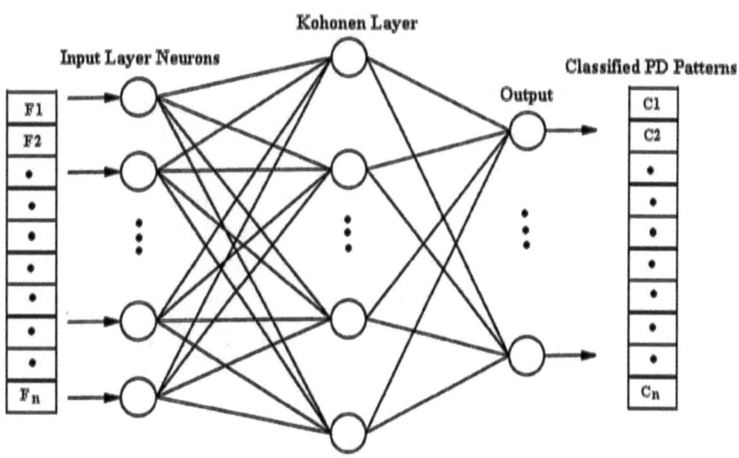

Fig. 3.30 Learning vector quantization network for classification of PD

A typical backpropagation network is shown in Fig. 3.29. A schematic of learning vector quantization method is also given in Fig. 3.30. Contin and Cavallini have used fuzzy logic-based classifiers for the same purpose [80].

In recent times Rough set-based classification technique has also been used for this purpose where cross-wavelet transform-based feature extraction is employed [92]. It is evident that there is no apriori knowledge regarding features of the cross-wavelet spectrum which will be suitable and should be taken for classification of PD patterns. So, the data table obtained after the extraction of the features may contain imprecise or superfluous information as in the case of impulse fault classification, explained in Chap. 2. As Rough sets are well-suited for this kind of problems [85–92], a classifier based on Rough sets may be used here.

3.5.5 Pulse Sequence Analysis

In classical approach, the partial discharge data is acquired as phase-resolved plots. This means, the partial discharge pulses are acquired as time-domain signals and are superimposed over the applied sinusoidal voltage. On the basis of this data, the pulse heights and the phase of the applied sinusoidal voltage at which these pulses occur are the two key elements from which several analytical methods are developed to determine the nature of the partial discharges that occur in a system.

However, the classical approach to analyze the partial discharge data has some limitations. Detailed study on the nature of the acquired data reveal that though the pulse heights can give a prediction about the severity of discharges that are taking place inside a system, the phase of the applied voltage at which the discharges occur is not a meaningful parameter to analyze the results. This statement is especially true in the case of liquid or solid dielectrics because, the inception of the partial discharge is determined by the local electric field inside the defect in the dielectric and not by the amplitude or phase of the applied sinusoidal voltage. Defect is a term used to denote the site of origin of partial discharge inside a dielectric. This is because it is expected that a perfect dielectric should not give rise to any sort of partial discharge in a system.

Modern analysis of partial discharge data rely on sequence correlated parameters such as the voltage differences of the applied voltage between two or three consecutive discharge pulses. These types of studies are far more decisive to contribute in the correct prediction about the test specimen because the studies incorporate one key phenomenon, the local electric field inside the discharge site that is responsible for the very next discharge. By means of the pulse sequence analysis, called PSA in short, the nature of defects can be characterized with a better precision and discharges occurring from single and multiple discharge sites can be clearly distinguished, which was out of the scope in classical approach.

3.5.5.1 Basic Principles of Standard Analysis Methods

In standard analysis methods, as stated earlier, involving analog to digital conversion process for storing data into the computer, data is acquired with respect to the power frequency cycle from the object or system under test. The entire measuring time with several number of complete power frequency cycles is then transformed into one reference cycle of the applied voltage as shown in Fig. 3.31. It should be noted here that the data acquisition remains synchronized with the applied power frequency cycle for the entire duration of capture to avoid any phase error in the acquired data. Thus, a single power frequency cycle exists and all the discharge data are placed within this span of 0° and 360° according to their angle of occurrence with respect to the beginning of the applied cycle. No correlation between pulses in the same or in different cycles of the applied voltage is taken into consideration.

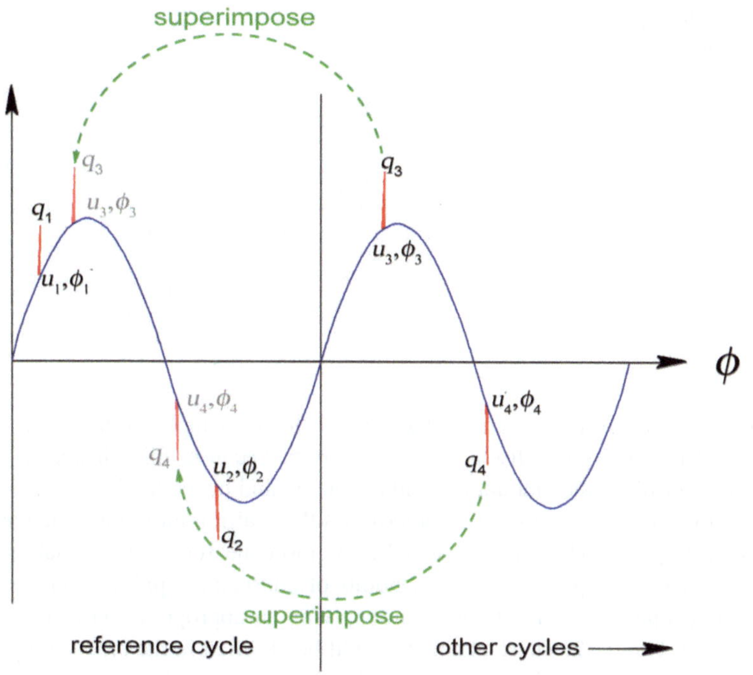

Fig. 3.31 Basic principles of standard PD analysis methods

The parameters of interest here are (i) pulse height (abbreviated as q_i in general), (ii) the applied voltage u_i at the instant of discharge or the phase of occurrence ϕ_i of each discharge pulse relative to the beginning of each voltage cycle, and (iii) the number of discharges (abbreviated as n_i in general) that occur at a particular phase angle of the reference cycle. Subsequent analysis is then done on the basis of so-called ϕ–q, ϕ–n, or three-dimensional ϕ–q–n characteristics with some more or less meaningful statistical parameters [93].

3.5.5.2 Basic Principles of PSA Method

From the preceding discussion about the onset of the partial discharge, it is obvious that the local space charge along with its associated electric field strongly influences the discharge process. Thus, very prominent correlations between consecutive discharge pulses must exist. In other words, during partial discharge, the successive pulses cannot be considered to be independent events without a correlation with the preceding discharges. This idea is the basis behind pulse sequence analysis. Here, more importance is given to the local electric field and its change or the elapsed time between consecutive discharges than the value of the external voltage or the phase angle where the discharge occurs. As far as the

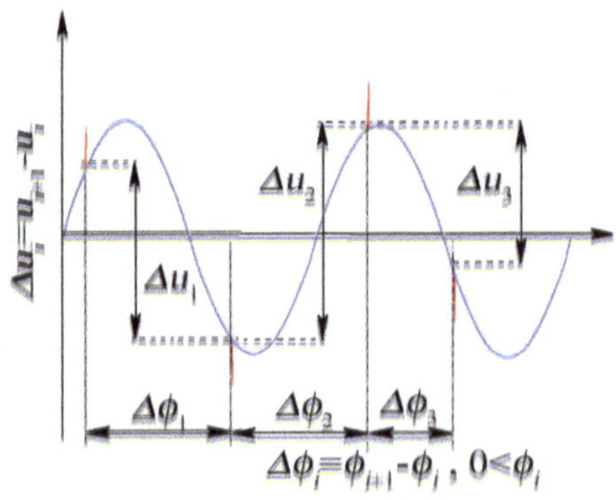

Fig. 3.32 Basic principles of PSA method

external parameter is concerned, the change in the supply voltage that occurs between two successive discharges is important because, this change in supply voltage determines the amount of residual charge and hence the local electric field involved in the discharge site that determines the ignition/initiation characteristics of the next pulse. The analysis will be far more nearer to the actual physical process involved if pulse/pulse correlations of consecutive pulses over the entire measuring time are taken into account, applying an appropriate set of parameters as shown in Fig. 3.32. These parameters will be clarified in the practical approach in the next section. Analyzing the mutual influence of discharge pulses and their non-stationary behavior with continuing insulation degradation provides a better understanding of physical processes involved in partial discharge phenomena. It reveals basic information on space charge buildup and decrease, the corresponding time constants and the field modifying influence of positive and negative space charges as well [94].

3.5.5.3 Practical Approach

Data Acquisition Method

For practical partial discharge data acquisition, the differences of instantaneously applied voltage magnitude that are taken between two consecutive pulses, called Δu features, are the key parameters to study PD sequence[94–97]. In PSA method, the partial discharge data are stored in phase-resolved pulse sequence (PRPS) data format synchronized with the power frequency cycle for the entire duration of data acquisition. These data are never superimposed into a single cycle. In such a way, the Δu features of consecutive pulses could be extracted to analyze and identify the nature of defect, i.e., the origin of partial discharge. It should be noted here that the origin of partial discharge, henceforth regarded as defect, can take many forms. There can be corona, surface, and void discharge within insulation as well as single and multiple sites of origin in all these cases which makes the analysis more complicated. From this Δu feature, two more patterns are extracted. These are called the n–Δu (read as n versus Δu) and $\Delta u_i - \Delta u_{i-1}$ patterns. These patterns are extracted by subtracting the values of instantaneous voltages at the time of occurrences of two consecutive discharge pulses divided to the peak of the applied high voltage. The magnitude of this feature can be changed from −2 up to +2 in the X-axis. This feature is supposed to be in a close relation with the changes of space and surface charges which remain in discharge source location. Remaining charges influence the sequence of pulses and cause the phenomenon which is called the memory effect in partial discharge activities [95–98]. In this way, pulse sequence analysis could be done by using Δu based patterns [93].

Typical Laboratory Setup

The schematic of data acquisition of partial discharge signals in a typical laboratory setup is given in Fig. 3.33. Test voltage is applied across the sample and the partial discharge that takes place inside the void is filtered from the applied high voltage through the coupling capacitor. A detailed description about the process is available in the literature [99] and in Sect. 3.3.1. The partial discharge signal appears across the inductor, where the role of the damping resistor is to prevent sustained oscillatory signal across the inductor in the case of a partial discharge pulse. C1 and C2 constitute the potential divider for the power frequency high voltage and the corresponding low voltage across C2 is used as a phase-reference of partial discharge pulses.

What Should We Expect During Analysis Using PSA Method?

In order to evaluate PD patterns, some test records are obtained on the basis of artificial defects like corona, surface, and void discharges. Since these discharges originate from different physical phenomena, they are expected to give results from which, the respective defects can be distinctly differentiated. Data

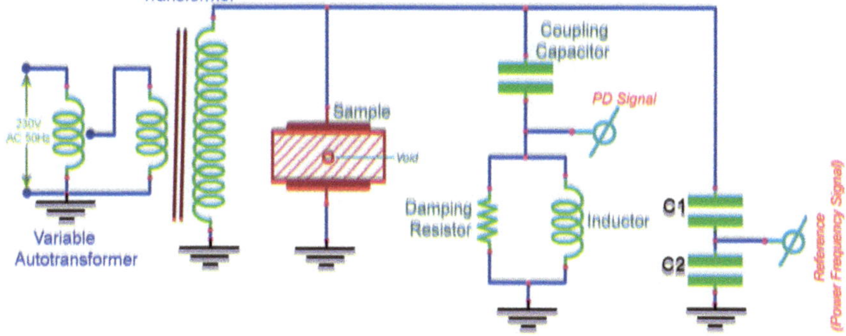

Fig. 3.33 Schematic of PD data acquisition from a sample

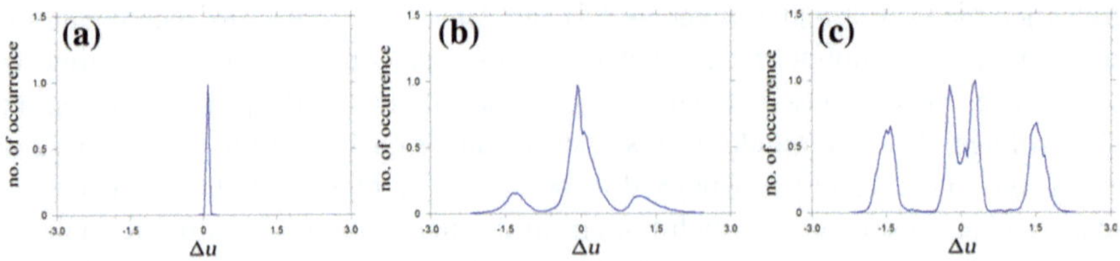

Fig. 3.34 Normalized n–Δu patterns of artificial PD defects: **a** corona, **b** surface and **c** void discharges

acquisitions were performed separately on three types of defects and were stored in phase-resolved pulse sequence (PRPS) data format.

In n–Δu pattern, the frequency of occurrences of Δu features in PD activities is displayed. The Δu feature is displayed on the X-axis and the corresponding frequency of occurrences (n) is displayed on the Y-axis. n is normalized between the values 0 and 1. The n–Δu patterns of partial discharge originating from various types of defects are shown in Fig. 3.34. Figure 3.34a shows the result due to a corona discharge, b shows the result due to a surface discharge, and c shows the result originating from partial discharge due to a void type of defect. From Figs. 3.34a–3.35c, it can be clearly seen that the patterns are distinct and have a good potential in discriminating between different defects. However, the n–Δu pattern is in cumulative form and the relation between Δu features of consecutive pulses is not taken into account [100].

In Δu_i–Δu_{i-1} patterns, the Δu of each discharge pulses is plotted versus its previous pulse Δu_{i-1}. For a comparative study, Fig. 3.35a–c is shown for the same partial discharges originating from the same set of defects from corona discharge, surface discharge, and void discharge, respectively. As shown in Fig. 3.35, the number of clusters, their location and relative densities are the major parameters of this pattern that could discriminate between different defects. Since in the Δu_i–Δu_{i-1} patterns, the Δu features of two consecutive pulses are used, they include more information about PD pulses sequence with respect to n–Δu patterns as can be seen from Fig. 3.35.

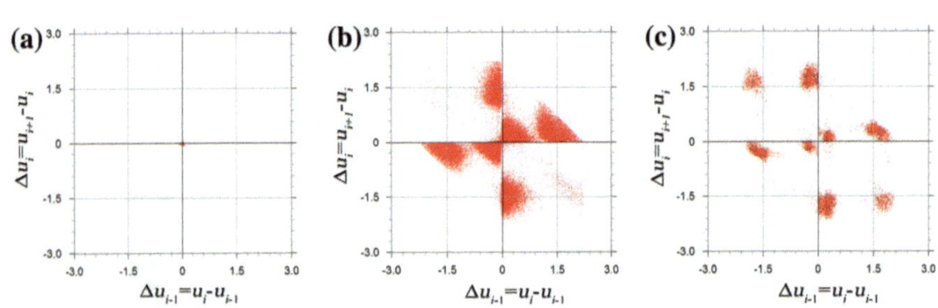

Fig. 3.35 The Δu_i–Δu_{i-1} patterns of artificial PD defects: **a** corona, **b** surface and **c** void discharges

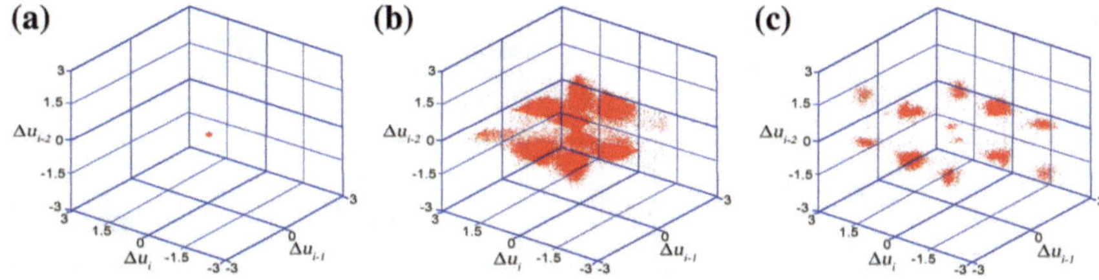

Fig. 3.36 The Δu_i–Δu_{i-1}–Δu_{i-2} patterns of artificial PD defects: **a** corona, **b** surface and **c** void discharges

It can be interpreted that inclusion of all the features by combining Figs. 3.34 and 3.35 will increase the discriminative pattern compared to individual n–Δu and Δu_i–Δu_{i-1} patterns. Thus, bringing more sequences in patterns will increase the ability to identify the type of partial discharges originating from various types of defects. An example of the above statements can be given by the patterns in Fig. 3.36, where the third sequences of Δu features are introduced in a 3D Δu_i–Δu_{i-1}–Δu_{i-2} pattern. It is clearly seen that in these patterns some extra discriminative clusters appear in respect to Δu_i–Δu_{i-1} patterns. So, the third sequences of Δu features could be used in defect identification to enhance its reliability. Now it becomes a choice for the reader to try various types of correlations between combinations of partial discharge pulses in his/her particular case to extract the most suitable features in identifying the particular nature of partial discharge related to the work.

Uniqueness of PSA Method: Prediction of Multiple Discharge Site

The unique feature of the PSA is the ability to find out multiple discharge sites, in contrast to the standard analysis that gives no possibility to separate these situations. A clear comparative study between ordinary pulse height versus phase angle plot and PSA shows that the former cannot identify multiple discharge sites inside insulation [101]. An example with such a comparative study is given in Fig. 3.37. In an artificially made sample, two defects are generated in insulation such that they have different inception voltages U_i and 1.31 U_i that can be stressed in parallel. When voltage is applied to this sample up to 1.3 U_i, only one of the defects generates partial discharges. At a voltage of 1.33 U_i both the defects produce discharges. So it is a known case that there must be a distinct difference in the physical phenomena on increasing the inception voltage from 1.3 U_i to 1.33 U_i. Figure 3.37a shows pulse heights (V) versus phase angles (ϕ) for the sample with inception voltages up to 1.33 U_i. Between 1.3 U_i and 1.33 U_i, no significant difference in the pattern exists that can identify between single and multiple site discharges.

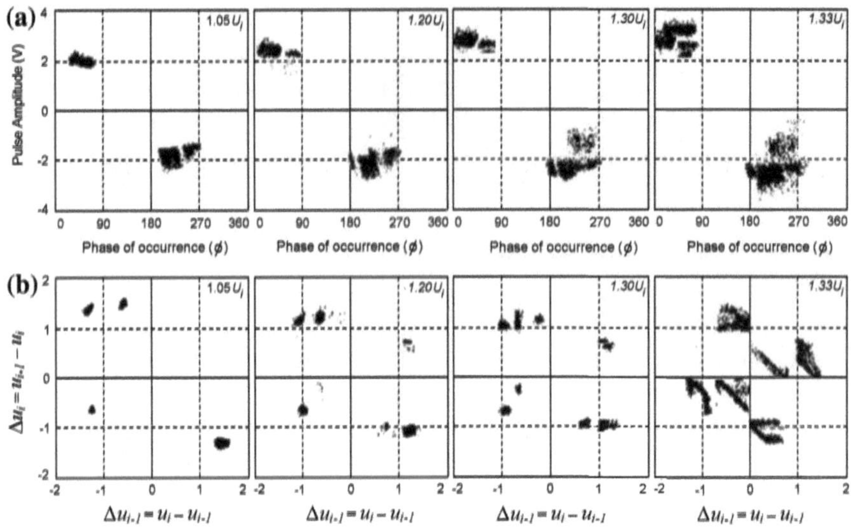

Fig. 3.37 **a** Pulse heights (V) versus phase angles (ϕ) for an experimental sample with inception voltages up to 1.33U_i [101]. Reproduced with kind permission ©IOP Publishing Ltd 2012. **b** The result of a PSA analysis for the same experimental sample of (**a**) with inception voltages up to 1.33U_i [101]. Reproduced with kind permission © IOP Publishing Ltd 2012

Now, Fig. 3.37b shows the result of a PSA analysis. A characteristic change in the Δu pattern occurs between 1.30 U_i and 1.33 U_i, as line-shaped regions, appear either parallel to the axes or at 135°. Thus, PSA analysis is able to identify the defects due to single and multiple discharge sites.

However, for automated operations of the scheme one may design an expert system based on the analysis algorithms as described earlier.

3.6 Future Scope and Challenges

In the case of acoustic sensors, internal mounting within an electrical equipment such as transformer is not permissible in most of the cases as they transmit an electrical signal that may interfere with the normal functioning of the transformer. Hence, efforts are now on to develop acoustic sensors suitable for internal mounting within high voltage equipment. One such possibility is to use fiber optic technology in tandem with acoustic sensors for internal mounting.

Real-life feedback from power utilities suggest that although the multisensor PD monitoring system would be a useful tool, key issues in its successful deployment are sensor design and installation. Practicing engineers are reluctant to modify the external housing for internal sensor installation. In such cases, it may be possible to install miniature internal sensors without modifying the equipment in any way. This could be achieved by inserting the sensor through small gaps in the equipment. For this purpose highly sensitive UHF sensors, which require entry dimensions of less than 1 mm diameter, are currently under development.

Experimental results suggest that for pulse injection testing of PD locating systems, a UHF sensor designed specifically for this purpose may be more suitable than the one designed for optimum broadband sensitivity. Efforts are now on to design UHF sensors for efficient pulse injection that will seek to optimize the rise-time of coupled UHF signals in order to quantify the accuracy of UHF-based PD location systems.

In the case of MV and HV switchgears, in addition to monitoring discharges, the wide bandwidth of a UHF monitoring system would allow it to accurately time arcing that occurs during the operation of switchgears. By relating these measurements to the timing of the relevant control signals, it may be possible to monitor other parameters of interest, such as contact wear.

References

1. IEC Standard (2000) High-voltage test techniques: partial discharge measurements. IEC-60270
2. Petersen W (1912) Archiv für Elektrotechnik 1:28
3. Schering H (1924) Die Isoliestoffe der Elektrotechnik. Verlag Julius Springer, Berlin
4. Lloyd WL, Starr EC (1927) Methods used in investigating corona loss by means of the cathode ray oscillograph. AIEE Trans 46:997–1024
5. Gemant A, von Philippoff W (1932) Die funken-strecke mit vorkondensator. Zeitschrift fur technische Physik 13:425–430
6. Tykociner JT, Brown HA, Paine EB (1933) Oscillations due to ionization in dielectrics and methods of their detection and measurement. University of Illinois Bulletin No 49 and 50, vol 30, Urbana, Illinois
7. Quinn GE (1940) Detection of the ionization point in electrical apparatus. AIEE Trans 59:680–682
8. Paine EB, Brown HA (1940) Corona discharge in rubber insulated cables. AIEE Trans 59:709–714
9. Austen AEW, Whitehead S (1941) Discharges in insulation under alternating current stresses. J IEE 88(11):18–22
10. Mason JH (1953) The deterioration and breakdown of dielectrics resulting from internal discharges. J IEE 100(I):149–158
11. Bartnikas R, d'Ombrain GL (1963) A method for determining the dissipation factor value from corona intensity and pulse count. IEEE Trans Power Apparatus Syst 82:366–375
12. Bartnikas R, d'Ombrain GL (1965) A study of the corona discharge rate and energy loss in spark-gaps. IEEE Trans Power Apparatus Syst 84:770–779
13. Bartnikas R (1966) Pulsed corona loss measurements inartificial voids and cables. In: Proceedings of the CIGRE, report 202, Paris
14. Bartnikas R, Levi J (1969) A simple pulse height analyzer for PD rate measurement. IEEE Trans Instrum Measur 18:311–345
15. Bartnikas R (1973) Use of multi-channel analyzer for corona pulse-height distribution measurements in cables and other electrical apparatus. IEEE Trans Instrum Measur 22:403–407

16. Bapt JC, Bui-Ai, Mayoux C (1974) Corona frequency analysis in artificial cavities in epoxy resins. In: 1973 annual report, conference on electrical insulation and dielectric phenomena, NAS/NRC, Washington DC, pp 282–288
17. van Glahn P, van Brunt RI (1995) Continuous recording and stochastic analysis of PD. IEEE Trans Dielectr Electr Insul 2:590–601
18. Ward BH (1992) Digital techniques for PD measurements. IEEE Trans Power Delivery 7:469–479
19. Osvath P (1995) Comment and discussion on digital processing of PD pulses. IEEE Trans Dielectr Electr Insul 5:685–699
20. Rizk F, Gervais Y, Luhrmann H (1975) Performance of electromagnetic shields in high voltage laboratories. IEEE Trans Power Apparatus Syst PAS-94(6):2077–2083
21. Casey KF (1988) Electromagnetic shielding behavior of wire mesh screens. IEEE Trans Electromagn Compat 30(3, part 2):298–306
22. Chatterjee B, Dey D, Chakravorti S, Roy CK (2008) Experimental investigation of the electromagnetic signal attenuation characteristics of the Partial Discharge Laboratory at Jadavpur University. Int J Emerg Electr Power Syst 9(2): Art 4 (BEPRESS Publication, CA, USA)
23. Sahoo NC, Salama MMA, Bartnikas R (2005) Trends in partial discharge pattern classification: a survey. IEEE Trans Dielectr Electr Insul 12(2):248–264
24. Gulski E (1995) Discharge pattern recognition in high voltage equipment. IEE Proc Sci Measur Technol 142:51–61
25. Cacciari M, Contin A, Rabach G, Montanari GC (1995) An approach to partial discharge investigation by height distribution analysis. IEE Proc Sci Measur Technol 142:102–108
26. Gulski E (1995) Digital analysis of partial discharges. IEEE Trans Dielectr Electr Insul 2(5):822–837
27. Abdel-Galil TK, Hegazy YG, Salama MMA, Bartnikas R (2005) Fast match-based vector quantization partial discharge pulse pattern recognition. IEEE Trans Instrum Measur 54(1):3–9
28. Lundgaard LE (1992) Partial discharge, part XIII: acoustic partial discharge detection—fundamental considerations. IEEE Electr Insul Mag 8(4):25–31
29. Boggs SA (1990) Partial discharge, part II: detection sensitivity. IEEE Electr Insul Mag 6(5):35
30. Schlemper HD, Feser K, Blaum H, Kirchesch P (1996) Sensitivity of acoustic PD detection in GIS Laboratory experiments and on-site experience. In: Proceedings of the IEEE international symposium on electrical insulation, Montreal, Canada, 16–19 June 1996, pp 99–102
31. IEEE PC57.127/D2 (1989) Trial use guide for the detection of acoustic emissions from partial discharges in oil-immersed transformers
32. Robles E (2000) Field diagnostic testing of power generator and transformers using modern techniques. In: Proceedings of the HV testing monitoring and diagnostic workshop, Alexandria, Virginia, paper no 12, Sept 2000
33. Lundgaard LE (1992) Partial discharge, part XIV: acoustic partial discharge detection—practical application. IEEE Electr Insul Mag 8(5):34–43
34. Lundgaard LE, Runde M, Skyberg B (1990) Acoustic diagnosis of gas insulated substations; a theoretical and experimental basis. IEEE Trans Power Delivery 5(4):1751–1758
35. Chen LJ, Lin WM, Tsao TP, Lin YH (2007) Study of partial discharge measurement in power equipment using acoustic technique and wavelet transform. IEEE Trans Power Delivery 22(3):1575–1580
36. Kreuger FH (1989) Partial discharge detection in high voltage equipment. Butterworth & Co, London. ISBN 0-408-02063-6
37. Howells E, Norton ET (1978) Detection of partial discharges in transformers using acoustic emission techniques. IEEE Trans PAS 97(5):1538–1549
38. Yamagiwa T et al (1991) Development of preventive maintenance system for highly reliable gas insulated switchgear. IEEE Trans Power Delivery 6(2):840–848

39. Hampton BF, Meats RJ (1988) Diagnostic measurements at UHF in gas insulated substations. IEE Proc 135(Pt. C 2):137–144
40. Meijer S, Agoris PD, Smit JJ (2005) UHF PD sensitivity check on power transformers. In: Proceedings of the international symposium on high voltage engineering (ISH), Beijing, China
41. de Kock N, Coric B, Pietsch R (1996) UHF PD detection in GIS: suitability and sensitivity of the UHF method in comparison with the IEC270 method. IEEE Electr Insul Mag 12(6):20–26
42. Markalous SM, Boltze M, Bolliger A, Wilson A (2009) On-line partial discharge diagnosis at power transformers. In: Proceedings of the 76th annual international Doble client conference
43. Klein T, Denissov D, Kohler W, Tenbohlen S (2008) Application of UHF method for on-line PD diagnostics of cable terminations. In: Proceedings of the CIGRE session, Paris, France, paper no B1-203
44. Orr JG, Reid AJ, Judd MD (2008) Sensor response characteristics for UHF location of PD sources. In: Proceedings of the international conference on condition monitoring and diagnosis, Beijing, China, 21–24 April 2008
45. Meijer S, Cichecki P, Agoris PD, Smit JJ, Reijnders HF (2007) UHF sensors for partial discharge diagnostics of power transformers. In: Proceedings of the international conference on electrical engineering and informatics, Indonesia, paper no F-03, 17–19 June 2007
46. IEC 62478: High voltage test techniques: measurement of partial discharges by electromagnetic and acoustic methods
47. CIGRE TF 15/33.03.05 (1999) PD detection system for GIS: sensitivity verification for the UHF method and the acoustic method. In: Proceedings of the ELECTRA, pp 75–87
48. Meijer S, Agoris PD, Smit JJ, Judd MD, Yang L (2006) Application of UHF diagnostics to detect PD during power transformer acceptance tests. In: Proceedings of the conference on record of the 2006 IEEE international symposium on electrical insulation, June 2006
49. CIGRE A2.18 (2003) Life management techniques for power transformers. In: Proceedings of the CIGRE brochure 227, Jan 2003
50. Muhr M, Strehl T, Gulski E, Feser K, Gockenbach E, Haushild W, Lemke E (2006) Sensors and sensing used for non-conventional PD detection. In: Proceedings of the CIGRE session paper D1.102
51. Ma XD, Zhou C, Kemp IJ (2000) DSP based partial discharge characterization by wavelet analysis. In: Proceedings of the IEEE 19th international symposium on discharge and electrical insultaion in vacuum, Xi'an, China, pp 780–783
52. Ma X, Zhou C, Kemp IJ (2002) Interpretation of wavelet analysis and its application in partial discharge detection. IEEE Trans Dielectr Electr Insul 9:446–457
53. Hang W, Kexiong T, Deheng Z (1997) Extraction of partial discharge signals using wavelet transform. In: Proceedings of the 5th international conference on properties and applications of dielectric materials, Seoul, Korea,vol 1, pp 322–325
54. Zhang H, Blackburn TR, Phung BT, Sen D (2007) A novel wavelet transform technique for on-line partial discharge measurements, part 1: WT de-noising algorithm. IEEE Trans Dielectr Electr Insul 14(1):3–14
55. Ming Y, Birlasekaran S (2002) Characterization of partial discharge signals using wavelet transform and statistical techniques. In: Proceedings of the IEEE international symposium on electrical insulation, Boston, MA, USA, pp 9–13
56. Satish L, Nazneen B (2003) Wavelet-based de-noising of partial discharge signals buried in excessive noise and interference. IEEE Trans Dielectr Electr Insul 10:354–367
57. Shim I, Soraghan JJ, Siew WH (2001) Detection of PD utilizing digital signal processing methods, part 3: open-loop noise reduction. IEEE Electr Insul Mag 17(1):6–13
58. Shim I, Soraghan JJ, Siew WH (2000) Digital signal processing applied to the detection of partial discharge: an overview. IEEE Electr Insul Mag 16(3):6–12
59. Tian Y, Lewin PL, Davies AE (2002) Comparison of on-line partial discharge detection methods for HV cable joints. IEEE Trans Dielectr Electr Insul 9:604–615

60. Zhang H, Blackburn TR, Phung BT, Sen D (2007) A novel wavelet transform technique for on-line partial discharge measurements, part 2: on-site noise rejection applicaton. IEEE Trans Dielectr Electr Insul 14(1):15–22
61. Bartnikas R, McMohan EJ (eds) (1979) Corona measurement and interpretation, engineering dielectrics, vol I. STP 669, ASTM, Philadelphia
62. Bartnikas R (1973) Note on multichannel corona pulse-height analysis. IEEE Trans Electr Insul 8:2–5
63. Bapt IC, Bui AI, Mayoux C (1974) Coronal frequency analysis in artificial cavities in epoxy resins. In: Proceedings of the conference on electrical insulation and dielectric phenomena, NASrNRC, Washington DC, pp 282–288
64. Mazroua A, Salama MMA, Bartnikas R (1993) PD pattern recognition with neural networks using the multilayer perceptron technique. IEEE Trans Electr Insul 28:1082–1089
65. Sahoo NC, Salama MMA, Bartnikas R (2005) Trends in partial discharge pattern classification: a survey. IEEE Trans Dielectr Electr Insul 12(2):248–264
66. Jang J-K, Kim S-H, Lee Y-S, Kim J-H (1999) Classification of partial discharge electrical signals using wavelet transforms. In: Proceedings of the IEEE 13th international conference on dielectric liquids, Nara, Japan, pp 552–555
67. Carminati E, Cristaldi L, Lazzaroni M, Monti A (2001) A neuro-fuzzy approach for the detection of partial discharge. IEEE Trans Instrum Measur 50:1413–1417
68. Satish L, Zaengl WS (1995) Can fractal features be used for recognizing 3-D partial discharge patterns? IEEE Trans Dielectr Electr Insul 2:352–359
69. Krivda A, Gulski E, Satish L, Zaengl WS (1995) The use of fractal features for recognition of 3-D discharge patterns. IEEE Trans Dielectr Electr Insul 2:889–892
70. Lalitha EM, Satish L (1998) Fractal image compression for classification of PD sources. IEEE Trans Dielectr Electr Insul 5:550–557
71. Mandelbrot B (1983) The fractal geometry of nature. Freeman, New York
72. Lovejoy S, Schertzer D (1986) Scale invariance, symmetries, fractals and stochastic simulations of atmospheric phenomena. Bull Am Meterol Soc 67(1):21–32
73. Niemeyer L, Pietronero L, Wisemann HJ (1984) Fractal dimension of dielectric breakdown. Phys Rev Lett 6(6):661–674
74. Fuji M, Watanabe M, Kitani I, Arii K, Yoshino K (1991) Fractal character of dc trees in Polymethylmethacrylate. IEEE Trans Electr Insul 26:159–162
75. Kranz H-G (1999) Diagnosis of partial discharge signals using neural networks and minimum distance classification. IEEE Trans Electr Insul 28:1016–1024
76. Hong T, Fang MTC, Hilder D (1996) PD classification by a modular neural network based on task decomposition. IEEE Trans Dielectr Electr Insul 3:207–212
77. Salama MMA, Bartnikas R (2002) Determination of neural network topology for partial discharge pulse pattern recognition. IEEE Trans Neural Netw 13:446–456
78. Satish L, Gururaj BI (1993) Partial discharge pattern classification using multilayer neural networks. IEE Proc Sci Measur Technol 140:323–329
79. Gulski E, Krivda A (1993) Neural networks as a tool for recognition of partial discharges. IEEE Trans Electr Insul 28:984–1001
80. Contin A, Cavallini A, Montanari GC, Pasini G (2002) Digital detection and fuzzy classification of partial discharge signals. IEEE Trans Dielectr Electr Insul 9:335–348
81. Ruessink BG, Coco G, Ranasinghe R, Turner IL (2006) A cross-wavelet study of alongshore non-uniform near shore sandbar behavior. In: Proceedings of the joint conference on neural networks, pp 4310–4317, Vancouver, Canada, July 2006
82. Torrence C, Compo GP (1998) A practical guide to wavelet analysis. J Am Meteorol Soc 79(1):61–79
83. Li X, Yao X, Jefferys JRG, Fox J (2005) Computational neuronal oscillations using morlet wavelet transform. In: Proceedings of the IEEE annual conference on engineering in medicine and biology, Shanghai, China, Sept 2005, pp 2009–2012
84. Plett MI (2007) Transient detection with cross wavelet transforms and wavelet coherence. IEEE Trans Signal Process 55(5):1605–1611

85. Pawlak Z (1991) Rough sets: theoretical aspects of reasoning about data. Kluwer, Boston
86. Peng JT, Chien CF, Tseng TLB (2004) Rough set theory for data mining for fault diagnosis on distribution feeder. IEE Proc Gener Transm Distrib 151:689–697
87. Cabral JE, Pinto JP, Gontijo EM, Filho JR (2004) Fraud detection in electrical energy consumers using rough sets. In: Proceedings of the IEEE conference on systems, man and cybernetics, pp 3625–3629
88. Kusiak A (2001) Rough set theory: a data mining tool for semiconductor manufacturing. IEEE Trans Electron Packag Manuf 24:44–50
89. Pal SK, Mitra P (2004) Case generation using rough sets with fuzzy representation. IEEE Trans Knowl Data Eng 16:292–300
90. Yao YY (1999) Rough sets, neighborhood systems and granular computing. In: Proceedings of the IEEE conference on electrical and computer engineering, pp 1553–1558, Canada, May 1999
91. Nguyen HS (1997) Discretization of real value attributes: a boolean reasoning approach. PhD thesis, Department of Mathematics, Warsaw University, Poland
92. Dey D, Chatterjee B, Chakravorti S, Munshi S (2008) Rough-granular approach for impulse fault classification of transformers using cross-wavelet transform. IEEE Trans Dielectr Electr Insul 15(5):1297–1304
93. Hoof M, Patsch R (1995) Pulse-sequence analysis: a new method for investigating the physics of PD-induced ageing. IEE Proc Sci Measur Technol 142(1):95–101
94. Patsch R, Hoof M (1994) The influence of space charges and gas pressure during tree initiation and growth. In: Proceedings of the 4th ICPADM-94, Brisbane, Australia
95. Patsch R, Hoof M, Reuter C (1993) Pulse-sequence analysis, a promising diagnostic tool. In: Proceedings of the 8th ISH93, Yokohama, Japan, vol 4, pp 157–160
96. Patsch R, Hoof M (1993) Pulse-sequence-analysis, the physics behind. In: Proceedings of the joint conference: international workshop on electrical insulation and 25th symposium on electrical insulation materials, Nagoya, Japan, pp 111–114
97. Patsch R, Hoof M (1993) Pulse-sequence-analysis, a way to get a better insight into the physics of discharges. In: Proceedings of the international conference on partial discharge, Canterbury, UK, pp 49–51
98. Gulski E (1991) Computer-aided recognition of partial discharges using statistical tools. Dissertation, University of Delft, Netherlands
99. Chatterjee B (2009) Studies on some novel aspects of measurement and data acquisition techniques in high voltage systems. PhD dissertation, Jadavpur University, Kolkata
100. Mirzaei HR, Akbari A, Allahbakhshi M (2009) Partial discharge identification using pulse sequence analysis and neural network. In: Proceedings of the 16th international symposium on high voltage engineering, Johannesburg, paper D-14, pp 1–4
101. Patsch R, Berton F (2002) Pulse sequence analysis: a diagnostic tool based on the physics behind partial discharges. J Phys D Appl Phys 35(1):25–32
102. Dey D (2009) Studies on intelligent digital signal processing tools for measurements in high voltage systems. PhD Thesis, Faculty of Engineering and Technology, Jadavpur University
103. Dey D, Chatterjee B, Chakravorti S, Munshi S (2010) Cross-wavelet transform as a new paradigm for feature extraction from noisy partial discharge pulses. IEEE Trans Dielectr Electr Insul 17(1):157–166

Chapter 4
Conventional Diagnostic Techniques

4.1 Introduction

A utility can cut costs well while managing aging and its effects, if it can distinguish between the equipment that needs maintenance from that which does not. This is possible through condition monitoring. Condition monitoring of major equipment such as transformers receive more attention in most of the utilities due to the fact that such failures lead to potentially widespread interruption of service and very high replacement cost. Cellulose paper and mineral oil insulation of a transformer degrade due to aging while in service. Three factors play major role in accelerating the aging of transformer insulation, viz. moisture, oxygen, and heat. While there are cases where the degradation of insulation could be sudden, it usually goes down gradually, giving plenty of warning if tested periodically. If advanced insulation degradation goes undetected, then failures will come as a surprise, unplanned, inconvenient, and quite possibly very expensive in time and resources and therefore money to rectify.

In general, measurement of a basic parameter of equipment with a threshold alarm is termed as monitoring. If suitable analysis is incorporated in the monitoring system so that an assessment of equipment condition along with suggestions on proactive steps is available, then it is called diagnostic technique. Purposes of diagnostic testing are threefold: (a) to identify increased aging, (b) to identify the cause of aging, and (c) to identify, if possible, the most appropriate corrective actions.

Common diagnostic techniques for transformers rely on periodic tests based on physical, chemical, and electrical parameters. Physical Measurements, in general, involve measurement of temperature, vibration, acoustic emission, etc. But the diagnostic methods that give useful information on transformation insulation condition are chemical and electrical tests.

S. Chakravorti et al., *Recent Trends in the Condition Monitoring of Transformers*, Power Systems, DOI: 10.1007/978-1-4471-5550-8_4, © Springer-Verlag London 2013

4.2 Chemical Techniques

For reliable assessment of overall condition of a transformer, utilities over the years are using chemical analysis of oil sample. This is due to the fact that decomposition by-products of transformer oil and paper or pressboard insulation are transported by the oil through the transformer. One major by-product of insulation decomposition is moisture which further accelerates insulation degradation with heat acting as catalyst. Some other by-products are low molecular weight gases that remain dissolved in oil, while others include furans, cresols, and phenols that indicate solid insulation degradation. Consequently, chemical techniques of transformer oil diagnostics, which are most commonly used in practice, are moisture analysis in oil, dissolved gas analysis, measurement of degree of polymerization, and furan analysis. This section discusses these commonly used chemical diagnostic techniques.

4.2.1 Moisture Analysis in Transformer Oil

The importance of moisture presence in transformer paper-oil insulation system has been recognized since the 1920s. It is now well-known that moisture has a harmful effect on the life of transformer insulation. The most readily identifiable effects of increase in moisture content in transformer insulation are increased in dielectric dissipation factor, decrease in electrical resistivity, and reduction in dielectric strength. Moreover, increase of moisture content in zones of high electrical stress can lead to partial discharge. Large amount of moisture in transformer oil can lead to bubble formation with dangerous consequences. Researchers have found that moisture content within solid insulation is one of the major factors for degradation of solid insulation. In 1942 Clark [1] reported that if the moisture content of paper insulation is doubled, then its life in terms of mechanical strength is halved. According to Fabre and Pichon [2], the rate of thermal aging of paper is proportional to its moisture content.

4.2.1.1 Sources of Moisture in Transformer Insulation

There are three major sources of moisture build up in transformer insulation: (a) residual moisture, (b) Ingress from atmosphere, and (c) aging decomposition of cellulose and oil [3].

When a transformer is manufactured in the factory, it is subjected to extensive drying before installation. But, even after that residual moisture can be present particularly in plastic insulation, which needs much longer drying time compared to paper or pressboard. Plastic insulation is used in supports of leads, support insulation of neutral of winding, core support insulation, etc. During service this moisture can gradually migrate into oil, thereby increasing the moisture content in oil insulation.

There are three mechanisms of moisture ingress in transformer: (i) moisture sorption takes place when transformer insulation is exposed to air during installation and repair works, (ii) difference in water concentration in atmosphere and oil in the tank causes molecular flow of moisture from atmosphere into the tank, e.g., when transformer is treated under vacuum, and (iii) viscous flow of wet air into the transformer under the action of pressure difference between atmosphere and inside of the tank due to inadequate sealing.

Decomposition of cellulose causes water generation within transformer insulation, typically three molecules of water for one elementary scission of cellulose chain. However, findings of different researchers differ from one another. Lampe and Spicar [4] suggested that depolymerization up to 5–6 scission leads to formation of 2 % water. This reported value was considered to be on the higher side. In a later paper, Shroff and Stannet [5] have shown that reduction of cellulose chain length to one-third of its initial value causes about 0.4 % of water generation. In 1962, Murphy [6] reported that water generation increases abruptly when temperature exceeds 135 °C, because at this temperature degradation of the main structure of cellulose is activated. Hence, the process of intensive water generation is expected to be localized in "hot spots," which comprise about 5 % of total insulation mass in a transformer.

4.2.1.2 Moisture Distribution in Transformer Insulation

Moisture in paper is typically expressed in %, which is the weight of moisture divided by the weight of dry oil-free paper. On the other hand, moisture in oil is typically expressed in parts per million (ppm), which is the weight of moisture divided by the weight of oil (μg/g).

A typical case study [7] shows that a 150 MVA, 400 kV transformer, which contains about seven tons of solid insulation, can have as much as 223 kg of water in paper for an average 3 % moisture content in solid insulation. On the other hand, for 20 ppm moisture content in oil, the total amount of moisture is only about 2 kg in oil as the oil volume of the transformer is about 80,000 l.

This is due to the fact that affinity to moisture of cellulose is several hundred times higher than the oil. Consequently, once moisture is in a transformer, most of it ends up in the cellulose insulation. Practical experience shows that most of the moisture is present in the components of "Thin Cold Insulation," e.g., pressboard barriers, end caps, etc., which operate at oil bulk temperature. About 10 % (by weight) of this group forms "wet zones" with moisture content having 1.0–1.5 % higher than the average. Components of this group are the main sources of oil contamination at elevated temperatures. Thin Hot Insulation components, e.g., turns, coil, layer insulation, etc., operate at the temperature close to the conductor temperature. Moisture content in the components of "Thin Hot Insulation" is less than that in "Thin Cold Insulation." However, its detrimental effect may be higher because of higher temperature at those locations.

Moisture in oil is present in dissolved form and also as a hydrate adsorbed by polar aging products (bonded water). Oil at large is moisture transferring medium.

Moisture distribution between cellulose insulation (paper, pressboard) and oil is not constant. Moisture migration from one insulation component to the other takes place depending on the working condition of a transformer. Moisture migration within transformer insulation takes place by three mechanisms: (a) After an onsite repairing or after an oil change by concentration gradient, (b) During drying of a transformer by pressure gradient, and (c) by temperature gradient.

Under normal operating condition, temperature gradient is the most important mechanism in moisture dynamics. When temperature increases, part of the moisture absorbed in solid insulation migrates to oil. Conversely, moisture returns to solid insulation when temperature decreases. This migration takes place throughout the transformer following the contours of temperature and fibers. The process of moisture migration from paper to oil is faster than the migration of moisture from oil to paper. Further, solubility of water in oil increases sharply with temperature. Hence, oil is able to absorb a larger amount of moisture coming from the solid insulation when the transformer is hotter.

When moisture in oil exceeds the saturation value, free water appears. Oil saturation can take place when the temperature of the transformer decreases abruptly after a period of operation at high temperature in which large amount of water has migrated from solid insulation to oil, e.g., when a highly loaded transformer trips on a cool day. In such a case, oil solubility reduces quickly and moisture in oil tries to return back to solid insulation. But the migration of water to solid insulation is slower than the decrease of solubility so that the saturation value in oil could be exceeded. As a result, free water could be found within the transformer [3]. This may give rise to a very critical situation when that transformer is restarted after the tripping, but before the moisture is reabsorbed into the solid insulation.

4.2.1.3 Moisture Equilibrium in Transformer Insulation

The equilibrium between water content of paper and oil has been widely studied by several researchers, viz. Fabre & Pichon [2], Norris [8], Oommen [9], Griffin [10], etc., who obtained equilibrium curves showing water content in paper versus water content in oil for different temperatures. Figure 4.1a, b show the Oommen's curves [9] and Griffins' curves [10], respectively, for moisture equilibrium of oil-paper insulation system. A detailed review on moisture equilibrium in oil-paper insulation is available in [11].

It is useful to know the moisture equilibrium curves of oil-paper insulation system of transformer. Under equilibrium operating condition of a transformer, these curves provide a quick way of estimating the moisture content in paper by measuring the moisture in oil. Three steps need to be taken for this purpose [3], [12]:

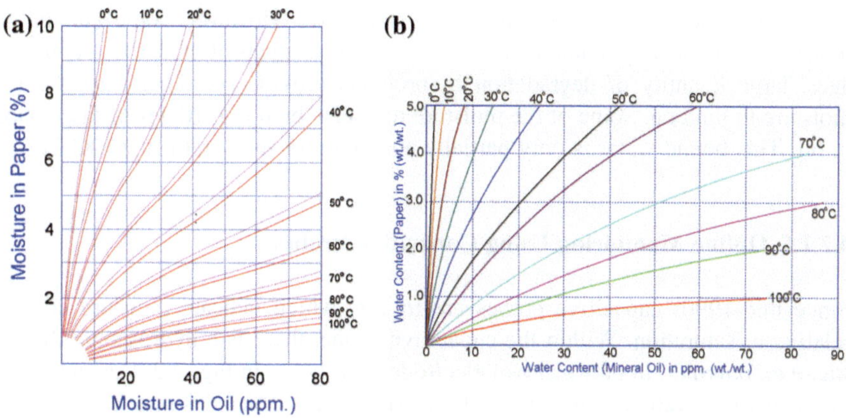

Fig. 4.1 Moisture equilibrium curves of oil-paper insulation system. The *solid lines* indicate the adsorption curves, whereas the *dashed lines* indicate desorption curves. **a** Oommen's curve. **b** Griffins' curve

(i) Taking oil samples from in-service transformers, (ii) Measurement of water content in oil sample by Karl Fischer titration technique, and (iii) estimating moisture content in paper insulation with the help of oil-paper moisture equilibrium curves.

But, this approach suffers from several errors that may result into poor accuracy. Equilibrium depends on moisture adsorption capacity of solid insulation and oil and aging of insulation significantly changes the moisture absorption capacity. There is also possibility of moisture ingress from the atmosphere to oil during onsite oil sampling, transportation, and sample preparation during laboratory testing as transformer oil is highly hygroscopic in nature. The steep gradient of the equilibrium curves in the low moisture region increases error. An important point to be noted here is that the validity of oil-paper moisture equilibrium curves is restricted to the original materials that were used to prepare these curves [3, 12].

4.2.1.4 Moisture in Oil Detection by Karl Fischer Titration

An accurate method of measuring very small amount of moisture in oil is the Karl Fischer Titration (KFT) technique. KFT uses a reagent that consists of iodine, sulfur dioxide, a base, and a solvent such as alcohol. Accurate measurement of moisture content is possible due to quantitative and selective reaction of the reagent with water. This method can be used in both volumetric and coulometric titration systems, the details of which are available in [3, 13]. Automatic coulometric Karl Fischer Titration technique can indicate moisture content as low as 1–2 ppm.

However, as the dynamic exchange of moisture between the paper and oil is very slow, KFT provides, at best, an indication of the insulation quality of the oil alone. KFT provides virtually no indication of the moisture content of the solid insulation system and measures the moisture dissolved in oil. However, some amount of moisture is chemically bound to agents such as by-products of oxidation. As the oil ages, large quantity of degradation by-products provide additional sites for the moisture to bind on. Some of the moisture may also bind to particles in suspension in oil. This bound water is only partially available for measurement by KFT.

4.2.1.5 Online Monitoring Using Capacitive Probes

Since mid-1990s capacitive probes are in use, which indicate moisture content relative to saturation. Within the capacitive probe, there is a hydroscopic dielectric, e.g., polymer, in between two electrodes, as shown in Fig. 4.2. Depending on the relative saturation of the ambient material, water molecules penetrate into the polymer dielectric, which changes the capacitance of the probe because of the high relative permittivity (≈ 80) of water. This variation of capacitance is measured and calibrated with respect to relative saturation [3, 12]. Advantages of capacitive probes are as follows: (i) very convenient for onsite and online applications, (ii) possibility of continuous measurement, (iii) no error due to oil sampling, transportation etc., and (iv) since the capacitive probe measures moisture content relative to saturation, the type of oil, and the oil condition does not affect the measurement [3].

Though the water content in a transformer does not vary rapidly, the problem arising out of temperature variations in transformers can be solved by the continuous measurement of moisture in oil. It is known that the moisture equilibrium curves for oil-paper insulation system are applicable only under stable temperature conditions. In reality most transformers experience daily variation of load and ambient temperature. In such cases, a single sampling of oil may cause erroneous evaluation of moisture content in solid insulation. With continuous monitoring, diffusion time can be taken into account and significant improvement in the estimation of moisture content can be achieved [3].

4.2.2 Dissolved Gas Analysis (DGA)

In 1928, Buchholz [14] published the method of diagnosis of fault in a transformer by testing the gases that evolve rapidly during the failure of oil-filled transformers.

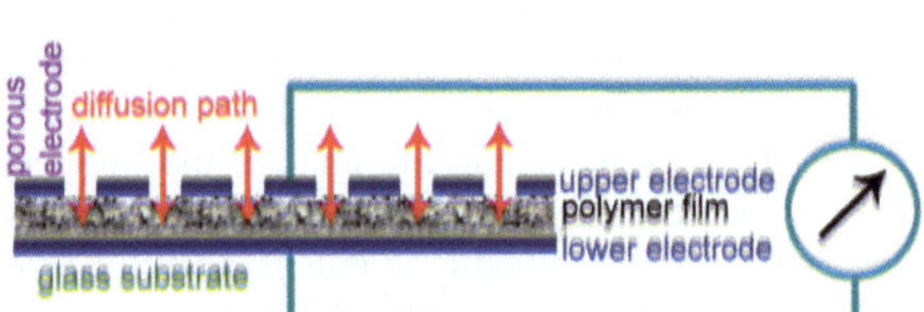

Fig. 4.2 Schematic diagram of a polymer thin film capacitive probe

Subsequently, it was realized that slowly developing faults within a transformer would produce gases which will remain dissolved in oil. As a result such faults could be detected by analyzing the gases dissolved in oil. Hence, DGA monitoring has become a common tool for condition monitoring of transformers and is being extensively practiced by the utilities/manufacturers over the decades. By means of DGA, it is possible to distinguish faults such as partial discharge, corona, thermal heating, and arcing in transformers as well as in other oil-filled equipment. It also helps to detect incipient or newly formed faults at much early stage than the other protective gadgets like Buchholz relay, thereby, saving the repair cost appreciably. This reduction in repair cost is further added with other less tangible benefits of early detection: (i) possible mitigation of oil fires and consequential damage, (ii) ability to schedule outages for units in the warning state (planning minimum shutdown in most convenient time), and (iii) documentation of first year in-warranty problems before out-of-warranty failure.

Mineral insulating oils are mixtures of hydrocarbon molecules in paraffinic or aromatic form comprising CH_3, CH_2 and CH chemical groups. Gas molecules such as Hydrogen (H–H), Methane (CH_3–H), Ethane (CH_3–CH_3), Ethylene (CH_2=CH_2), and Acetylene (CH≡CH) are generated due to decomposition of mineral oil. A theoretical thermodynamic assessment of the formation of decomposition gases in mineral oil was published by Halstead in 1973 [15]. In this work it was suggested that the proportion of each fault gas varies with the temperature of the point of decomposition. On the basis of Halstead equilibrium pressures at various temperatures, it can be shown that the maxima of gas generation would be methane, ethane, ethylene, and acetylene, respectively, with increasing temperature.

At low energy level Hydrogen (H_2), Methane (CH_4), and Ethane (C_2H_6) are formed, e.g., due to partial discharge or heating at low temperature (<500 °C). Due to heating at intermediate temperature Ethylene (C_2H_4) and at very high temperature (>1000 °C), e.g., in arcs, acetylene (C_2H_2) are formed [16].

On the other hand, paper insulation is made of cellulose molecules in cyclic form comprising CH_2, CH, and CO chemical groups. Out of these three groups C–O bonds are weak and result in gas formation at a temperature as low as 100 °C. Thermal degradation of paper insulation due to thermal heating at low temperature produces CO_2 and at high temperature produces CO [16].

Apart from the above-mentioned seven gases (H_2, CH_4, C_2H_6, C_2H_4, C_2H_2, CO_2, and CO), there are two non-fault gases which are commonly present in oil, viz. Oxygen (O_2), and Nitrogen (N_2). Relative values of CO_2 and CO under normal as well as abnormal conditions are dependent upon the amount of oxygen available to cause thermal degradation of cellulose insulation. A decrease in oxygen content usually means abnormal temperature rise within the transformer [16]. The relative magnitudes of all these gases can therefore be related to the type of fault and the rate of gas generation can indicate the severity of the fault.

The gases after generation remain dissolved in oil till they reach solubility limit (which of course is temperature and pressure dependent). These gases come out of oil and get collected in Buchholz relay when their quantity exceeds the solubility

Table 4.1 Solubility limit of different gases at a pressure of 760 mm Hg and a temperature of 25 °C

Gas	H_2	CO	CH_4	CO_2	C_2H_6	C_2H_4	C_2H_2
Solubility limit (by volume)	7 %	9 %	30 %	120 %	280 %	280 %	400 %

limit. The solubility limit of different gases at typical temperature and pressure are given in Table 4.1.

There are three steps involved in DGA analysis. The first step is oil sampling and extraction as well as identification of the dissolved gases. Only small volume of oil is required from the transformer for the DGA. The most popular practice is oil sampling by glass syringe. The recommended methods for oil sampling, extraction of the gases and methods of analysis are given in details in IEC 60567 [17]. Commonly oil samples are taken from bottom drain valve of the transformer. But for specific purposes, oil samples may also be taken from top drain valve or radiator. Well-documented gas extraction techniques [17] are used to get the dissolved gases separated from oil. Then gas chromatography is done to identify the extracted gases.

The second step is to establish whether or not a fault exists in the transformer. Transformers in service always have some fault gases dissolved in oil [18]. A fault is detected only when the levels of fault gases exceed the threshold levels. The threshold levels are set by statistical assessment of DGA results in such a way that about 90 % of transformers in operation have gas levels below these set values, which indicate normal operating condition. The transformers which have gas levels beyond these threshold levels are stated to have incipient faults likely to cause trouble in future if allowed to go unnoticed.

The third and final step is to determine the type of fault. Currently there are several methods for interpretation of DGA results. It is important to note here that there is no interpretation method, which can pin point the location and the type of fault. The different interpretation methods only provide guidelines to take a judicious decision.

Moreover, in many cases DGA result by itself does not provide sufficient information for the evaluation of the insulation integrity of a transformer. Other related information such as the history of transformer in terms of its loading, maintenance, previous faults, etc., form an integral part of the information based on which a meaningful evaluation can be made [18].

Some common DGA interpretation techniques, which are used in practice, are discussed in following subsections.

4.2.2.1 Total Dissolved Combustible Gases in a Transformer

Possibility of an incipient fault can be anticipated from the total concentration of dissolved combustible gases, i.e., by adding the concentrations of H_2, CH_4, C_2H_6,

Table 4.2 Preset limits of total combustible dissolved gases

Total dissolved combustible gas (ppm)	Transformer insulation condition
0–500	Satisfactory
500–1000	Decomposition of oil in excess of normal aging; needs monitoring
>1000	Significant decomposition of oil. Needs close monitoring
>2500	Substantial decomposition of oil. Needs inspection to detect fault

C_2H_4, C_2H_2, and CO. By determining the TDCG of dissolved gas within the oil and referencing these numbers to preset limits, it is possible to determine whether there is an abnormal level of gas in the oil that suggests a fault is occurring. Typical preset limits are given in Table 4.2.

4.2.2.2 Permissible Concentration of Gases in a Transformer

The normal concentrations of gases for healthy transformers of various ages are based on the rich experience of the leading utilities on the globe. In this context, it is to be noted that the combustible gas concentration in oil depends on: (i) the volume of oil involved, (ii) the age of the transformer, and (iii) the type of transformer, i.e., sealed or free breathing, construction of tap changer, etc. Suggestive values are given in Table 4.3, which are obtained from statistical analysis of DGA results of large number of healthy transformers as well as transformers having incipient faults.

IEEE Standard C57.104-1991 [19] also specifies acceptable limits for individual gas concentrations and Total Dissolved Combustible Gas (TDCG) levels.

4.2.2.3 Key Gas Method

This method is based on the correlation of key gases generated with the fault type. Table 4.4 summarizes the key gas interpretation method.

Table 4.3 Permissible concentration of gases in a transformer

Gas (ppm)	<4 years of service	4–10 years of service	>10 years of service
Hydrogen (H_2)	100–150	200–300	200–300
Methane (CH_4)	50–70	100–150	200–300
Ethane (C_2H_6)	30–50	100–150	800–1000
Ethylene (C_2H_4)	100–150	150–200	200–400
Acetylene (C_2H_2)	20–30	30–50	100–150
Carbon monoxide (CO)	200–300	400–500	600–700
Carbon dioxide (CO_2)	3000–3500	4000–5000	9000–12000

Table 4.4 Key gas interpretation method

Key gases	Nature of fault
Methane (CH_4) and Ethane (C_2H_6)	Gradual overheating
CO_2 or CO or both	Transformer overloaded or operating hot
Ethylene (C_2H_4)	Hot spots in overheated joints, core bolts, etc
Carbon monoxide (CO)	Overheating involving cellulose insulation
H_2	Corona discharge, electrolysis of water or rusting
H_2, CO_2 or CO	Corona discharge involving cellulose or severe overloading
Hydrogen (H_2) and Methane (CH_4)	Corona/Partial discharge in oil
H_2, CH_4 with small amount of C_2H_6 and C_2H_4	Sparking or other minor fault, causing oil breakdown
H_2, CH_4 with CO_2, CO, and small amount of other hydrocarbons except C_2H_2	Sparking or other minor fault in presence of cellulose
Hydrogen (H_2) and Acetylene (C_2H_2)	Arcing in oil
High amount of H_2 and other hydrocarbons including C_2H_2	High energy arc causing deterioration of oil
High amount of H_2, CH_4, high C_2H_4, and some C_2H_2	High temperature arcing of oil but in a confined area
High amount of H_2, CH_4, high C_2H_4, and some C_2H_2 and CO_2 and CO	High temperature arcing of oil in combination with cellulose

4.2.2.4 Gas Ratio Methods

Gas ratio methods are based on correlation of ratio of fault gas concentrations with incipient fault types. These methods are convenient, reliable for fault diagnosis, independent of transformer capacity, and can be computer programmed. Moreover, the choice of unit of gas concentration is insignificant as these methods consider one gas as per unit of another gas. Care should always be taken while using ratio methods because dividing one small value of a fault gas by another small value of another fault gas will give a significant ratio, but the magnitudes of the fault gases in such cases are too small to be significant for practical purposes. The disadvantages of ratio methods are that they may not always yield an analysis or may yield an incorrect one. It is, therefore, mostly used in conjunction with other interpretation techniques such as key gas method.

Dornenburg Ratio Method

Dornenburg and Strittmater in 1974 [20] first introduced the gas ratio method. This method is capable of determining mainly three fault types: viz. thermal decomposition, corona or low intensity partial discharge, and arcing or high intensity partial discharge. This method uses four gas ratios R1(CH_4/H_2), R2(C_2H_2/C_2H_4), R3(C_2H_2/CH_4), and R4(C_2H_6/C_2H_2). At first, the concentrations of fault gases in oil are determined to see whether these values are sufficiently

Table 4.5 Acceptable limits *L1* for Dornenburg ratio method

Main gases	Concentration *L1* (ppm)
Carbon monoxide (CO)	350
Methane (CH_4)	120
Hydrogen (H_2)	100
Ethane (C_2H_6)	65
Ethylene (C_2H_4)	50
Acetylene (C_2H_2)	35

above acceptable limits *L1* as given in Table 4.5 [20]. The transformer is considered to be faulty if at least one of the gas concentrations for H_2, CH_4, C_2H_4, and C_2H_2 exceeds twice the values for limit *L1* and one of the other three gases exceeds the values for limit *L1* as given in Table 4.5.

Each of the four ratios, i.e., R1, R2, R3, and R4, is compared to the values given in Table 4.6 [20]. If all the ratios for a specific transformer fall within the values given in one particular row of Table 4.6, then the fault is identified as the one given in the first column of Table 4.6.

It should be noted here that when the gas ratios completely fit into the intervals of Table 4.6, Doernenburg method performs diagnosis with 100 % confidence. However, gas chromatography results rarely fit completely into all ratio intervals which lead to the need for a revised technique. In this respect Rogers' method is often more assertive and accurate.

Roger's Ratio Method

Roger's ratio method is an improvement over Dornenburg method. It has been found to provide accuracy often higher than 80 % for dissolved gas analysis. This technique considers two of the four ratios introduced by Dornenburg, viz. CH_4/H_2, C_2H_2/C_2H_4, and in addition considers two new ratios, viz. C_2H_4/C_2H_6, and C_2H_6/CH_4. Fault diagnosis is done using the ratio codes given in Table 4.7 [21].

Twelve different types of faults as presented in Table 4.8 could be identified using the gas ratio codes given in Table 4.7.

IEC Ratio Method

The ratio C_2H_6/CH_4 is deleted because it only indicates a limited temperature range of decomposition. The remaining three gas ratios have different ranges as

Table 4.6 Fault diagnosis by Dornenburg ratio method

Fault type	Ratio R1 (CH_4/H_2)	Ratio R2 (C_2H_2/C_2H_4)	Ratio R3 (C_2H_2/CH_4)	Ratio R4 (C_2H_6/C_2H_2)
Thermal decomposition	>1.0	<0.75	<0.3	>0.4
Corona	<0.1	Insignificant	<0.3	>0.4
Arcing	>0.1 & <1.0	>0.75	>0.3	<0.4

Table 4.7 Gas ratio codes for Roger's ratio method

Gas ratio	Range	Code	Range	Code
C_2H_6/CH_4	<1.0	0	≥1.0	1
C_2H_4/C_2H_6	<1.0	0	≥1.0 and <3.0	1
	≥3.0	2		
C_2H_2/C_2H_4	<0.5	0	≥0.5 and <3.0	1
	≥3.0	2		
CH_4/H_2	≤0.1	5	>0.1 and <1.0	0
	≥1.0 and <3.0	1	≥3.0	2

Table 4.8 Fault classification using Roger's ratio codes

Fault type	CH_4/H_2	C_2H_6/CH_4	C_2H_4/C_2H_6	C_2H_2/C_2H_4
Normal deterioration	0	0	0	0
Partial discharge	5	0	0	0
Slight overheating below 150 °C	1/2	0	0	0
Slight overheating 150–200 °C	1/2	1	0	0
Slight overheating 200–300 °C	0	1	0	0
General conductor overheating	0	0	1	0
Winding circulating currents	1	0	1	0
Core and Tank circulating currents, overheated joints	1	0	2	0
Flashover without power follow through	0	0	0	1
Arc with power follow through	0	0	1/2	1/2
Continuous sparking to floating potential	0	0	2	2
Partial discharge with tracking	5	0	0	1/2

compared to Roger's method. The IEC gas ratio codes are given in Table 4.9 and the faults are classified into nine categories as shown in Table 4.10 [22].

For specific fault diagnosis, two other gas ratios are introduced in IEC 60599. To detect possible contamination from the on-load tap changer compartment, the value of C_2H_2/H_2 ratio greater than 3 is recommended. Similarly, abnormal oil heating or oxidation is detected by finding the ratio of O_2/N_2 to be less than 0.3 [22–24]. It is important to note here that all these tables are intended to demonstrate the range of phenomena, which are expected to occur within a transformer in service, e.g., normal aging, incipient electrical faults, or thermal overheating that

Table 4.9 Gas ratio codes for IEC ratio method

Gas ratio	Range	Code	Range	Code
C_2H_2/C_2H_4	<0.1	0	≥0.1 & <3.0	1
	≥3.0	2		
CH_4/H_2	<0.1	1	≥0.1 & <1.0	0
	≥1.0	2		
C_2H_4/C_2H_6	<1.0	0	≥1.0 & <3.0	1
	≥3.0	2		

Table 4.10 Fault classification using IEC ratio codes

Fault Type	C_2H_2/C_2H_4	CH_4/H_2	C_2H_4/C_2H_6
Normal aging, no fault	0	0	0
Partial discharge of low energy density	Insignificant	1	0
Partial discharge of high energy density	1	1	0
Discharges of low energy	1 → 2	0	1 → 2
Discharges of high energy	1	0	2
Thermal fault of <150 °C	0	0	1
Thermal fault of range ≥150 °C and ≤300 °C	0	2	0
Thermal fault of range >300 °C and ≤700 °C	0	2	1
Thermal fault of >700 °C	0	2	2

are likely to reduce the functional life of the transformer as well as major faults or thermal overheating that requires emergency action to be taken [21].

IEC TC 10 database of IEC Publication 599, 1978, contains the DGA results obtained from testing several tens of thousand of transformers operating in fifteen different power networks spread all over the world. These transformers were removed from service and the faults, if any, were clearly identified, so that DGA results and the fault types could be properly correlated. This has been further revised in IEC Publication 60599 [22], in which the gas ratio limits are made more precise using the augmented database, so that the cases in which the faults could not be identified were reduced to negligible percentage.

It is to be noted here that most of the transformers in service are healthy and the dissolved gas concentrations are low. When a fault occurs in a transformer, the level of gas concentration starts increasing depending on the type and severity of fault. In this context, as a compromise between reduction in costs due to premature failure and the increase in costs due to preventive maintenance, a normality percentage is used in IEC 60599. Typical or normal values of gas concentrations are computed using the following steps [24]: (i) all the DGA results in relation to transformers in a network are to be collated, (ii) for each characteristic gas, the cumulative number of transformers for which the gas concentration is below a given value is calculated, and (iii) the gas concentration for the cumulative number of transformers lying below a certain value, say X %, is then called X % typical or normal value of that gas. In other words, if X % typical value of CH_4 is Y ppm in the transformer population of a power network, then for X % transformers of that network the concentration of CH_4 will be less than Y ppm. If for a transformer, concentrations of all the gases are found to be below the typical values, then the probability of failure is low, but not zero. For transformers the typical values are generally set at 90 %, while for cables and bushings these are set at 95 %.

4.2.2.5 Duval's Triangle

In the 1970s it was identified that the gas ratio methods has a disadvantage that some DGA results may not fall within the ratio codes and hence the diagnosis

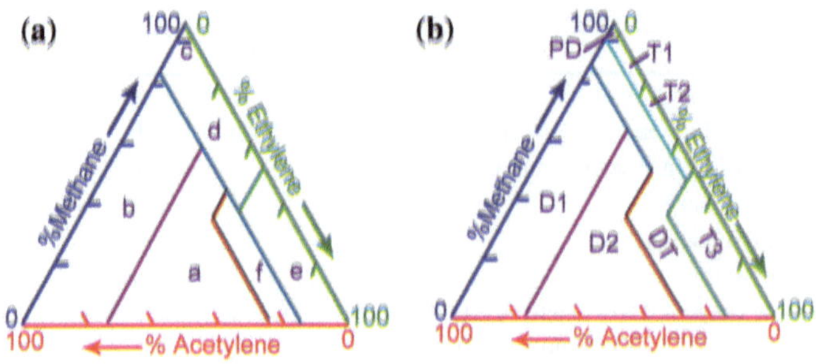

Fig. 4.3 **a** Original Duval's triangle. **b** Revised Duval's triangle

could remain unresolved. To overcome this problem, a graphical method was proposed by Duval in 1974 [25]. Since the triangle method as proposed by Duval is a closed system rather than an open one, the above-mentioned difficulty in fault diagnosis does not arise in this graphical method, which is now popularly known as Duval's triangle. The original Duval's triangle [25] is shown in Fig. 4.3a and the revised version is shown in Fig. 4.3b [16]. Six different fault classes were diagnosed using the original triangle method as given in Table 4.11. In the revised form, seven different fault classes could be diagnosed, which are also given in Table 4.11 following the nomenclature given in IEC 60599 [22]. Six fault classes are listed in IEC 60599, viz. PD, D1, D2, T1, T2, and T3. In the revised Duval's triangle a seventh zone DT is introduced corresponding to a mixture of electrical and thermal faults. As Duval's triangle represents DGA results in user-friendly graphical form, it is very useful for visualization of different faults and also for following the evolution of faults with time, e.g., evolution of an innocuous thermal fault into a potentially dangerous electrical fault.

Duval's triangle uses three characteristics gases, viz. CH_4, C_2H_4, and C_2H_2, which correspond to increasing energy levels that are required to generate gases in a transformer. The three sides of the triangle are expressed in triangular coordinates (x, y, z), where x, y, z are the relative concentrations in percentage of CH_4,

Table 4.11 Fault classification by Duval's triangle

Original Duval's triangle [25]		Revised Duval's triangle [16]	
Fault code	*Fault type*	*Fault code*	*Fault type*
a	High energy arcing	PD	Partial discharge
b	Low energy arcing, tracking	D1	Low energy discharge
c	Corona discharge	D2	High energy discharge
d	Hot spots, T < 200 °C	DT	Mixture of electrical and thermal faults
e	Hot spots, 200 °C < T < 400 °C	T1	Thermal faults, T < 300 °C
f	Hot spots, T > 400 °C	T2	Thermal faults, 300 °C < T < 700 °C
		T3	Thermal faults, T > 700 °C

C_2H_4, and C_2H_2, respectively. Let A, B, and C be the concentrations of CH_4, C_2H_4, and C_2H_2, respectively, in ppm. Then x = relative concentrations of CH_4 in % = A × 100/(A + B + C), y = relative concentrations of C_2H_4 in % = B × 100/ (A + B + C), and z = relative concentrations of C_2H_2 in % = B × 100/ (A + B + C). It is to be noted here that x, y, and z lie within 0–100 % and (X + Y + Z) is always 100 %. Plotting x, y, and z in the triangle provide only one point within the triangle as shown in Fig. 4.3b. The fault type is indicated by the zone in which the point (x, y, z) lies within the triangle.

4.2.2.6 Effect of Stray Gassing in DGA

Most of the mineral oils that are in use in transformers do not produce measurable amount of gases at temperature below 300 °C. However, some insulating oils are stated to be "stray gassing" type [26]. Such oils produce significant amount of H_2 and CH_4 at temperatures as low as 100 °C at the beginning of their service life. It occurs in the first year of the service life of such oil and is in general a non-recurrent process. The gas concentrations of H_2 and CH_4 reach a plateau after some time in service. These values must be taken into account for such oil types to avoid misinterpretation of DGA results.

4.2.2.7 Partial Discharge Identification using DGA

Low Energy partial discharges produce small amount of gases and causes little damage to the transformer insulation, e.g., low energy PDs often produce only about 5 μl of gas or about 0.1 ppm in the oil volume of a transformer, which in many cases is the detection limit for DGA. Thus low energy PDs may not be detected by DGA unless such PD activity is intense and/or occurring over a long period of time [16]. As a result, DGA is more suited for the determination of onset of the next stage of PD activity, i.e., when PD starts to cause damage to paper insulation. Damage or carbonization of paper insulation requires relatively higher energy and hence detectable amount of gases in DGA analysis are found for high energy PDs. Other monitoring techniques such as electrical or acoustic methods are more efficient in detecting PDs of low energy levels, before they become onerous to the paper insulation and produces measurable amount of gases.

4.2.2.8 Influence of Accuracy of DGA Results on Fault Diagnosis

Accuracy of fault diagnosis by DGA depends greatly on the accuracy and reliability of the DGA results as obtained from the chemical laboratories. CIGRE TF11 has evaluated that the analytical error of laboratories worldwide, where DGA is done, is about ±15 % on an average [27]. At low values of gas concentration this error increases to about 35 % and may become 100 % or more as concentration level become close to the analytical detection level of DGA. In order to get reliable and accurate fault diagnosis by DGA, analytical error of laboratory should be lower than ±10 %. Between ±10 and ±40 % uncertainty in fault diagnosis increases significantly and above ±40 % fault diagnosis becomes completely erroneous [27].

4.2.3 Measurement of Degree of Polymerization

Paper and pressboard, which are used as solid insulation in transformers, are materials made in the form of sheet from vegetable cellulose fibers, the main source of which is wood. Insulation grade paper is made from wood pulp by Kraft process, which is not only a method of delignification of wood pulp, but also removes other constituents of the living plant like carbohydrates, waxes, etc., to leave the cellulose fibers. Insulation grade paper contains about 90 % cellulose, 6–7 % hemicellulose, and 3–4 % lignin. Natural moisture content of paper is 4–5 % and the insulation is dried to less than 0.5 % for use in high voltage equipment. As the major constituent of paper is cellulose, hence the strength of dried paper primarily depends on the condition and strength of the fibers themselves and the physio-chemical bonding between the cellulose molecules. Cellulose is a linear polymer comprising anhydro D-glucopyranose units held together at the first and fourth carbon atoms through a glycosidic-linkage [5] as shown in Fig. 4.4. The monomer units are combined in long straight chains in the paper insulation and the number of monomer units in a polymer is known as Degree of Polymerization (DP).

The properties of paper are critically dependent on the DP of the cellulose. The DP of insulation grade paper after manufacturing varies between 1000 and 1300 [5, 28]. It is reduced to around 950 in the drying process of the transformer winding and it reduces further due to aging of paper while in use. The mechanism of degradation of cellulose could be broadly classified into three categories as detailed below:

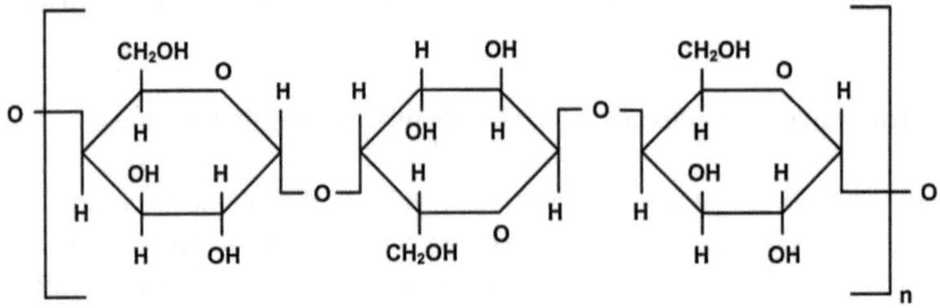

Fig. 4.4 Structure of cellulose

(a) Thermal Degradation: If cellulose is heated to around 200 °C, in the absence of oxidizing agents and moisture, the glycosidic bonds tend to break and open the glucose rings. Because of such reaction free glucose, moisture, carbon oxides, and organic acids are produced.
(b) Oxidative Degradation: Cellulose is very prone to oxidation. The products of oxidation of cellulose depend on the type oxidizing agent and the pH. In general, the hydroxyl groups are oxidized to carbonyl (aldehydic) and carboxyl (acidic) groups. This can lead to chain scission and moisture is produced during oxidative degradation.
(c) Hydrolytic Degradation: The cleavage at the glycosidic bond is caused by water and acids, which yield free glucose. Hydrolytic degradation is aided by the moisture produced due to thermal and oxidative degradation. As a matter of fact, moisture is the most powerful degrading agent of paper insulation. The nature of relationship between moisture content of paper and the degree of polymerization has been given in [5] as shown in Fig. 4.5.

It should be mentioned here that, unlike the tensile strength of paper, the dielectric properties of paper is not severely affected by the degradation mechanisms of paper as discussed above. It has been reported that the dielectric strength and dielectric loss factor do not deteriorate enough to cause dielectric failure [5]. Mechanical failure of paper due to degradation is the primary reason behind major equipment failures.

4.2.3.1 DP Measurement by Viscometric Method

Studies on solution viscometry have proved that the DP of a polymer can be determined from the viscosity of dilute polymer solutions. The number of monomers present in a cellulose polymer, as measured by viscometric method, is denoted by DP_v. There are international standards for the measurement of DP of cellulose in Cupriethylene-diamine (Cuen) solvent, e.g., ASTM D1795-90 [29] and BSI 6306 [30]. Other solvents such as Cadmium Oxide-Ethylenediamine (Cadoxen) have been used but have not been incorporated in the standards. ASTM D4243-86 [31] and IEC 60450 [32] describe viscometric DP measurement methods specific to insulating papers used in transformers.

4.2.3.2 Relationship Between Viscosity and Molecular Weight

The viscosity of a dilute polymer solution and the molecular weight of the polymer is related by the Mark Houwink Sakurada (MHS) equation as given in Eq. (4.1).

$$\eta = kM_w^{\alpha} \tag{4.1}$$

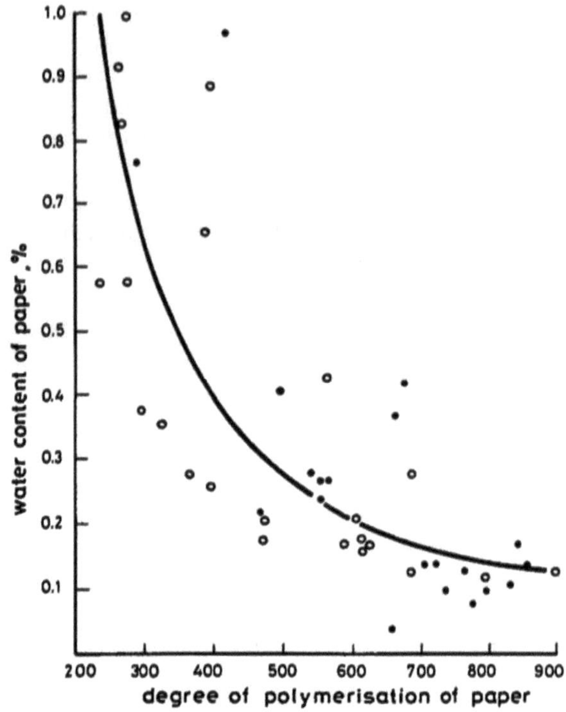

Fig. 4.5 Moisture content versus DP of paper [5]. Reproduced with kind permission ©IET 2012

where, η is the intrinsic viscosity (expressed normally in dl/g), M_w is the molecular weight, k, and α are constants related to solvent and polymer types.

However, Eq. (4.1) is valid only for dilute solutions ($\approx$0.1–1.0 %). Hence, ASTM 1795-90 [29] and BSI 6306 [30] suggest that the product of intrinsic viscosity (η in dl/g) and concentration (c in g/dl) should remain within 1–1.5. Whereas, ASTM D4243-86 [31] state that the value of ηc should be less than 1. This is because of the fact that at ηc values greater than 1, complete dissolution of paper is difficult to obtain [33].

4.2.3.3 Sources of Error in DP Measurement by Viscometric Method

In [33], researchers have very clearly identified the five main sources of error in DP measurement by viscometric method as mentioned below.

(i) Ambient temperature during dissolution: If the temperature during dissolution of paper is above 20 °C, then complete dissolution does not occur.
(ii) Exposure to air: The solution is sensitive to oxygen. The DP_v value decreases as the solution is exposed to air.

(iii) Effects of shear rate: A polymer undergo mechanical degradation when shaken over a long period of time. The measured viscosity is thus dependent on the rate of shear in the viscometer. The calculated DP_v values are lower for higher shear rates.
(iv) Temperature of viscosity measurement: In [33] it has been reported that the calculated DP_v decreases by 1 % per 3 °C increase in temperature. Consequently, ASTM D4243-86 specifies 20 °C for Kraft paper, while ASTM 1795-90 and BSI 6306 specify 25 °C. However, both these temperatures are difficult to maintain accurately during the measurement unless a water bath is used which utilizes a heater working in tandem with a cooler [33].
(v) MHS equation constants: There are more than six values of MHS equation constants as reported in literature. The calculated DP_v values vary significantly depending upon the MHS constants used. ASTM D4243-86 mentions the values of $\alpha = 1$ and $k = 0.0075$, while ASTM 1795-90 makes an approximate conversion of intrinsic viscosity to DP_v by multiplying 190 with intrinsic viscosity. BSI 6306 does not provide any conversion rule.

In addition to the above-mentioned sources of error, there is one more factor that affects the DP_v measurement, which is removal of oil from the paper sample. As the paper insulation in transformer is impregnated with oil, it is necessary to remove any oil from the paper sample to obtain the dry weight of paper accurately. In addition the moisture content of the paper should also be measured accurately for the same purpose.

Thus to obtain consistent DP_v values, it is essential that [33]

(a) the solvent be pre-cooled to 4 °C
(b) the paper be dissolved under nitrogen at an ambient temperature of 20 °C.
(c) the viscosity is measured immediately upon exposure to air
(d) the temperature should be maintained constant to ±0.2 °C of the temperature specified in the standard during the measurement.
(e) the value of ηc should be kept nearly equal to 1
(f) a consistent set of MHS equation constant are used, e.g., $\alpha = 1$ and $k = 0.0075$, for Kraft paper.

4.2.3.4 Interpretation of Measured DP_v Value

Researchers have shown that the tensile strength of paper remains practically constant for DP_v from 950 to 500. Thereafter in the range of 500–200, it decreases linearly with DP_v. The strength of paper is reduced to 20 % of its initial magnitude at a DP_v of 150 and below 150 the paper has no strength for all practical purposes [5] and may be taken as end of life for paper insulation. Table 4.12 presents typical interpretations of DP_v values for transformers as prevalent in practice [34].

Several investigators have attempted to relate paper strength to its DP_v and hence to define a DP_v value as end of life criterion. In [35] Bozzini used EOL

Table 4.12 Interpretation of measured DP_v

DP_v	Interpretation
>900	No paper degradation detected
601–900	Very little paper aging
451–600	Notable paper degradation but not critical
351–450	Paper degradation approaching critical condition. Re-sampling suggested within 1–2 years time
251–350	Paper degradation very near to critical condition. Re-sampling suggested within 1 year
200–250	Paper degradation has reached critical condition. Transformer should be taken out of service as soon as possible and inspected thoroughly
<200	Paper degradation has crossed critical condition. Transformer must be taken out of service immediately and inspected thoroughly

criterion as 100–150 while in Lampe et al. [36] suggested a value of 200. Fabre and Pichon [2] used 100–200 as EOL. It should however be noted here that many transformers, which are functioning well in service, have been found to have DP_v value of less than 150 [2]. McNutt [37] in his paper described the significance of EOL criterion for paper insulation and used his own EOL value as 200.

The measurement of degree of polymerization is therefore used as a diagnostic tool for determination of the condition of the solid insulation within a transformer.

4.2.3.5 Collection of Paper Sample for DP_v Measurement

DP measurement of paper insulation requires removal of a few strips of paper from appropriate site within the transformer in such a way that the transformer is not damaged. In many cases the sample collection locations are re-taped with new paper insulation to avoid any damage. Sometimes dummy samples are introduced at strategic locations into the design of the transformer itself, which may be removed for DP measurement as and when necessary. In many cases, the samples are taken from bushing leads or winding leads, which are at bulk oil temperature and not at hot-spot temperature.

4.2.3.6 Studies on Molecular Weight Distribution by Size Exclusion Chromatography

The DP values measured by viscometric method give only an approximate estimate of the average chain length of cellulose. The variation of DP_v during aging thus gives nothing more than a coarse measure of the changes occurring in paper insulation. To get a full view of the changes occurring at the macromolecular level, size exclusion chromatography (SEC) of paper could be used. The most commonly used technique is Gel Permeation Chromatography (GPC). GPC separates the paper molecule into decreasing order of size, i.e., hydrodynamic volume, and thus provides a detailed molecular weight distribution of paper. Consequently, GPC is useful for studying paper aging as molecular weight distribution of paper insulation changes significantly due to degradation [23, 33].

4.2.3.7 Advantages and Disadvantages of DP Measurement

The advantages of using DP measurement in assessing insulation condition are: (i) It is easy to make DP measurements and (ii) DP values can be empirically related to insulation condition.

On the other hand, the disadvantages are manifold and are discussed in detail in [28]. These are: (i) The mechanisms and kinetics of paper degradation processes are not properly defined, (ii) The type of paper and its final chemical treatment significantly influence the rate of degradation, (iii) For temperatures above 140 °C, the rate of degradation increases discontinuously with increasing temperature, (iv) Accessibility of the reactants to the interiors of transformer insulation affects the rate of hydrolytic and oxidative degradation. It has been reported in literature that the rate of degradation varies across the thickness of insulation [38]. It is highest in the outermost and innermost layers [39], (v) The DP_v values and the strength of the paper are not linearly related except in the range of DP_v from 200 to 500, and (vi) The reproducibility of DP measurement of paper insulation is very poor.

4.2.4 Furan in Oil Analysis

Determination of DP value of paper insulation requires removal of a few strips of paper from within the transformer, which is not practical for in-service transformers. Such procedure can be carried out during the repairing of transformers and also during strip down of redundant units, which can be used to perform audit of transformers of the same design in use. As a result the number of DP measurements for transformers are always limited. Hence, the problem is to relate the few DP measurements to general condition of paper insulation. This calls for a method of determination of condition of paper insulation without involving the paper sample. In other words, an alternative method of assessing paper degradation preferably by performing tests on transformer oil is needed.

The by-products of degradation of paper insulation are not only carbon oxides but also derivatives of the aromatic compound called furan, which are soluble in transformer oil. Thus the amount of furans present in oil could be a good indicator of the condition of cellulose insulation.

4.2.4.1 Generation of Furans

All three forms of paper degradation as mentioned in Sect. 4.2.3 yield glucose or degraded forms of glucose. Hence, it is logical that monitoring of glucose level in transformer oil should give a fair idea about the extent of paper degradation. But the problem is that the solubility of glucose in transformer oil is very poor. Moreover, glucose tends to remain associated with the polar molecular structures within the paper.

Glucose degrades under the influence of moisture and acids, which are produced during paper degradation, to produce 5-hydroxymethyl-2-furaldehyde. The mechanism is given in [40] and is shown in Fig. 4.6. However, this is an unstable compound and decomposes further to generate other furaldehydes. The degraded forms of glucose take part in reactions to produce some other furans. The major furanic compounds that are released into the oil are as follows: 2-furfuraldehyde (2FAL), 5-hydroxy-methyl 2-furfuraldehyde (5H2F), 2-acetylfuran (2ACF), 5-methyl 2-furfuraldehyde (5M2F), and 2-furfuryl-alcohol (2FOL).

4.2.4.2 Furanic Concentration in Oil

Experimental studies [41] on the production of furanic compounds in oil due to paper aging report that furanic concentrations increase initially, as shown in the inset of Fig. 4.7, followed by a slow rate of furan compound formation. After that there is an approximately exponential rise in furan concentration with aging duration. Finally, the rate of increase slows down and the concentrations reach a maximum and eventually decrease, particularly at high temperatures. Similar trends were seen at all temperatures, but with different maximum concentrations of furanic products at different temperatures. The concentrations of 2-furfuraldehyde (2FAL) in the oil were found to be always the highest [41].

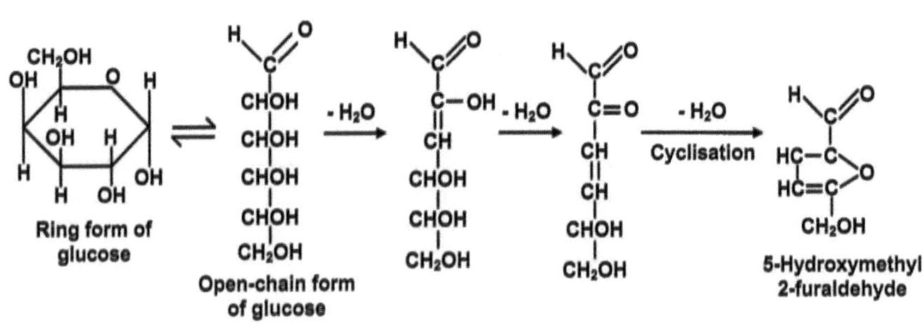

Fig. 4.6 Degradation mechanism of glucose to 5-hydroxymethyl 2-furaldehyde

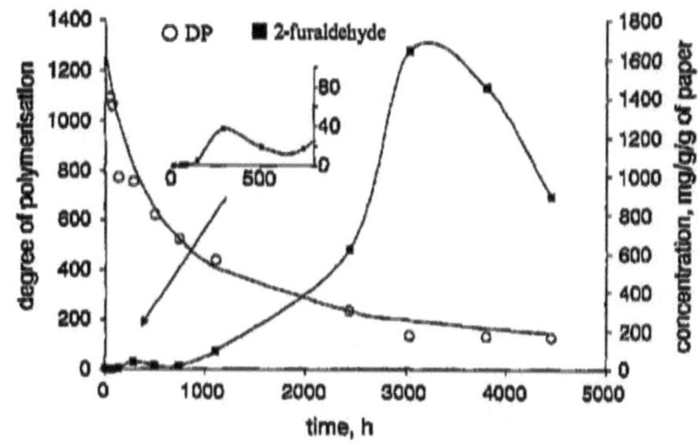

Fig. 4.7 Comparison of increases in 2FAL and DP of paper during aging [41] Reproduced with kind permission ©IET 2012

4.2.4.3 Significance of Furanic Compounds in Oil

The furan in oil analysis for in-service transformers is a powerful tool for condition monitoring of oil-paper insulating systems. Attempts have also been made to set acceptable limits on furans in transformer oil, which can be summarized as follows: (i) less than 0.1 mg/l—acceptable, (ii) greater than 0.1 mg/l but less than 0.25 mg/l—questionable, and (iii) greater than 0.25 mg/l—unacceptable [42, 43]. However, paper in a transformer does not degrade uniformly and hence the furan levels in oil relate to the average deterioration of the paper insulation. The extent of paper deterioration at the hot-spot within the transformer will be higher than that indicated by the furan levels in oil.

Furthermore, furan in oil analysis can provide complementary information to the dissolved gas analysis, particularly when the presence of carbon mono-oxide in oil indicates problems related to paper insulation. As reported in [44], the types and concentration of furanic compounds in oil can also indicate abnormal stresses in a transformer. The following indicative observations depending upon the presence of furanic compounds have been mentioned in [44]: (a) 2FAL—Overheating, old faults, (b) 5H2F—Oxidation, (c) 5M2F—Severe local overheating, and (d) 2FOL—High Moisture content.

4.2.4.4 Stability of Furanic Compounds in Oil

Experimental results [41] show that the furans degrade themselves in oil, but the rate of decomposition is very low with time, with the exception of 2-furfuryl-alcohol (2FOL), at all temperatures up to 140 °C above which the rate of decomposition increases significantly. However, the decomposition rates are usually too small and it cannot account for the loss of furans that has been observed during paper aging experiments. Such loss of furans is most probably due to evaporation since the boiling points of furans are in the range of 160–200 °C. In other words, furanic compounds are relatively stable up to 140 °C except 2FOL. Thus furans should be detectable in an aging paper-oil insulating system.

However, existing literature [41, 45] shows that out of the five furanic compounds mentioned in Sect. 4.2.4.1, only 2FAL, 2FOL, and 5H2F are mostly detected in practice. Moreover, the concentrations of furanic compounds in oil are not consistent except for 2FAL. Hence, 2-furfuraldehyde (2FAL) is considered to be the most stable furanic compound in oil and is commonly used as the indicator of paper degradation.

It has also been reported [45] that the furanic compounds have the ability to migrate back and forth between oil and paper insulation. As a result the concentrations of furans vary within ± 15 % when the actual sample collection for experimentation is carried out over a period of time spanning several months. Furans are polar compounds and hence require time and specific conditions for release into the oil, which also adds the variation in results.

4.2.4.5 Measurement of Furan Concentrations by HPLC

The method of measurement of furan concentrations in oil have been described in [40, 41]. The measurement is done in two phases: solid stationary phase and mobile phase. In the solid stationary phase, 1–5 ml of oil is diluted with 5 ml of hexane and injected in a steel tube packed with particle of solid material. Commonly used solid in the stationary phase is octadecyl groups bonded to silica particles. It is then washed with 10 ml of hexane to remove the oil. Nitrogen is applied for 5 min to evaporate the hexane. The furans are then eluted with 1.5 ml acetonitrile. About 20 μl of the elute solution is then injected into the High Performance Liquid Chromatography (HPLC) column. A solvent called the mobile phase is then pumped through the column. The solvent is a mixture of methanol and water. The flow of the mobile phase separates the furan components. The different extent to which the different furan components interact with the solid stationary phase and the mobile phase determines the extent of separation of furan components. The detector senses the components as they are eluted from the column. The detector may be an ultra-violet detector or a photo diode array detector.

4.2.4.6 Correlation of Furan Concentration with DP and Moisture Content of Paper

The initial attempt to correlate the furfuraldehyde content with the DP of paper was done in [5]. The relationship as obtained in [5] is presented in Fig. 4.8a, which shows that the furfuraldehyde content of the oil is related well with the DP of the paper. Subsequently, in [41] a comprehensive study has been made on the correlation of DP and 2-furfuraldehyde (2FAL), which is the most prominent furan component generated by paper degradation. The correlation as obtained in [41] for Kraft paper is shown in Fig. 4.8b, which confirms the relationship reported in [5]. From Fig. 4.8b it may be seen that the most significant rise in furfural level occurs

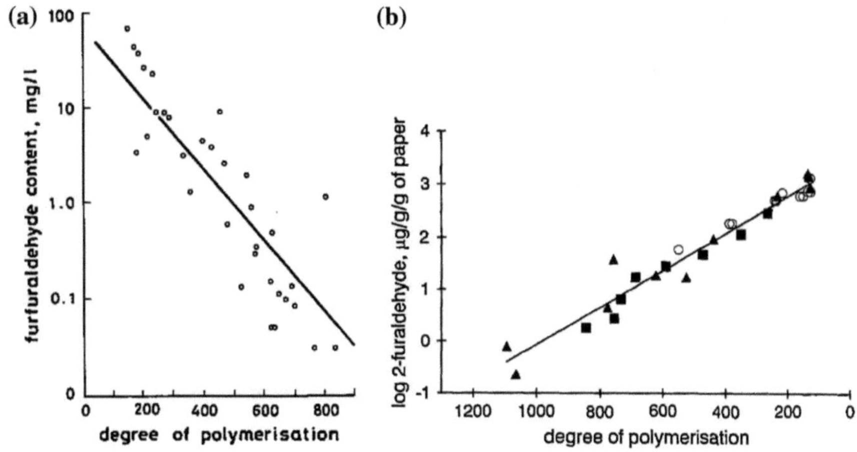

Fig. 4.8 **a** Correlation of the furfuraldehyde content with the DP of paper [5]. Reproduced with kind permission ©IET 2012. **b** Correlation of DP and 2-furfuraldehyde (2FAL) for Kraft paper [41]. Reproduced with kind permission ©IET 2012

below DP_v value of 400 as it approaches the critical value of 200, at which the paper looses its mechanical strength and becomes susceptible to mechanical damage leading to equipment failure. Hence, furan analysis will be most suitable as an indicator of rapid rates of paper aging rather than a monitor of normal aging [41]. It has been reported in literature [44] that a furan production rate of 25 ppb/month is a cause for concern.

Chendong [45] has developed a widely accepted relationship between 2-furfuraldehyde (2FAL) in oil and DP of paper as given in Eq. (4.2)

$$\log_{10}(2FAL_{ppm}) = 1.51 - 0.0035 \times DP \tag{4.2}$$

Subsequent studies on thermally upgraded insulation led to the modified Chendong's equation as proposed by Stebbins et al. in [46], which is given in Eq. (4.3)

$$\log_{10}(2FAL_{ppm}) = 1.5655 - 0.0035 \times DP \tag{4.3}$$

It is to be mentioned here that during oil treatment by vacuum filtering, gases and moisture in oil are removed, but furans remain in oil. Hence, if the paper insulation of a transformer got degraded before the treatment of oil, then the furans will remain in oil even after oil treatment and a high level of furan may be obtained in oil even after oil treatment as the same transformer oil is used after treatment.

Attempts have also been made to correlate moisture content in paper with 2-furfuraldehyde (2FAL) in oil. Such a relationship, as reported in [47], is shown in Fig. 4.9.

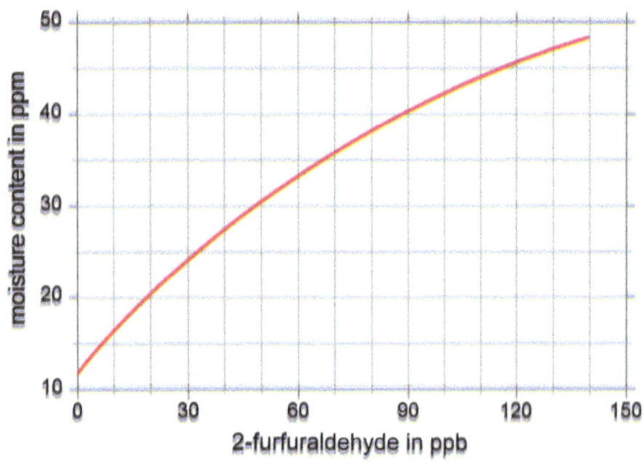

Fig. 4.9 Relationship between DP and 2-furfuraldehyde (2FAL)

4.3 Electrical Techniques

The useful life of the oil-paper insulation system of a transformer is practically ended when the paper insulation has become brittle enough to develop cracks under the mechanical stresses to which it is subjected. Cracking in paper insulation results from mechanical stresses, which are caused by (a) short-circuit currents, (b) thermal expansion and contraction, and (c) vibration. Experimental results reported in [5] show that although the electrical properties of paper and oil degrade during aging, they do not do so sufficiently to promote transformer failure in most of the cases. The dielectric strength of insulation is not significantly reduced by brittleness alone; however, electrical breakdown will quickly follow the development of ensuing cracks. For this reason, close inspection for insulation should be made at frequent intervals, and repairs made as necessary.

The conventionally used electrical techniques for insulation diagnostics, which are most common in practice, are: (i) Insulation Resistance (IR) measurement and (ii) dielectric loss factor (tanδ) measurement.

4.3.1 Insulation Resistance Measurement

The types of insulation used in high voltage transformers have continued to evolve and cause changes in insulation resistance measurement techniques over the years. One particular insulation resistance test result cannot accurately predict the breakdown voltage of insulation. The results must be compared with values when the insulation was new and of known voltage withstand capability. Therefore, insulation resistance measurements must be made periodically, beginning when the equipment is new and in good condition, and the trend of the test results are to be noted.

When a dc voltage from a high voltage dc source is suddenly applied to an insulation system, the current through the insulation starts at a relatively high value, gradually decreases with time, and finally levels off to a low yet stable value. The insulation resistance at any given instant of time is the quotient of the voltage applied over the current flowing through the insulation at that time instant. The low initial insulation resistance is partly caused by the high initial capacitive charging current. This capacitive current quickly decreases to a negligible value, usually within 15 s, as the geometric capacitance of the insulation gets charged. The low initial insulation resistance is also partly caused by the high initial dielectric absorption current. This current also decreases with time, but more gradually, requiring from 10 min to several hours to decay to a negligible value. However, for the purpose of conventional insulation resistance tests, the change in dielectric absorption current after 10 min is commonly disregarded. Basically, the insulation system consists of dipoles that line up (polarize) with the applied DC voltage. As the dipoles get polarized, the absorption current decreases and the insulation resistance increases. The leakage current through the insulation is due to the volume resistance of the insulation and does not change with time of voltage application, if the voltage remains constant.

For insulation resistance measurement of transformers, the tank and the core should always be grounded. Each winding of the transformer should be short-circuited at the bushing terminals. Resistance measurements are to be made between each winding and all other windings grounded or between two windings. Windings should never be left floating for insulation resistance measurements.

One factor that affects the insulation resistance measurements is the presence of a previous charge within the insulation system. The charge may come from normal operation of a transformer, or from a previous measurement of insulation resistance. Time may be saved if the transformer winding is kept grounded until the test on the winding is made. A winding should also remain solidly grounded long enough after insulation resistance testfor the absorbed charge to be completely drained off. Insulation systems vary greatly in the length of time to which the absorbed charge will remain, but in the absence of specific knowledge, it is recommended that the winding be grounded for at least one hour.

The principal methods used today for insulation resistance measurements are: (i) Spot reading test, (ii) Time versus Resistance test, and (iii) Stepped Voltage test. Each insulation resistance test gives a different view, or window, into the condition of the insulation; the whole picture is only available when all required tests are performed.

4.3.1.1 Spot Reading Test

For this test the insulation resistance meter is connected across the insulation system and a constant dc test voltage is applied for a fixed amount of time, usually 60 s, and an IR reading is taken. The spot reading test should only be carried out if the temperature during testing is above the dew point. This helps in converting the measured IR to a base value. For proper interpretation of spot reading test results, it is necessary to have previous readings, which have been taken at the same temperature and humidity for the same test voltage and same duration of voltage application. In such comparative analysis, a downward trend indicates unfavorable insulation condition and a sharp decrease indicates insulation failure. A general rule of the thumb is that the minimum IR value in spot reading test should be greater than 1 MΩ for every 1 kV rating of the machine.

Correction of Insulation Resistance for Temperature

Temperature variation affects all the components of current flowing through an insulation system, except the geometric capacitive current. As a result, the value of insulation resistance for a given system varies inversely, on an exponential basis, with the temperature. In order to avoid the effects of temperature in trend analysis, all the insulation resistance values should be corrected to the same temperature. IEEE Std. 43-2000 [48] recommends that all insulation resistance values be corrected to a common base temperature of 40 °C using Eq. (4.4).

$$R_C = K_T R_T \tag{4.4}$$

where,

R_C = Insulation resistance in MΩ corrected to 40 °C
R_T = Insulation resistance in MΩ corrected to T °C
K_T = Insulation resistance temperature coefficient at T °C and is given by Eq. (4.5)

$$K_T = (0.5)^{(40-T)/10} \tag{4.5}$$

4.3.1.2 Time Versus Resistance Test

The concept of this test is to take successive readings of insulation resistance at specified intervals keeping the applied voltage same. Good insulation shows a continual increase in resistance over time, because for good insulation as the absorption current decreases with time the insulation resistance increases. On the other hand, in the case of moist or degraded insulation, the absorption current is masked to a great extent by a relatively high leakage current which remain constant with time and hence the increase in insulation resistance with time becomes small. This method of testing is also independent of the rating of the equipment, as the nature of change in the current through the insulation system of same quality will be the same irrespective of the equipment being large or small.

The steepness of the dielectric absorption curve taken at a given temperature indicates the relative dryness of the insulation. This steepness may be expressed as

Polarization Index (PI). Prior to 1974, the power industry relied on non-standardized systems to evaluate the polarization of the insulation system, which led the IEEE to form a standards committee to address this situation [49]. It was found that the time required for insulation systems to polarize vary anywhere from 10 min to several hours. So it was recommended to produce a ratio using the insulation resistance readings at the 1 and 10 min marks [48], i.e.,

Polarization Index (PI) = Insulation Resistance at 10 min/Insulation Resistance at 1 min

Since it is a ratio, it does not have any unit.

IEEE Std. 43-2000 [48] states the recommended minimum values of polarization index for ac rotating machines as follows: (i) For class-A insulation—1.5, (ii) For class-B insulation—2.0, and (iii) For class-F insulation—2.0.

The following are practical guidelines for evaluating transformer insulation using polarization index values: (i) Less than 1.0: Unacceptable, (ii) 1.0–1.1: Poor, (iii) 1.1–1.25: Questionable, (iv) 1.25–2.0: Fair, and (v) Above 2.0: Good.

Results of PI test performed on a hot insulation system could be quite different than that of a PI test performed on a cold one. A test on hot insulation system is carried out immediately after the shutdown from full-load operation of at least 4 h. It has been found that the difference in polarization index between dry and moist insulation is more pronounced at higher temperatures and therefore, the hot test is more sensitive. Cold test is carried out 4–8 hours after the hot resistance test or when equipment has cooled to approximately ambient temperature.

The Polarization Index test is particularly valuable in revealing moisture ingress, oil soaks, and similar pervasive contamination.

4.3.1.3 Stepped Voltage Test

In this method, test voltage is increased in five equal steps at one-minute increments and the insulation resistance at one-minute is recorded at each voltage level. Any unusual resistance reduction is an indication of incipient insulation weakness.

There are insulation defects that pass a PI or simple spot reading test by yielding high resistance values at a given voltage. Such defects include localized physical damage like pinholes or dry, brittle insulation in aged equipment. Stepped Voltage tests reveal such problems as the insulation defects will pass increasingly higher current as the voltage is increased and will be reflected in a decreasing insulation resistance at higher voltage. Moreover, higher voltage will pull arcs across small air-gaps, thereby providing an early warning of incipient faults. This test can reveal aging and physical damage even in relatively dry insulation system which would not have been apparent in PI test.

4.3.1.4 Limitations of Insulation Resistance Test

Insulation resistance test data is useful in evaluating the presence of some insulation problems such as contamination, absorbed moisture, or severe cracking; however, there are several major limitations as mentioned below:

(a) Unless the insulation defect is concentrated, it is impossible to specify the value of insulation resistance at which the insulation system will fail.
(b) A single insulation resistance measurement at one particular voltage does not indicate whether foreign matter is concentrated or distributed throughout the insulation system.
(c) Direct voltage measurements, such as IR and PI tests, may not detect voids present within the insulation system.

4.3.2 Dielectric Loss-Factor (tanδ) Measurement

Under the influence of an alternating field, the current through an insulation system has a capacitive component and a resistive component, which are 90° out of phase. The angle between the capacitive current and the resultant current phasors is called the dielectric loss factor angle (δ). The factor tanδ is defined as the dielectric dissipation factor and represents the dielectric loss in the insulation system.

Dielectric dissipation factor is also directly related to the permittivity of the insulating material. If the permittivity is expressed as complex quantity as in Eq. (4.6),

$$\bar{\varepsilon} = \varepsilon' - j\varepsilon'' \tag{4.6}$$

then, the ratio of the imaginary to the real part of the permittivity is equal to tanδ.

The dielectric dissipation factor relates to the orientation of dipoles in the insulating medium with an alternating electric field. The ability of the dipoles to orient themselves is dependent on the temperature of the insulation, the size of the dipoles involved, and their polarity. It is also dependent on the frequency of the alternating field. The permittivity and the dielectric dissipation factor are both affected by the molecular size, composition, and relative orientation of functional groups within the molecules. So permittivity and, hence, dielectric dissipation factor can be used to monitor the quality of the insulation with regard to deterioration in use and for the presence of contaminants.

In this test, the test voltage is applied from a power frequency voltage source and the tan δ is measured. The test voltage is increased in steps to a level that is high enough (typically 10–30 kV) to overcome any local interference at the test site. The tan δ readings are plotted against the applied voltage and the trend is studied. There are no standard formulae or benchmarks to interpret the results of tan δ test. The health of the insulation is ascertained by observing the nature of the variation of tan δ. A steady, linear trend would indicate a healthy insulation, while a rising trend would indicate an insulation that has been contaminated with moisture and other impurities. Dielectric dissipation factor test is often carried out at different temperatures in an attempt to identify the concentration of contaminants as the presence of these contaminants including moisture influences the tan δ differently at different temperatures.

The IEC 60422 [50] has recommended limits depending on the type of transformer based on tan δ at 90 °C determined as per the IEC Test Method 60247 [51]. These action limits are presented in Table 4.13.

Category-O: Power transformers/reactors with a nominal system voltage of 400 kV and above.

Category-A: Power transformers/reactors with a nominal system voltage above 170 kV and below 400 kV.

Category-B: Power transformers/reactors with a nominal system voltage above 72.5 kV and up to and including 170 kV.

Category-C: Power transformers/reactors for MV/LV application, e.g., nominal system voltages up to and including 72.5 kV.

Recommended Actions: (i) Good: Continue normal sampling, (ii) Fair: More frequent sampling, and (iii) Poor: Reclaim or replace the oil.

In [5] an attempt has been made to check the possibility of using electrical tests on oil to monitor paper degradation. For this purpose, tan δ of oil were measured as against the DP of paper. Apart from indicating that the tan δ increases in a general way as aging proceeds, Fig. 4.10 shows that there is poor correlation between tan δ of the oil and DP of the paper. This is presumed to be because other factors intervened such as oil degradation.

Researchers have found that the dielectric dissipation factor is very much dependent on frequency and the studies on variation of permittivity and tanδ with frequency has given rise to another branch of transformer condition monitoring, which will be discussed in more details in Chap. 6.

4.4 Possible Ways to Overcome the Shortcomings

In relation to transformer operation for power utilities, there is often a problem to reconcile operational needs to have the unit in service with the assessment of risk of failure. An on-line gas monitor could be useful, because some faults are load related and gas generation may be used to indicate maximum safe load. It is important to note that some faults may take less than one year to progress from onset to failure, while some others may remain in a stable state for a much longer

Table 4.13 Recommended action limits for transformers as per IEC 60422

Category of transformer	Good	Fair	Poor
O, A	<0.10	0.10–0.20	>0.20
B, C	<0.10	0.10–0.50	>0.50

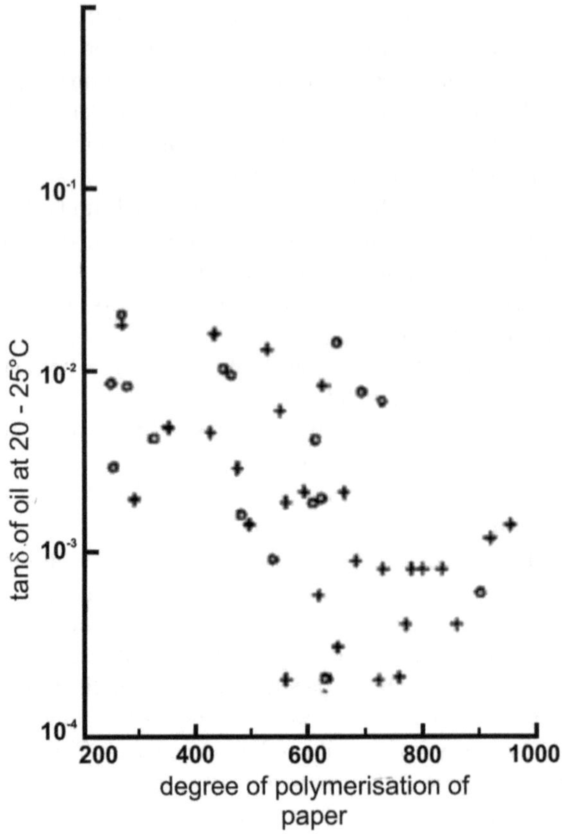

Fig. 4.10 Correlation between DP of paper and tan δ of oil [5]. Reproduced with kind permission ©IET 2012

period but have the potential for a rapid increase. A very widely used system has a membrane which allows preferentially lighter molecules to pass through it, which can then be detected in a gas reaction cell (HYDRAN). More recently, various researchers have developed Fourier Transform Infra Red (FTIR) detectors which will detect most of the gases of interest and quantify their amounts. These methods are capable of providing a greater understanding of the load, temperature, and time dependencies of generated gases.

In DGA more information is needed on limit values taking into consideration the equipment subtype, age, and fault type. In order to reach agreement on methods of analysis and interpretation that will yield reliable DGA results and diagnoses, it is very important to have international cooperation for sharing data and experience. For factory acceptance tests, the gas concentrations are very low, typically 1–10 ppm. Hence, if agreement is to be reached between the manufacturer and power utilities, the precision and accuracy requirements would have to be higher.

Since, conducting direct tests on paper samples taken from transformers which are in-service is very difficult; attempts have been made to correlate electrical properties of insulation system directly to paper degradation. So far such attempts have met limited success. However, new techniques have been developed based on dielectric response measurements, which are claimed to be able to predict paper degradation in general, and moisture content in paper, in particular.

Many of the limitations of power frequency dielectric dissipation factor measurements could be resolved with the help of Frequency Domain Spectroscopy (FDS), which is emerging as a very promising method of non-invasive condition monitoring of transformers.

References

1. Clark FM (1942) Factors affecting the mechanical deterioration of cellulose insulation. Trans Electr Eng 61:742–749
2. Fabre J, Pichon A (1960) Deteriorating processes and products of paper in oil—application to transformers. In: Proceedings of international conference on large high voltage electric system, CIGRE, Paris, Paper 137
3. CIGRE Working Group A2.30 (2008) Moisture Equilibrium and Moisture Migration within Transformer Insulation Systems. CIGRE Report ISBN: 978-2-85873-036-0
4. Lampe W, Spicar E (1976) The oxygen-free transformer, reduced aging by continuous degassing. CIGRE Paper 12-05
5. Shroff DH, Stannett AW (1985) A review of paper aging in power transformers. IEE Proc, Part C 132:312–319
6. Murphy EJ (1962) Thermal decomposition of natural cellulose in vacuum. J Polymer Sci 58:649–665
7. Beer G, Gasparini G, Osimo F, Rossi F (1966) Experimental data on the drying-out of insulation samples and test coil for transformers. CIGRE Paper No. 135
8. Norris ET (1963) High voltage power transformer insulation. IEE Proc 110(2):433
9. Oommen TV (1983) Moisture equilibrium in paper-oil systems. In: Proceedings of the electrical/electronics insulation conference, Chicago, IL, pp 162–166
10. Griffin PJ, Bruce CM, Christie JD (1988) Comparison of water equilibrium in silicone and mineral oil transformers. In: Minutes of the 55th annual international conference of double clients, Sec 10-9.1
11. Du Y, Zahn M, Lesieutre BC, Mamishev AV, Lindgren SR (1999) Moisture equilibrium in transformer paper-oil systems. IEEE Electr Insul Mag 15(1):11–20
12. Koch M, Krueger M, Tenbohlen S (2009) Comparing various moisture determination methods for power transformers. In: Southern Africa regional CIGRE conference
13. IEC 60814 (1997) Insulating liquids—oil-impregnated paper and pressboard—determination of water by automatic coulometric Karl Fischer titration
14. Buchholz M (1928) The Buchholz protection system and its application in practice. E.T.Z. 49, pp 1257–1262
15. Halstead WD (1973) A thermodynamic assessment of the formation of gaseous hydrocarbons in faulty transformers. J Inst Petrol 59:239–241
16. Duval M (2002) A review of faults detectable by gas-in-oil analysis in transformers. IEEE Electr Insul Mag 18(3):8
17. IEC Publication 60567 (2005) Oil-filled electrical equipment-sampling of gases and of oil for analysis of free and dissolved gases—guidance

18. Wang M, Vandermaar AJ, Srivastava KD (2002) Review of condition assessment of power transformers in service. IEEE Electr Insul Mag 18(6):12–25
19. IEEE Standard C57.104-1991 (1991) IEEE guide for the interpretation of gases generated in oil-immersed transformers
20. Dornenburg E, Strittmatter W (1974) Monitoring oil cooled transformers by gas analysis. Brown Boveri Rev 61(5):238–247
21. Rogers RR (1978) IEEE and IEC codes to interpret incipient faults in transformers, using gas in oil analysis. IEEE Trans Electr Insul 13(5):349–354
22. IEC Publication 599 (1978) Interpretation of the analysis of gases in transformers and other oil-filled electrical equipment in service
23. Saha TK (2003) Review of modern diagnostic techniques for assessing insulation condition in aged transformers. IEEE Trans Dielectr Electr Insul 10(5):903–917
24. Duval M, dePablo A (2001) Interpretation of gas-in-oil analysis using new IEC publication 60599 and IEC TC 10 databases. IEEE Electr Insul Mag 17(2):31–41
25. Duval M (1974) Fault gases formed in oil-filled breathing ehv power transformers: the interpretation of gas analysis data. In: IEEE-PES conference, Paper no. C74-476-8
26. dePablo A (2001) Gassing in mineral insulating oils. Abnormal hydrogen production. CIGRE TF 15/12-01-11
27. Duval M (2006) Dissolved gas analysis and the Duval triangle. http://www.avo.co.nz/technical-papers/transformers/109-dissolved-gas-analysis-and-the-duval-triangle
28. Emsley AM, Stevens GC (1994) Review of chemical indicators of degradation of cellulosic electrical paper insulation in oil-filled transformers. IEE Proc Sci Meas Technol 141(5):324–334
29. ASTM D1795-90 (1990) Standard test method for intrinsic viscosity of cellulose
30. BS 6306—Part 1 (1982) Methods for determination of limiting viscosity number of cellulose in dilute solutions. Cupriethylenediamine (CED)
31. ASTM D4243-86 (1986) Standard method for measurement of average viscometric degree of polymerization of new and aged electrical papers and boards
32. IEC 60450 (2007) Measurement of the average viscometric degree of polymerization of new and aged cellulosic electrically insulating materials
33. Ali M, Eley C, Emsley AM, Heywood R, Xaio X (1996) Measuring and understanding the ageing of kraft insulating paper in power transformers. IEEE Electr Insul Mag 12(3):28–34
34. Gray IAR (2009) A guide to transformer oil analysis. http://www.satcs.co.za/Transformer_Oil_Analysis.pdf
35. Bozzini CA (1968) Transformer ageing diagnosis by means of measurements of the degree of polymerization—results of new experiments. CIGRE, Paris, France, Paper No. 12-08
36. Lampe W, Spicar E, Carrander K (1978) Continuous purification and supervision of transformer insulation system in service. In: IEEE winter power meeting, pp 111–117
37. McNutt WJ (1992) Insulation thermal life considerations for transformer loading guides. IEEE Trans Power Deliv 7:392–401
38. Zhilyaev TB, Linova LS, Granovskaya VN, Golovan NI (1983) Investigation of ageing of cellulose electrical insulation materials in transformer oil. Electrotechnika (USSR) 54(10):74–77
39. Mcnutt WJ, Kaufmann GH, Vitols AP, Mcdonald JD (1980) Short-time failure mode considerations associated with power transformer overloading. IEEE Trans PAS 99(3):1186
40. Unsworth J, Michell F (1990) Degradation of electrical insulation paper monitored by high performance liquid chromatography. IEEE Trans Electr Insul 25:737–746
41. Emsley AM, Xiao X, Heywood RJ, Ali M (2000) Degradation of cellulosic insulation in power transformers. Part 2: formation of furan products in insulating oil. IEE Proc Sci Meas Technol 147(3):110–114
42. Burton PJ, Carballiera M, Duval M, Fuller J, Graham CW, De Pablo Samat J, Spicar E (1988) Applications of liquid chromatography to the analysis of electrical insulating materials. In: CIGRE international conference on large high voltage electrical systems, Paris, Paper no. 15-08

43. Myers DP, Sans JR, Myers SD (1992) Field studies of furan formation in transformer fluids as an indicator of damage to paper insulation and insulating fluids. In: Proceedings of 59th Doble clients conference, vol 10, pp 1–14
44. Gray IAR Evaluation of transformer solid insulation. http://www.satcs.co.za/Evaluation_of_Transformer_Solid_Insulation-rev2.pdf
45. Chendong I (1991) Monitoring paper insulation ageing by measuring furfural contents in oil. In: 7th international symposium on HV engineering, Dresden
46. Stebbins RD, Myers DS, Shkolnik AB (2003) Furanic compounds in dielectric liquid samples: review and update of diagnostic interpretation and estimation of insulation ageing. In: Proceedings of 7th international conference on properties and applications of dielectric materials (ICPADM), vol 3, pp 921–925
47. Poniran Z, Malek ZA (2007) Life assessment of power transformers via paper ageing analysis. In: Proceedings of POWERENG 2007, Setubal, Portugal, pp 460–465
48. IEEE Std 43-2000 (2000) IEEE recommended practice for testing insulation resistance of rotating machinery
49. Penrose HW (2000) A review of polarization index and IEEE Std. 43-2000. http://ecmweb.com/mag/electric_review_polarization_index/
50. IEC 60422 (2005) Mineral insulating oils in electrical equipment—supervision and maintenance guidance
51. IEC 60247 (2004) Insulating liquids—measurement of relative permittivity, dielectric dissipation factor (tanδ) and dc resistivity

Chapter 5
Time Domain Dielectric Response Measurements

5.1 Introduction

As stated in Chap. 1, in recent times the catchword in utility companies is condition based or predictive testing and maintenance, or life management, as opposed to time-based or preventive maintenance only. Costs can be reduced, first of all, by a transition from time-based maintenance (TBM) to condition-based maintenance (CBM), as well as reliability can be increased if the actual conditions of the expensive high voltage components within the electric power transmission systems are reliably known [1, 2]. A CBM based on reliable diagnostic tools should thus be applied in the present scenario. In this respect, Dielectric response measurements for insulation systems have gained importance. The basic advantages of such methods are that these methods are non-invasive or non-destructive and are also terminal measurements. Polarization and Depolarization Current (PDC) and Recovery Voltage (RV) measurements are the two major techniques in the field of time domain dielectric response measurement for insulation condition monitoring of power apparatus which are discussed in detail in this chapter.

If we recall Chap. 1, it is already known that, when a dielectric is subjected to an electric field, a relative shift of positive and negative charges take place within the dielectric giving rise to dielectric polarization. There are essentially four kinds of polarization mechanisms: (a) Electronic polarization, (b) Ionic polarization, (c) Dipole Polarization, and (d) Interfacial Polarization, out of which interfacial polarization takes place in insulation system comprising multi-dielectric media, e.g., oil-paper insulation system of transformers. Under the action of the driving electric field, movable positive and negative charges may get deposited at the interface of two dissimilar dielectric media forming some kind of dipoles. This mechanism is slow and is effective in power frequency range or lower. Dipole and interfacial polarizations are inelastic phenomena and do involve power loss.

As discussed earlier, some or all of these mechanisms may act simultaneously. For example, electronic polarization, is always present in any dielectric and thus becomes superimposed on other active mechanism, if any. Electronic, ionic, and dipole mechanisms are amenable to basic considerations and calculations.

S. Chakravorti et al., *Recent Trends in the Condition Monitoring of Transformers*, Power Systems, DOI: 10.1007/978-1-4471-5550-8_5, © Springer-Verlag London 2013

On the other hand, no basic treatment is applicable for interface polarization. There is no general way to calculate the charges on interfaces or their contribution to the total polarization of a dielectric. This interfacial polarization mechanism has a significant role in determining the condition of the insulation by dielectric response measurements.

5.2 Basic Theory of Time Domain Dielectric Response Measurement

For proper utilization of time domain dielectric response measurement, one should understand the physics behind the processes. However, the physical mechanism of dielectric response of oil-paper insulation of transformer is very complex, as it includes moisture migration through cellulose depending on moisture potential and moisture equilibrium within transformer, adsorption of water vapor in cellulose material, and mechanisms of bubbles evolution [3, 4], etc. Hence, in this section we shall discuss the topic from an engineering point of view; omitting complicated mathematical derivations of the physical processes from a microscopic point of view. This chapter will try to answer the questions which are often asked by a power engineer or draw interest from the management of a power utility regarding transformer insulation condition.

Any dielectric material, under the influence of an electric field, results in a relative shift of positive and negative charges in the material. This phenomenon is called dielectric polarization which gives rise to what is called "displacement current" in addition to normal conduction current. In other words, the nature of this displacement current is dependent on polarization of the dielectric material which in turn depends on the condition of the material under consideration. The study of the variation of aforesaid nature of current with time by application of electric field constitutes time domain dielectric response measurements. Zaengl, in [1] explained the mathematical background behind the dielectric response analysis in detail.

Any dielectric material, under electric field $E(t)$ generates a total current density $j(t)$ which is the summation of conduction current and displacement current [1, 27].

$$j(t) = \sigma_0 E(t) + \frac{dD}{dt} \tag{5.1}$$

Here, σ_0 is the dc conductivity and $D(t)$ is the electric displacement given by $D(t) = \varepsilon_0 E(t) + P(t)$, related to macroscopic polarization $P(t)$ of the material. ε_0 is the permittivity of vacuum.

Equation 5.1 can be modified as

$$j(t) = \sigma_0 E(t) + \varepsilon_0 \frac{\partial E(t)}{\partial t} + \frac{\partial P(t)}{\partial t} \tag{5.2}$$

The term $P(t)$ is related to dielectric response function $f(t)$ by the convolution integral

$$P(t) = \varepsilon_0 \chi_\infty E(t) + \varepsilon_0 \int_0^t f(t-\tau)E(\tau)d\tau \tag{5.3}$$

χ is the electric susceptibility of the material, where $\varepsilon_\infty = 1 + \chi_\infty$. The index ∞ denotes very fast polarization processes and cannot be measured in time domain.

For $E(t)$ = constant, Eqs. 5.2 and 5.3 can be reduced to

$$j(t) = \sigma_0 E(t) + \varepsilon_0\{\varepsilon_\infty \delta(t) + f(t)\}E(t) \tag{5.4}$$

Equation 5.4 can be further modified to give the expression for polarization current in the dielectric material [1].

$$i_{pol}(t) = C_o U_0 \left(\frac{\sigma_0}{\varepsilon_o} + \varepsilon_\infty \delta(t) + f(t) \right) \tag{5.5}$$

Here C_O is called the geometric capacitance (i.e., vacuum capacitance of the electrode system between which the dielectric is sandwiched or the high frequency capacitance of the dielectric at staring time t_O of current measurement) and U_O is the step voltage applied.

As seen in Eq. 5.5, the charging current contains three terms: The first one is related to the intrinsic conductivity of the dielectric under test which is independent of any polarization processes, the last one represents a net reflection of all the active polarization processes during the span of applied U_O and the middle part with the delta function cannot be recorded in practice due to the large amplitude of current related to the very fast polarization processes occurring within an extremely short duration of time.

A polarization current measurement is usually stopped if the current becomes either stable due to σ_O or becomes very low. If the test object is now short-circuited at $t = t_c$, the depolarization or discharging current flows as $i_{depol}(t)$.

According to the superposition principle, sudden reduction of the voltage U_O to zero can be regarded as a negative step voltage at time $t = t_c$. Then we get for t $(t_O + t_c)$

$$i_{depol}(t) = -C_0 U_0 [\varepsilon_\infty \delta(t) + \{f(t) - f(t+t_c)\}] \tag{5.6}$$

$\delta(t)$, like polarization current, is again an extreme short duration current of very high magnitude that is to be avoided during the measurement processes. How this can be done is discussed in Sect. 5.3.7. Figure 5.1 shows typical nature of polarization and depolarization current [1, 27].

A further investigation on the above PDC phenomena, i.e., the "after effects" of the PDC measurement, involves what is called Recovery Voltage Measurement (RVM) [2]. A step voltage is used to charge the sample under RV measurement for a definite period of time (t_{ch}) and then earthed for discharging through a period, which is half of the charging period (½t_{ch}). Thereafter the earthing connections of the sample are withdrawn and the voltage across the unearthed sample is recorded

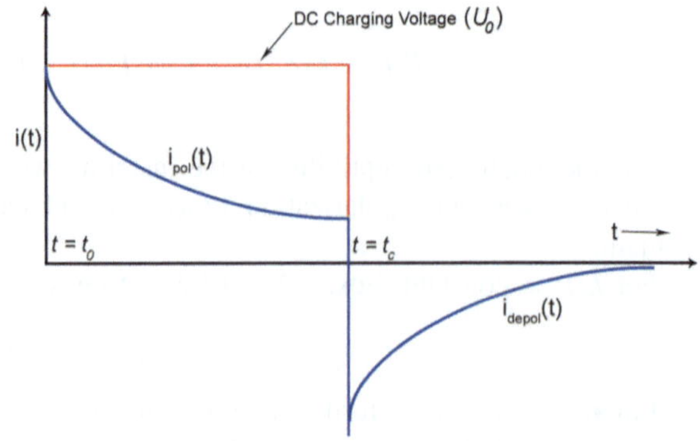

Fig. 5.1 Typical nature of polarization and depolarization currents

as the RV. The active relaxation processes inside the dielectric material gives rise to this voltage as these processes did not get enough time to relax fully during the insufficient discharging period. Thus, this RV is a significant characteristic of the insulating property of the dielectric material.

During recovery voltage measurement, the current density is zero as this voltage needs to be measured by open circuit condition [5]. Thus Eq. 5.4 can be solved with $j(t) = 0$. If the conductivity σ_0, permittivity ε_0 are known, then the recovery voltage depends on the response function *f(t,)* which in turn depends on the characteristics of the dielectric material. Hence recovery voltage measurement is another important parameter in time domain dielectric spectroscopy.

The curve in Fig. 5.2 shows the nature of typical Recovery Voltage waveform.

5.3 Time Domain Response Measurements

5.3.1 Polarization and Depolarization Current Measurements

In practice, the polarization and depolarization currents are measured with a "two electrode" technique [1] as shown in Fig. 5.3.

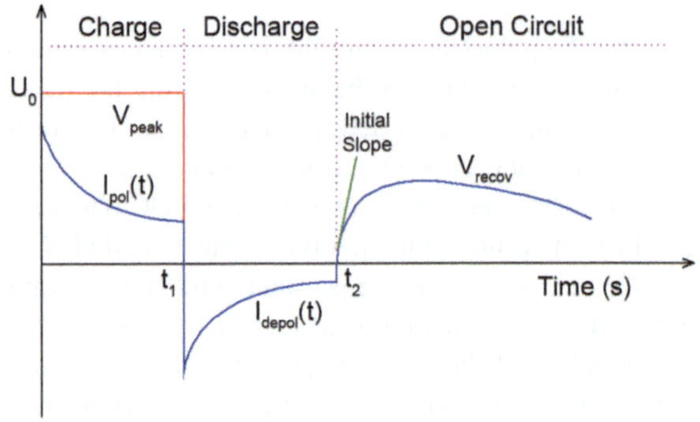

Fig. 5.2 Typical nature of recovery voltage waveform

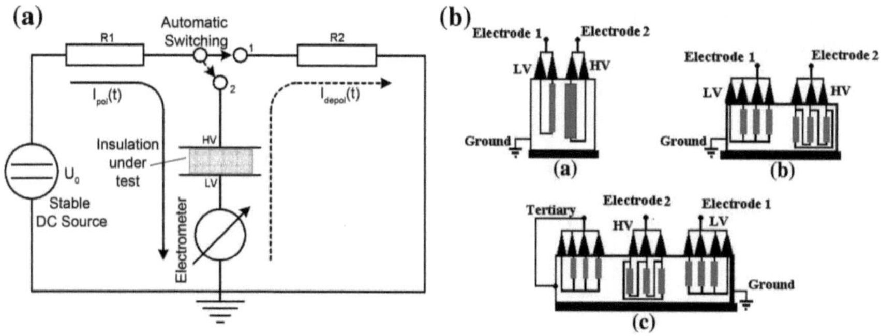

Fig. 5.3 **a** Two electrode technique for time domain dielectric response measurements, **b** actual terminal connections for different transformers

Two resistors of this circuit represent the protection resistances, which will not influence the recorded currents. The sample under test has two electrodes marked HV and LV. LV is permanently connected with electrometer, a type of precision ammeter (and also voltmeter on requirement). During charging or polarization, the switch *S* connects at position 2 due to which the polarization current $i_{pol}(t)$ flows under the influence of the applied step voltage U_0 as shown in the Fig. 5.3. The electrometer records this current. During discharging or depolarization, switch *S* connects to 1 and the sample under test is short circuited through the electrometer. The electrometer then records the depolarization current $i_{depol}(t)$.

5.3.2 Recovery Voltage Measurement

As per the basic theory of recovery voltage measurement, the charging (polarization) and discharging (depolarization) process can be performed using the same circuit as that of Fig. 5.3. The difference is in the time where the discharging time is much less than the charging time according to the requirement. A typical case may be discharging time ($t_{discharge}$) = half of charging time (i.e., $t_{ch}/2$). To measure the voltage, the electrometer should be able to change its mode from ammeter to voltmeter at the proper instant of time. This does not pose a problem, as modern electrometers are all configurable through a computer which adjusts the timing sequences. The curve in Fig. 5.4 shows the nature of a typical recovery voltage waveform. The important parameters of the recovery voltage waveform are the peak value $V_{recovery}$(peak) and the initial slope.

5.3.3 Concept of RV Spectra and the Central Time Constant

According to the methodology of RV measurement, as stated in Sect. 5.3.2, the charging time is increased gradually with the corresponding increase in discharging time from a small initial value and a different peak value of recovery

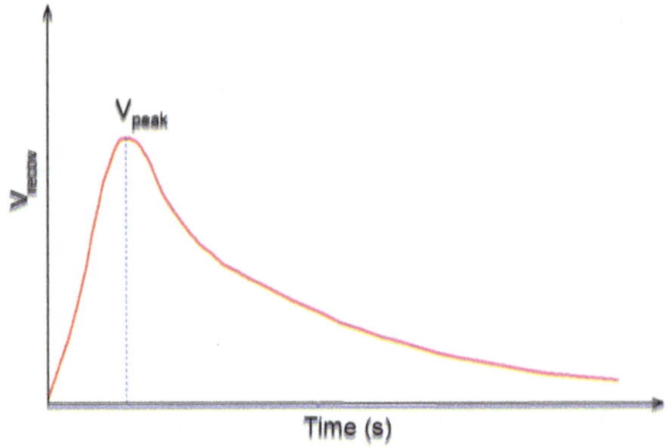

Fig. 5.4 A typical recovery voltage waveform

voltage, $V_{recovery}(peak)$ is obtained in each step. Peak values obtained for different charging times (t_{ch}) can be plotted to yield a $V_{recovery}(peak)$ versus t_{ch} curve, which is called *recovery voltage spectrum*. This recovery voltage spectrum is also analyzed for condition monitoring. The charging time corresponding to the peak of RV spectra is called *central time constant* (CTC). Figure 5.5 shows a representative diagram of recovery voltage spectrum [1–5].

5.3.4 Need for Using Electrometer in PDC and RVM

As stated earlier, the overall condition and the aging of the transformer insulation could be predicted using PDC and RV measurements. They can give a close prediction of the overall amount of moisture trapped in paper and pressboards that is very difficult to envisage by other non-invasive tests [7]. But to get such a close prediction, the measurements must be done with high precision without disturbing the internal processes occurring in transformer. For this purpose, two aspects of these measurements are very significant. First, the currents from the processes that

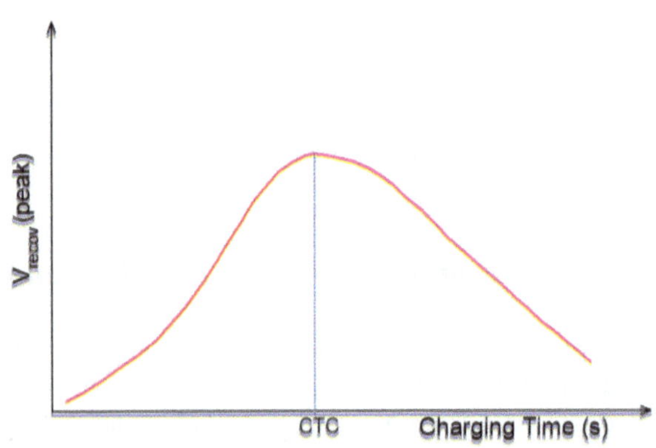

Fig. 5.5 A representative recovery voltage spectrum

are required to be measured are of the order of nanoamperes and secondly, the recovery voltage originates from a high insulation resistance that is equivalent to a voltage source with extremely high internal impedance. Thus recovery voltage needs to be measured by a voltmeter with even higher input impedance compared to the source to prevent any voltage drooping. Thus for PDC and RV measurement high precision electrometer is used instead of any ordinary meter. The electrometers are computer controlled, have a resolution as low as 100 aA during current measurement and have a very high impedance >200 TΩ during voltage measurement.

5.3.5 PDC Measurement Algorithm

The sequence of operation in PDC measurement is given in the following steps:

Step1: The electrometer is configured as an ammeter in auto mode through the controlling computer. In this mode, the meter is capable of auto ranging according to the current input;

Step2: One terminal of the sample is subjected to U_0, typically 1 kV ripple free dc, by means of the switch S in Fig. 5.3a;

Step3: The Polarization Current is acquired through the meter as current versus time in the computer. To do so, the electrometer has inbuilt ADC. Here, it is important to mention that during all these types of data acquisition, the rate and the integrating cycles for A/D conversion is to be set very carefully according to the total time for acquisition. This is because, the main goal is not only to perform precision measurement, but also to limit the acquired data size for storage and analysis. Too high a data is always cumbersome for transfer. For example, if the measurement is for 3 h, data at a rate of 50 S/s means almost infinite data for such a portable device. In that case the rate is calculated as the required number of data that is feasible for the system divided by the total time of measurement. Data acquisition is usually stopped when the response current becomes stable;

Step4: Both the terminals of the test samples are earthed, one directly and other through the electrometer, because the meter offers low input impedance during current measurement. In this configuration the depolarization process takes place;

Step5: The Depolarization Current is acquired in a manner similar to that of the polarization current through the electrometer. However, in this case the direction of current is reversed.

The flowchart of the PDC measurement is given in Fig. 5.6.

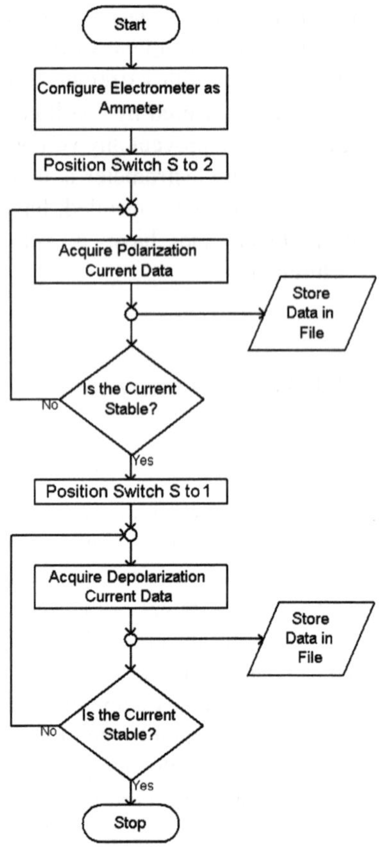

Fig. 5.6 Flowchart for PDC measurement

5.3.6 RV Measurement Algorithm

The sequence of operation during RV measurement is given in the following steps:

Step1: The Electrometer is configured as low impedance ammeter through the computer. The switch in Fig. 5.3a is in the position such that both the terminals of the sample are earthed, one directly and other through the electrometer. Thus the sample under test are shorted for a specified period of time. This time is set by the expert at the field, considering the overall condition and type of sample used;

Step2: One electrode of the test sample is charged for a definite period of time with a step voltage of 1 kV dc, i.e., U_0 for a definite period of time. Typical starting time may range from 1 ms to 1 s, depending on the hardware used;

Step3: The switch again discharges the test sample through a period, half that of charging. This discharging has same configuration like depolarization process of PDC measurement;

Step4: The electrometer is then configured as a high input impedance voltmeter as mentioned in Sect. 5.3.4;

Step5: The recovery voltage data is acquired through the meter and is programmed to be stored as voltage versus time data in a file in the computer.

In this algorithm, charging time is increased in steps by a factor of two with the corresponding increase in discharging time. In each step, the ratio of charging and discharging time is kept the same, typically 2. The process is repeated for a number of cycles as per RVM methodology. The flowchart of the RV measurement is shown in Fig. 5.7.

5.3.7 Avoiding the Transient Phenomena During PDC and RVM; A Practical Case

Expression of polarization and depolarization current in Eqs. 5.5 and 5.6 contains the $\delta(t)$ term, which is an extreme short duration current of very high magnitude that is already explained in Sect. 5.2. Inclusion of this part during PDC data acquisition has two distinct difficulties. First, to capture such impulse type data will require very fast ADC which will unnecessarily add to the cost of the instrument. Second, capturing such a high magnitude current will need a special type of attenuator that will be active only for that very short interval of time, otherwise the low magnitude data will be lost. Construction and control of such an attenuator seems to be very difficult. Since acquisition of $\delta(t)$ term does not add any extra information about the condition of the dielectric under test, effort to include this part with aforesaid complexity in instrument is avoided.

Both polarization and depolarization currents are measured during specified time periods starting at not less than about 0.1 to 1 s after the switching events [1]. This deliberate gap during data acquisition avoids the transients. t_0 and t_c respectively in Fig. 5.1 are actually the starting point of acquiring polarization and depolarization currents, respectively, and not the time of actual switching events. If measurement periods up to hours are made, the measurements can easily be performed by means of standard equipment [1] with a slow ADC speed to maintain data acquisition for hours.

5.4 Time Domain Response of Transformer Insulation

In electric utilities around the world, significant proportion of power transformers are being operating beyond their design life. Sudden failure of such transformers operating at a pivotal position may cause total power disruption resulting in

Fig. 5.7 Flowchart of the RV measurement

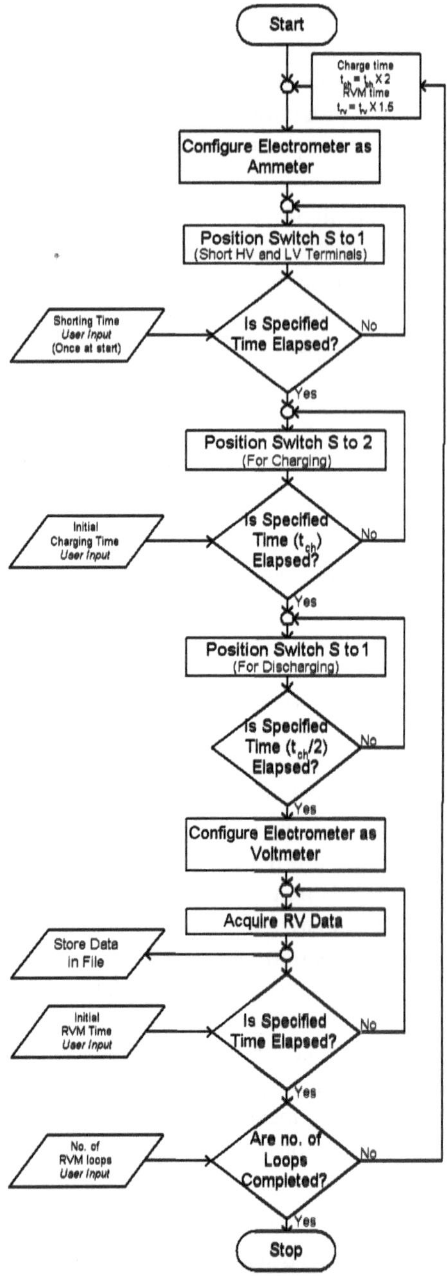

substantial loss of money and time. Thus in places round the globe, there have been urge to direct priority attention of research into improved diagnostic techniques for determining the condition of the insulation in aged transformers. Aging of a transformer is primarily due to the aging of its insulation. So, condition assessment of the insulation in a transformer is required to interpret its aging status. At present, emphases are given on non-invasive methods for condition assessment of oil-paper insulation of power transformers. As explained in the aforesaid sections, the methodology of PDC and RVM is best suited in assessment of condition in transformer insulation by time domain response measurement in dielectric spectroscopy. Henceforth, in the subsequent sections, results of PDC and RVM will be dealt with composite oil-paper insulation as dielectric under test.

In case test object contains an insulation system, like power transformers, the terminals of the transformer are appropriately shorted such that ultimately two electrodes, similar to that of HV and LV is obtained as already depicted in Fig. 5.3. The insulation between HV and LV is the bulk insulation of the transformer for which the dielectric spectroscopy is performed according to the setup. Before performing the test, HV and LV is short circuited for a substantial period of time such that the test object is completely "discharged." Otherwise some polarization will exist in the insulation that may adversely affect the obtained results. To prove the effectiveness of the PDC and RVM methodologies, transformer models are constructed in the laboratory using CIGRE standards [8, 9]. Various parameters like temperature, oil conductivity, moisture content in paper, artificial aging effect are varied in the model for a comprehensive study of the phenomena involved.

5.4.1 *Effect of Oil Conductivity*

It has been already established in the literature that the initial part of dielectric response in the case of time domain measurement is controlled primarily by the parameters of the liquid dielectric [10]. Thus in the case of oil-paper composite insulation, initial part of the polarization and depolarization currents and the initial slope of the recovery voltage will reflect the oil conductivity. Figures 5.8a and 5.8b show the variation of polarization and depolarization currents with oil conductivity.

In the case of polarization and depolarization currents, from Figs. 5.8a and 5.8b, it may be seen that with increase in oil conductivity, the initial parts of the curves show a higher value along the y-axis. This fact can be explained by using the concept stated in [1]. "Interfacial Polarization" is predominantly effective in insulating materials composed of different dielectric materials, such as oil impregnated paper/cellulose. The mismatch of the product of permittivity times resistivity for the different dielectrics causes, under the influence of an electric field, movable positive and negative charges to become deposited on the (micro- or macro-) interfaces of the different materials, also forming some kinds of dipoles. This phenomenon is often very slow and in general active in the power frequency range and below. Trapping and hopping of charge carriers between localized charge sites may occur that also enhances localized polarization [1]. Now with increase in oil conductivity, these interfacial and trapped charges get a chance to

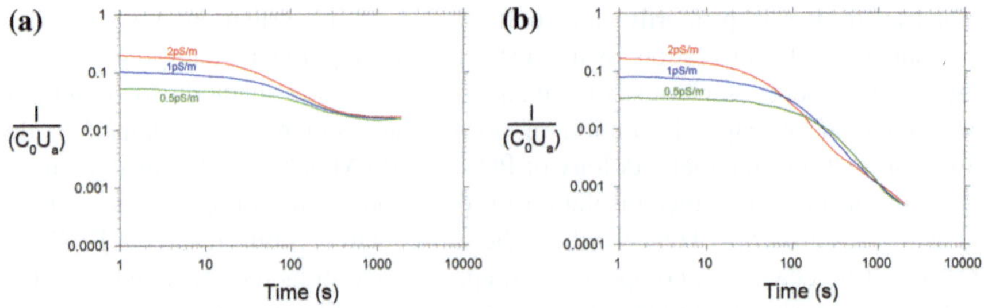

Fig. 5.8 Effect of oil conductivity on the nature of time domain response measurement: **a** polarization current; **b** depolarization current

migrate quickly resulting in easier dipole formation during polarization and quicker relaxation of the dipoles during depolarization. Thus with increase in oil conductivity the overall polarizability of the insulation increases. Due to this easier polarization and depolarization, current of higher magnitude is reflected in the initial part of the plotted current curves.

In the case of RVM, it is known that, the discharge time is not sufficient to cause full relaxation of the polarized material within the dielectric. These unrelaxed interfacial charges, on their course of relaxation, are responsible for rise of voltage in the case of RVM. Recalling the explanation in the case of PDC, more conductivity of oil causes easier polarizability in the overall insulation material. Easy polarizability enhances quicker relaxation of the formed dipoles to its initial stage. This causes faster voltage rise up to the peak and consequently the initial slope is higher in RVM as evidenced from Fig. 5.9.

5.4.2 *Effect of Temperature*

Figures 5.10a and 5.10b show polarization and depolarization currents obtained at different steady operating temperatures in a transformer model to visualize the effect of temperature in the case of a composite oil-paper insulation [6]. In each

Fig. 5.9 Effect of oil conductivity on the nature of recovery voltage

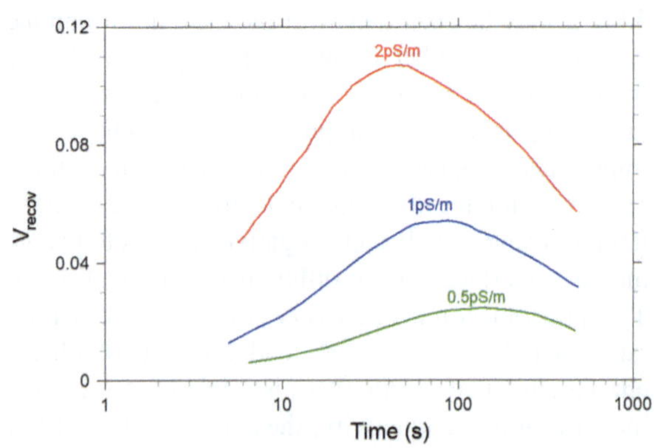

case, the transformer model was charged (polarized) with 300 V for a long time and then discharged (depolarized) for equal time period. It can be seen that variations in temperature cause significant shifting of both the polarization and depolarization current curves. It can be also seen from Figs. 5.10a and 5.10b that magnitude of the polarization and depolarization currents tends to shift to higher values with increasing temperature. It is further noted that deviation between polarization current at different temperatures are not really linear.

Polarization current is composed of two parts—one is the decaying current due to the polarization mechanisms, and the other part is the steady conduction current related to the conductivity in the oil paper insulation [11, 12]. In the case of depolarization, the conduction current does not exist due to absence of applied voltage. It should be noted that an increase in oil conductivity at same temperature and increase in oil conductivity due to increase in temperature does not affect the PDC curve in the same way. With increase in temperature, an overall increase in oil conductivity takes place in such a way that an overall uplift of the curve occurs during polarization. During depolarization, the applied voltage is absent, so there occurs no overall shifting of the curve. Instead, the overall increase in oil conductivity at elevated temperature invites faster depolarization resulting in quicker decay rate of current. Accordingly, it can be found from Fig. 5.10a that, with rise of temperature, the overall curve shifts upwards in the case of polarization current. In the case of depolarization current, as may be seen in Fig. 5.10b, the overall uplift of the curve is absent. Instead, with increase in temperature, the decay rate is faster in Fig. 5.10b. In Fig. 5.10a, the overall uplift of the curves with increase in temperature is due to the increase in the overall conductivity in the insulation as mentioned before. This later effect is much more pronounced in the polarization current data and supersedes the decay rate due to polarization mechanisms in most of the cases as seen from Fig. 5.10a. In the case of Fig. 5.10b, with increase in temperature, the overall conductivity of the insulation system increases due to migration of electrons from the valence band to the conduction band. Hence, as explained in Sect. 5.4.1, with increase in conductivity, the overall polarizability in the insulation increases resulting in faster decay rate of depolarization current.

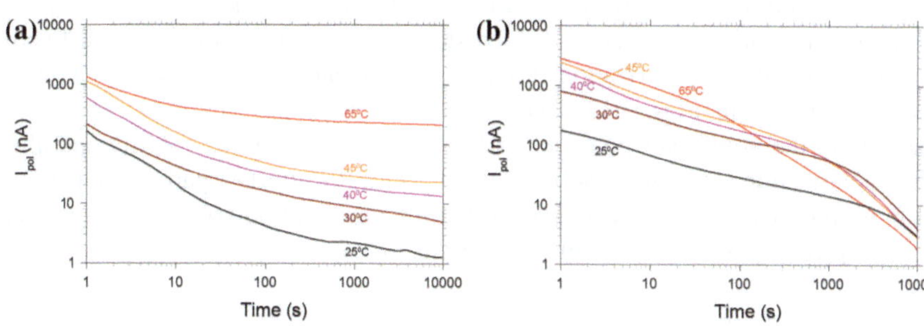

Fig. 5.10 Effect of temperature on the nature of time domain response measurement: **a** Polarization current; **b** depolarization Current

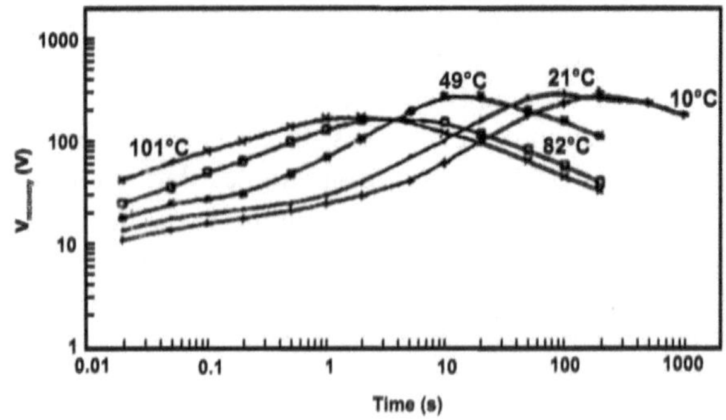

Fig. 5.11 Effect of temperature on the nature of recovery voltage [13]. Reproduced with kind permission ©IET 2012

As shown in Fig. 5.11, at elevated temperatures, the RV curve is found to peak at increasingly smaller times. This is due to the fact that the initial rise of the recovery voltage (i.e., initial slope) is determined by relaxation of interfacial charge through the oil [10]. In the case of short charging and discharging times during the recovery voltage measurement, surface charges completely dominate the magnitude of the recovery voltage. With increase in temperature, these charges are quickly released so that the RV attains the peak at smaller times.

5.4.3 Effect of Paper (Solid Insulation) Conductivity and Moisture

One of the significant features of application of PDC and RVM is to find the paper conductivity (i.e., conductivity of solid insulation), and consequently paper moisture in composite oil-paper insulation. It can be understood that finding the condition of solid insulation, i.e., paper, is almost impossible without complete dismantling of the equipment under test. Thus a methodology that can predict the conductivity and the amount of moisture absorbed in the solid insulation will be of immense help to the engineers. Figures 5.12a and 5.12b show the influence of paper conductivity on the results of PDC and RV measurement, respectively.

In contrast to the effect of oil conductivity, paper conductivity distinctly influences the tail of polarization current measurement as well as recovery voltage as evidenced from Figs. 5.12a and 5.12b. This happens because in composite insulation, the liquid part polarizes earlier due to its easy polarizability and after that, polarization starts in the solid material. Thus, the effect of paper conductivity is more pronounced in the later part of the time scale in polarization current and recovery voltage than that of oil conductivity.

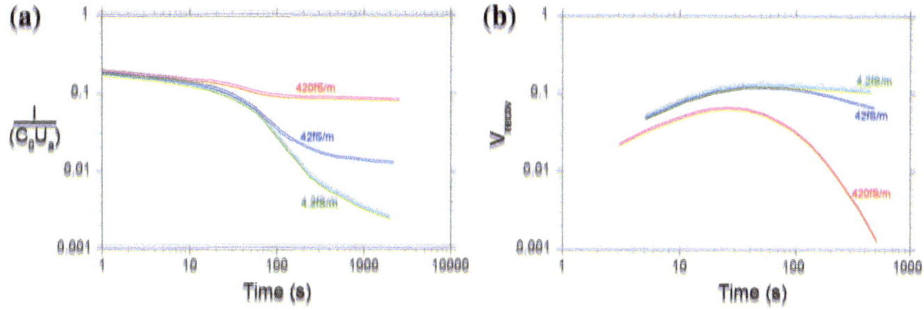

Fig. 5.12 Effect of paper conductivity on the nature of time domain response measurement: **a** polarization current; **b** recovery voltage

However, the effect of paper conductivity does not influence much in the depolarization current as may be seen from Fig. 5.13. This can be explained from the expression of depolarization current in Eq. 5.6, which is independent of the conductivity term.

Absorption of moisture in the paper increases its conductivity and due to the reason discussed earlier, polarization current and recovery voltage curves are similar in nature to that of increase in paper conductivity. In either cases of increase in paper conductivity or increase in paper moisture, the CTC of RV spectra decreases, i.e., peak of RV spectra occurs earlier. Figure 5.14a shows the effect of paper moisture in polarization current measurement and Fig. 5.14b shows the effect of paper moisture in recovery voltage measurement.

It is worth mentioning here that, according to the findings of the CIGRE Task Force 15.01.09 [14] there may be two or more peaks of the RV spectra, of which the additional or subsidiary peak is away from the dominant time constant depending upon the oil-paper insulation condition. In such a case, the dominant time constant is governed by the condition of the oil and any subdominant maximum would stem from polarization phenomena in the solid insulation.

Fig. 5.13 Effect of paper conductivity on the nature of depolarization current

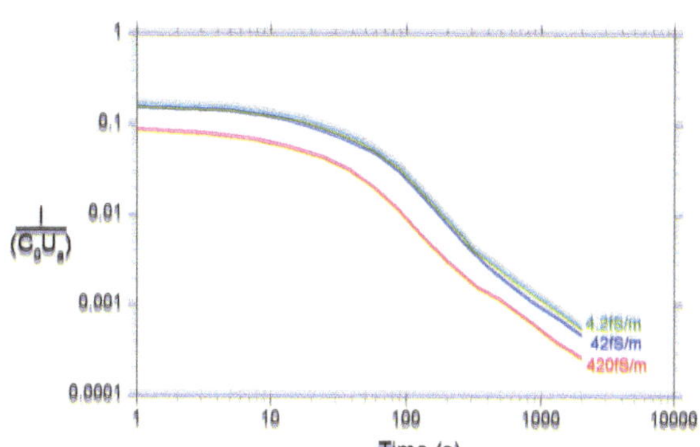

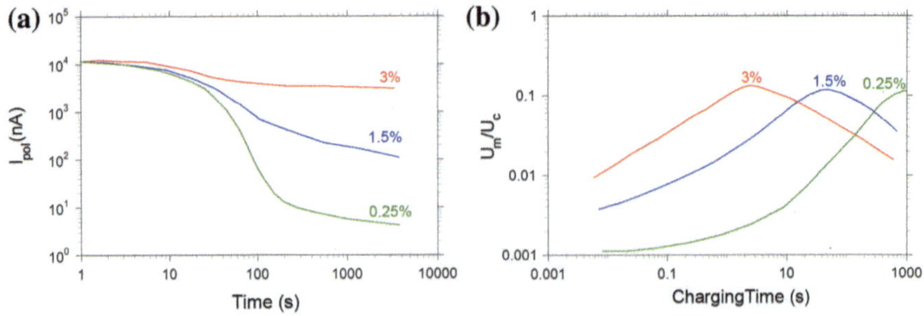

Fig. 5.14 Effect of paper moisture on the nature of time domain response measurement: **a** Polarization current; **b** recovery voltage

5.4.4 Effect of Insulation Aging

Nature of polarization and depolarization current varies significantly with overall aging of the oil-paper insulation. J. L. Wei et al. studied the aging characteristics of oil-paper insulation by preparing oil impregnated pressboard samples with accelerated aging. Results were recorded at various temperatures both before and after aging for a comparative study [15]. Figures 5.15a and 5.15b show the effect of insulation aging on polarization current. Similarly, Figs. 5.16a and 5.16b show the effect of aging on depolarization current of PDC measurements under different aging conditions.

A comparative study of Figs. 5.15 and 5.16 reveals that, due to aging, the slope of the initial part of both polarization and depolarization current becomes lower compared to unaged insulation. One of the explanations may be that, due to aging, the migration of interfacial charges becomes sluggish through the oil and consequently, the overall rate of polarization becomes slower. This results in lower slope in the obtained current.

In the case of RVM, the influence of overall aging in insulation is demonstrated by the curves as shown in Fig. 5.17. Curves for transformers are presented that widely differ in ages. An approximate assessment of the dielectric condition can be

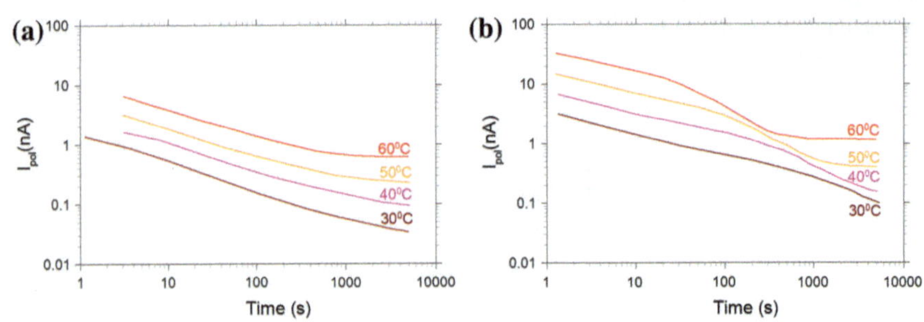

Fig. 5.15 Effect of temperature on polarization current; **a** unaged condition; **b** aged condition

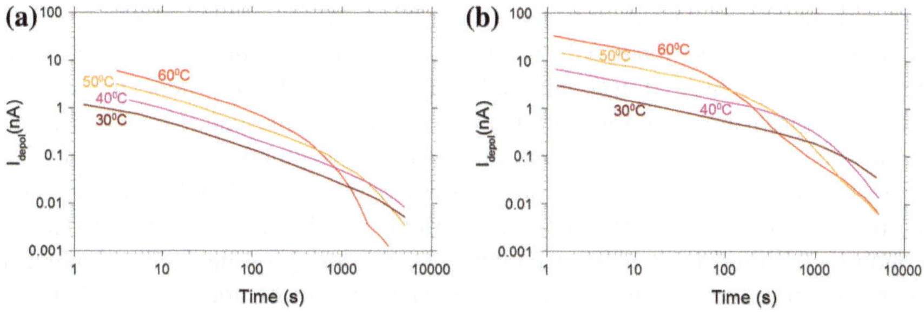

Fig. 5.16 Effect of temperature on depolarization current; **a** unaged condition; **b** aged condition

made by visualization of the curves. The closer the curve peak (CTC) is toward the small time constant, the worse is the condition of the dielectric [13]. It may be noted here that, the degeneration of cellulose insulation produce water as an end product of the chemical reaction, which contributes to the moisture content of the cellulose insulation (mainly paper). Hence, the paper moisture content reflects the aging condition of insulation.

5.5 Effect of Noise on Dielectric Response Measurements in Field Conditions

PDC and RVM are off-line diagnostic tests in the case of a transformer. Considering the ideal case, the transformer should be isolated from the electricity network. But in practice, at site, it is difficult to remove the transformer completely from the network in majority of the cases. Presence of undue long connecting lines and its associated supporting insulators and structures affect the measurements and sometimes produce adverse results. Such undesired additional networks along with other energized equipment located near the test site cause electromagnetic interference. In the case of insulation condition monitoring using PDC and RV

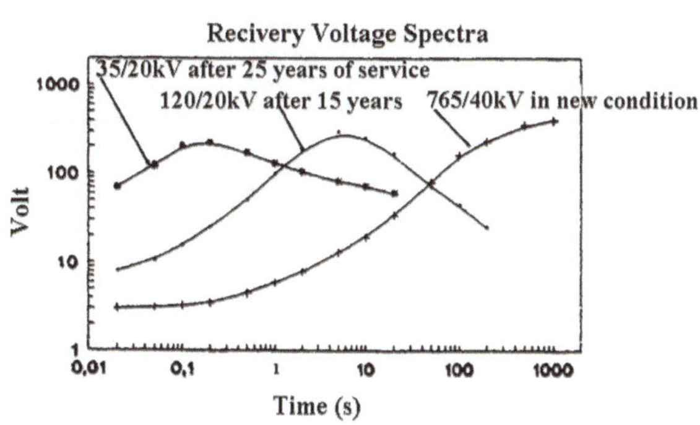

Fig. 5.17 Polarization spectra for various ages of transformers [13]. Reproduced with kind permission ©IET 2012

measurements it is necessary to have at least a fairly denoized recording. Otherwise wrong interpretation may be result from the noisy data during post-processing for analyzing the insulation condition. So, it is better to devise a denoizing scheme that can eliminate the noise during the recording of the data through the data acquisition system. Moreover, the signals to be acquired during time domain dielectric response measurement are of low magnitude level and hence are more prone to be affected by external noises; especially in the case of acquiring data for condition monitoring of outdoor equipments. So during on-field measurement of these condition-monitoring signals, significant amount of noise may pollute the actual waveforms. Therefore to extract important information from the recorded data, a suitable denoizing scheme should be adopted. Considerable amount of noise in the signals is evident from the real life recordings of PDC and RV waveforms, shown in Figs. 5.18 and 5.19, respectively [16].

5.5.1 Use of Denoizing Filter

As mentioned earlier, due to outdoor environmental condition and several hours of experimentation, various noises may contaminate the actual signal, such as, random noise with some unknown distribution, periodic noises or aperiodic noises with stationary or non-stationary properties, etc. In Fig. 5.18 the PDC signal is contaminated with random high frequency noise and also with impulses (spikes). Whereas in Fig. 5.19 the RV waveform is contaminated with random noise but the time variation of the noise is quite different in nature from that present in the PDC waveform shown in Fig. 5.18.

Fig. 5.18 Real-life noisy PDC waveform

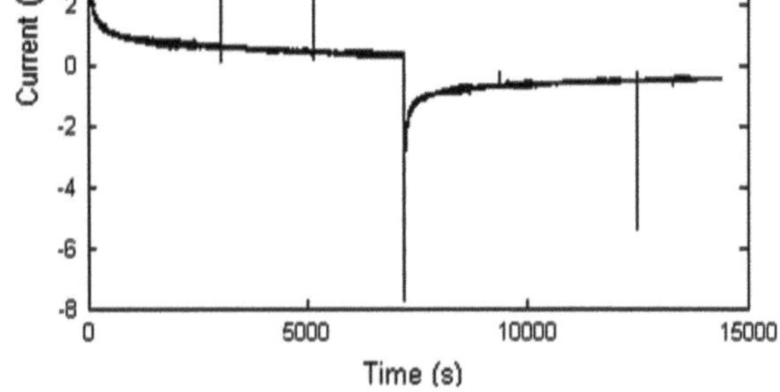

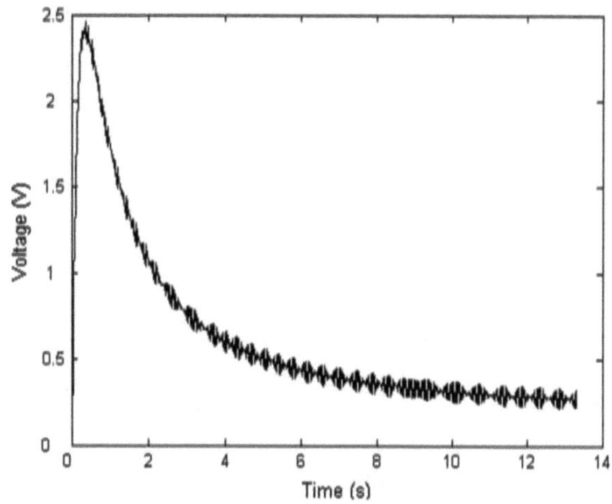

Fig. 5.19 Real-life noisy RV waveform

In PDC measurement the slope of different recorded PDC waveforms (i.e., nature of variation with time) are used as a discriminating factor for different insulation conditions. However, the noise present in the recorded signal may affect the precision with which this parameter (i.e., slope of the curve, or nature of variation) can be detected. It is also evident from the above argument that any filtering technique that changes the slope of the curves of PDC must not be used for denoizing of such signals. The typical PDC waveform, shown in Fig. 5.1, also indicates that there occurs a sudden change in the waveform during the transition from polarization current measurement to that of depolarization current. Such sudden change is not a noise but is inherent in the measurement procedure. Hence, such jump-discontinuity in the waveform must be preserved while filtering the waveform in real time. No linear filtering technique is capable of performing this task; so nonlinear filtering is a possible choice.

Similarly for RV measurement technique, properly denoized recordings of recovery voltage waveform are necessary. The determination of peak values of recovery voltage from periodic charging and discharging is of great importance. Therefore, a spike or any kind of spurious noise can lead to a wrong interpretation such as incorrect evaluation of the peak magnitude and the time of occurrence of the peak. Therefore RV signals also require a judiciously chosen filtering scheme.

There may be a number of ways to design such a filer to achieve the required performance. However, in this section one of the possible filtering schemes is discussed.

The scheme presents a simple, fast, and versatile hybrid-filtering method that is very effective for on-line denoizing of different signals of dielectric response measurement.

5.5.2 Detailed Description of Developed Hybrid Filter

The developed hybrid-filtering scheme incorporates weighted median (WM) filter, which belongs to the class of nonlinear filters, along with a low pass digital infinite impulse response (IIR) filter [16].

5.5.2.1 Weighted Median Filter

The weighted median (WM) filter is a generalized form of the conventional median filter, and it has several applications in image and signal processing [17–19]. A brief description of the WM filters is given below.

Let **S** = $[S_1, S_2, \ldots, S_N]$ is the discrete time real valued input vector and Y is the output of the WM filter. The integer weights of the WM filter having window length N are given by the weight vector **W** = $[W_1, W_2, \ldots, W_N]$. The association of X, W, and Y gives the filtering action as follows,

$$Y = MED\left[W_1 \diamond S_1,\ W_2 \diamond S_2, \ldots,\ W_N \diamond S_N\right] \tag{5.7}$$

Here *MED [•]* indicates median operator and $\diamond$ indicates replication operation. So, the filtering procedure can be stated as follows:

Sort the samples inside the filter window in ascending order after replicating each sample S to the number of the corresponding weight W, and choose the median value from the new sequence.

Figure 5.20 shows a typical structure of WM filter. For real positive weights this technique can be extended to find out θ, the weighted median value of S, by minimizing the expression,

$$L(\theta) = \sum_{i=1}^{N} W_i |S_i - \theta| \tag{5.8}$$

θ will definitely be one of the samples of **S**, as $L(\theta)$ is piecewise linear and convex for $W_i \geq 0$, for all i, [17]. So the output of weighted median filter for real positive weights is calculated as follows:

Step1: Rearrange the sample values within the filter window in ascending order;

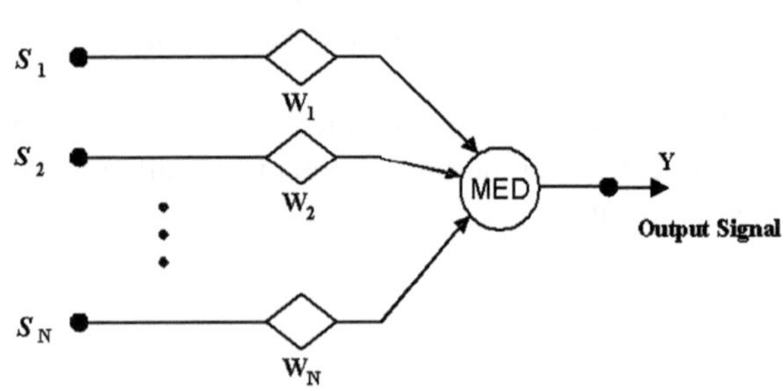

Fig. 5.20 Weighted median filter

Step2: Sum up the respective weights of the samples from the highest end of the sorted sample window;

Step3: Stop at that sample when the sum just becomes greater than or equal to half of the total sum of the weights (i.e., $\geq {}^1\!/_2 \sum_{i=1}^{N} W_i$);

Step4: The sample is the respective output of the WM filter.

In a more general way, for real valued weights the filter output is given in Eq. (5.9),

$$Y = MED\left[|W_i| \Diamond \operatorname{sgn}(W_i) S_i|_{i=1}^{N}\right] \tag{5.9}$$

where, sgn(x) = 1 for x > 0, sgn(x) = 0 for x = 0, and sgn(x) = −1 for x < 0.

5.5.2.2 Infinite Impulse Response Filter

IIR filters have an impulse response function which is nonzero over an infinite length of time. In digital IIR filters, the output feedback is immediately apparent in the equations defining the output. IIR filters may be implemented as either analog or digital filters. IIR filter operation can be represented in the mathematical form as given below:

General form of a Laplace transfer function,

$$G(s) = \frac{A_0 + A_1 s + A_2 s^2 + \ldots A_m s^m}{B_0 + B_1 s + B_2 s^2 + \ldots B_n s^n} = \frac{Y(s)}{S(s)} \tag{5.10}$$

Corresponding z transform,

$$H(z) = \frac{a_0 + a_1 z^{-1} + a_2 z^{-2} + \ldots a_m z^{-m}}{1 + b_1 z^{-1} + b_2 z^{-2} + \ldots + b_n z^{-n}} = \frac{Y(z)}{S(z)} \tag{5.11}$$

Corresponding recursive difference equation

$$y_i = a_0 s_i + a_1 s_{i-1} + a_2 s_{i-2} + \ldots + a_m s_{i-m} - b_1 y_{i-1} - b_2 y_{i-2} \ldots - b_n y_{i-n} \tag{5.12}$$

where

y_i = present value of output
y_{i-1} = previous output value (one time increment ago)
s_i = present value of input data and so on…
a_k and b_k = filter coefficients.

A representative diagram of IIR filter operation is shown in Fig. 5.21. As the designing of IIR filters is very common it has not been discussed in detail here. In

Fig. 5.21 A typical schematic of digital IIR filter

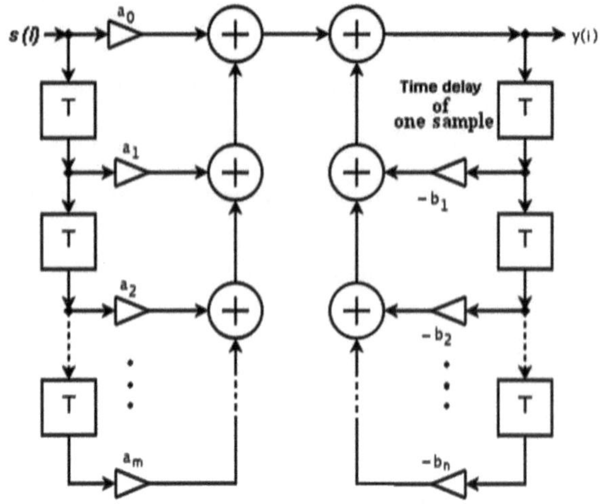

the present hybrid-filtering scheme a lowpass Butterworth digital IIR filter may be used.

5.5.2.3 Hybridization of WM and IIR Filter

The hybrid filter works with a novel switching algorithm. The algorithm dictates switching of the filtering technique between WM filter and lowpass digital IIR filter depending upon the nature of the input signal at that instant. If a sudden step change or an impulse signal occurs at the input then the switching algorithm selects WM filter to obtain the filtered output at that instant and for slow variation of input signal, contaminated only with random high frequency noise, lowpass digital IIR filter is selected through the switching algorithm [16]. The schematic of such a hybrid filter is shown in Fig. 5.22.

The switching algorithm used for the hybrid filter is explained below:

Step1:

Fig. 5.22 Schematic of switching based hybrid filter

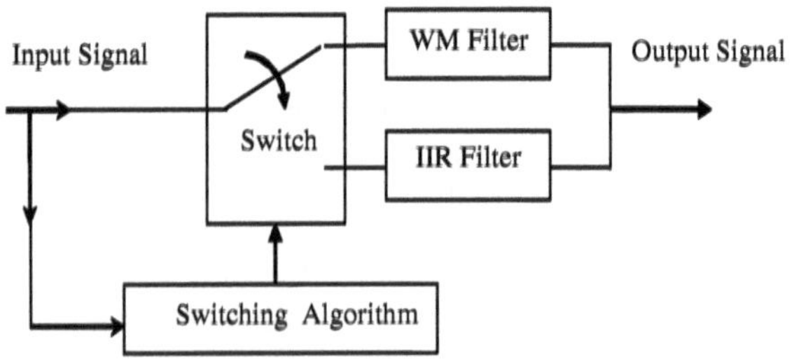

Number of recent data samples taken into account is equal to the window length of the WM filter, (it may be taken as 5 or 6 in the case of PDC waveforms);

Step2: Check whether any impulse signal or any step change occurs within this data sequence;

Step3: If YES then output is taken from WM filter.

(i.e., switch to WM filter).
If NO then output is taken from IIR Filter
(i.e., switch to IIR filter).

It is clear from the scheme that the computational complexity of the filtering technique is minimal, so it requires very little CPU time, which makes the scheme suitable for real-time applications.

5.5.2.4 Denoizing Performance of the Hybrid Filter

For obtaining the information related to the selection of the cutoff frequency of the IIR filter, the Power Spectral Density (PSD) of the recorded signals for different environmental conditions and also for different insulation states should be studied. Different experimentation for PDC measurement carried out by the authors show that the PSDs of recorded signal beyond 0.2 Hz (approximately) do not contain significant frequency components but noises. A typical PSD of a real life PDC waveform is shown in the Fig. 5.23 to ascertain this fact. Therefore the lowpass Butterworth IIR filter can be designed on the basis of a cutoff frequency of 0.2 Hz [20]. One can choose appropriate order of the filter to meet the desired performance, at the cost of increased computational load [16]. As a rule of thumb the order can be taken as 5 or 6.

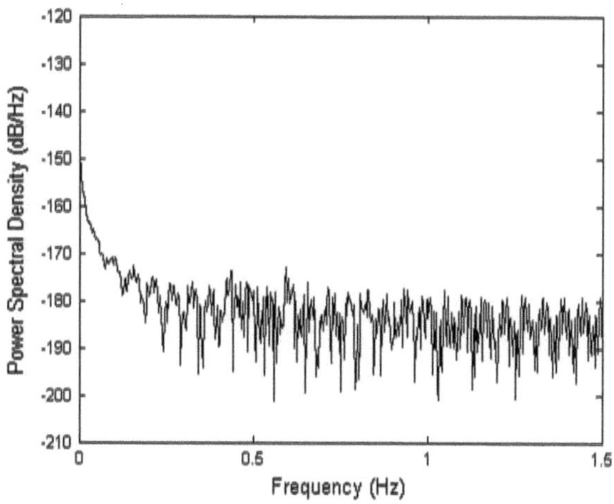

Fig. 5.23 Power spectral density of a typical PDC waveform

Figure 5.24 shows the denoized version of a real life noisy PDC waveform, given in Fig. 5.18, using WM filtering. It is established from the filter output that impulses present in the actual noisy data are removed completely by the WM filter and also the sudden change in the waveform during the changing from polarization to depolarization is preserved. The salient features of the parent waveform, such as wave shape and the nature of variation of the curve are also conserved. Nonetheless in the slowly varying portions of the waveform the random noises are present in good proportion, as evident from the magnified view of the selected segment in Fig. 5.24. Thus, low pass IIR filtering is applied in these portions of the waveform. However, IIR filter has its own disadvantages while acting on impulse noise or jump-discontinuity. Figures 5.25a and 5.25b show that during initial stage of polarization and depolarization current waveform, respectively, the IIR filter output not only alters the wave shape but also changes the slope of the curve, which is undesirable. For clarity, only a part of the waveforms are shown in Fig. 5.25.

Moreover, presence of any impulse noise in the signal makes the output of the IIR filter oscillatory in this case and therefore wave shape of the original signal gets altered. For example, due to the presence of an impulse noise in the original recorded data, shown in Fig. 5.26a, the output of the IIR filter, shown in Fig. 5.26b, has become distorted. Despite these facts Figs. 5.25 and 5.26 make it clear that the outputs of the IIR filter are smooth enough to ensure that the random high frequency noises are removed well by this filtering method.

Outputs of the hybrid filter are shown in Fig. 5.27. The region around the step change and that around the impulse noise, are magnified for better understanding. In RV measurements, as mentioned earlier, the peak values of recovery voltages obtained from consecutive cycles are important signatures for studying insulation

Fig. 5.24 WM filter output of a real-time noisy PDC waveform

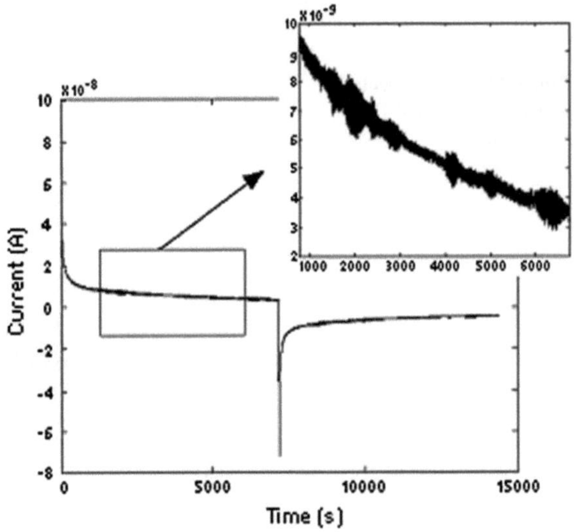

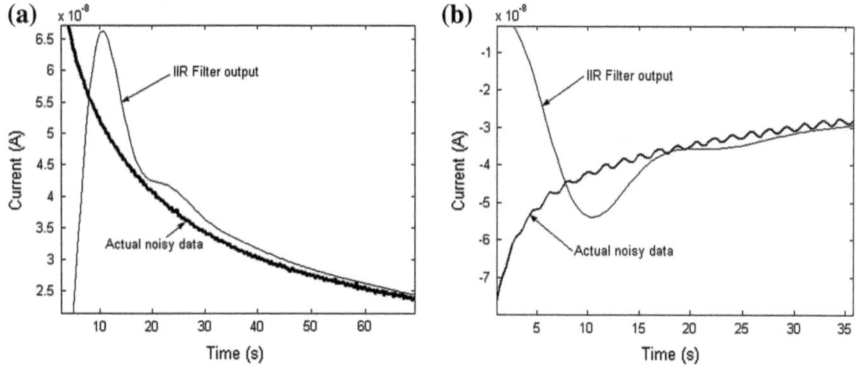

Fig. 5.25 Real-time noisy recorded data and lowpass IIR filter output in the initial stage of: **a** polarization current; **b** depolarization current

condition. As shown in Fig. 5.28, the impulse noise present in curve 2 makes improper peak value determination. The denoized waveforms obtained using parallel topology of the developed hybrid-filtering scheme are shown in Fig. 5.29, which depict that the actual peak values can be obtained from different curves.

This filtering does not change the slope or wave shape of the recorded data, which is an important parameter for detecting the condition of the insulation. This is a significant property of this hybrid-filtering scheme.

Fig. 5.26 **a** An impulse noise in recorded signal; **b** distortion in lowpass IIR filter output due to the impulse noise

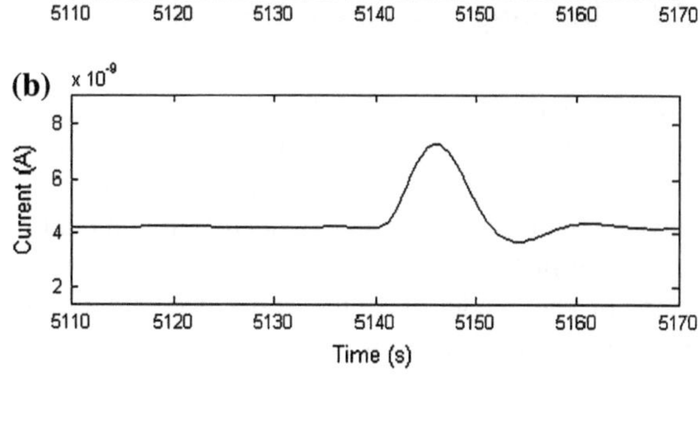

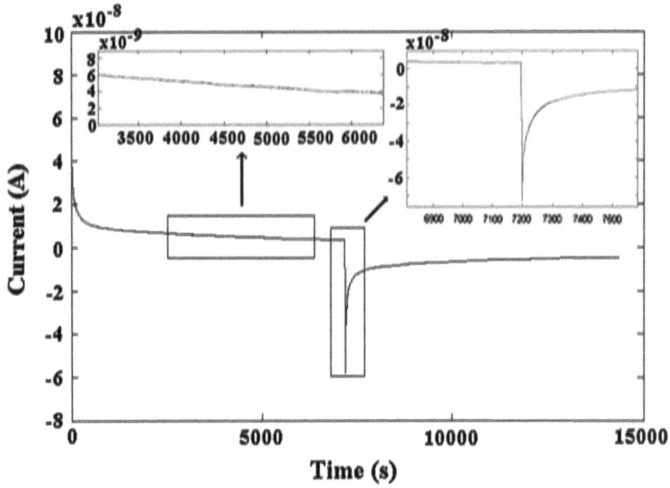

Fig. 5.27 Output of the hybrid filter for noisy PDC waveform

Fig. 5.28 Online recorded RVM data with impulse noise

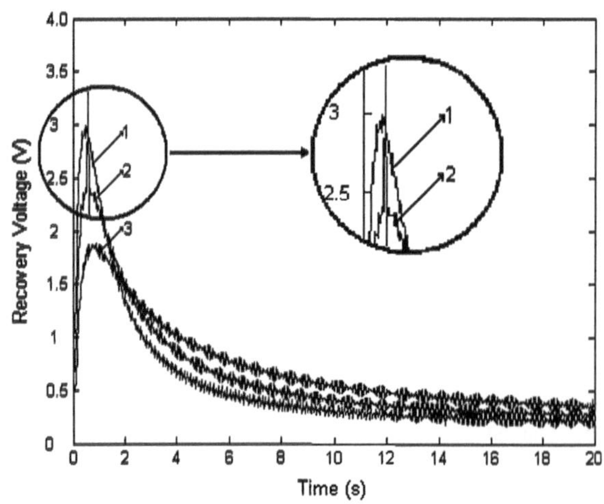

5.6 Modeling Dielectric Response

Over the past decade, several publications [1, 3, 21–25] have proposed a number of equivalent circuits for modeling the transformer oil-paper insulation system for better understanding of the dielectric response. All the models proposed so far have been derived from an extended Debye model based on a simple RC network as shown in Fig. 5.30. Depending on the condition of the insulation, the parameters of the model (i.e., the resistance and capacitance values in the parallel branches) vary. Therefore, the advantage of such model is that the values of the

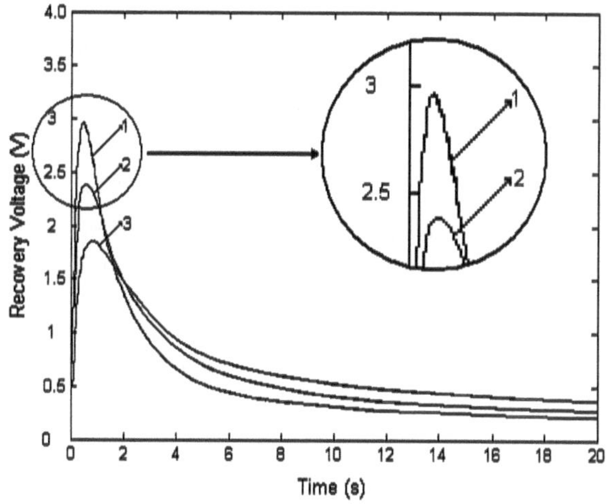

Fig. 5.29 RVM curves with real-time denoizing using hybrid filter

circuit parameters carry significant information about the condition of the insulation. In the case of designing an Expert System (ES) for insulation condition monitoring, as reported in [26], these parameter values may be considered as one of the inputs. Hence, to predict the insulation condition from time domain dielectric response measurement, the equivalent circuit parameters are required to be determined.

5.6.1 *Determination of Equivalent Circuit Parameters from Time Domain Measurements*

5.6.1.1 From PDC Data

For oil-paper insulation system of transformers, within the time periods of interest, all polarization and depolarization currents can be described by monotonously

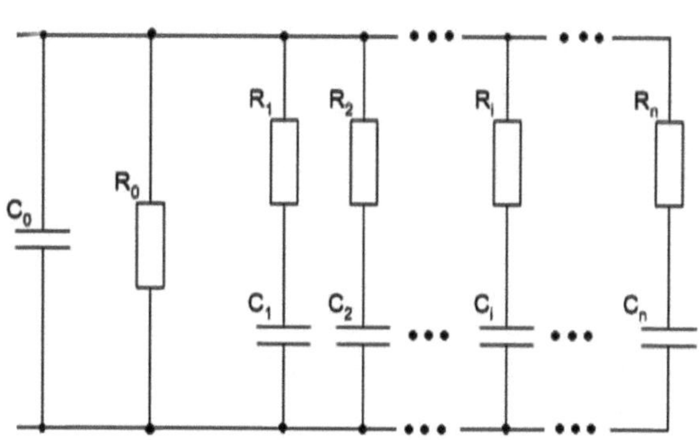

Fig. 5.30 Equivalent circuit model of oil-paper insulation in transformer (Debye model)

decreasing functions. If both polarization and depolarization currents are measured for long enough time, then their sum can be used to identify the DC resistance R_0 of the test object as follows:

$$R_0 = \frac{C_0\,\sigma_0}{\varepsilon_0} \tag{5.13}$$

where, C_0 is the geometric capacitance and σ_0 is the dc conductivity.

The other unknown elements of the different branches of the equivalent circuit, with their individual time constants, represent different polarization processes within the insulation system. These elements are to be determined by fitting the depolarization current with the equation

$$i_{\text{depol}}(t) = \sum_{i-1}^{n} A_i\, e^{-\frac{t}{\tau_i}} \tag{5.14}$$

where,

$$A_i = U_c\left[1 - e^{-\frac{T_c}{\tau_i}}\right]/R_i, \quad for\ i = 1,\ldots.,n \tag{5.15}$$

In the above equation U_c is the magnitude of the step voltage applied during charging and T_c is the time during which the charging voltage is applied, $\tau_i = R_iC_i$ is the time constant of the ith branch having R_i and C_i as its elements.

The time constants are often chosen equidistantly in logarithmic time scale [27], with two or three time constants per time decade to fit the measured depolarization currents with sufficiently good accuracy. After the circuit elements are thus determined, it can be used for any transformation that would be necessary.

5.6.1.2 From RVM Data

In the first step of RVM test, the insulation is charged with a constant dc voltage of magnitude U_c for a time period of T_c. Then the voltage U_c^i at the capacitance C_i in the ith branch of the equivalent circuit is given by

$$U_c^i = U_c\left(1 - e^{-\frac{t}{\tau_i}}\right),\ i = 1,\ldots,n;\ 0 \le t \le T_c \tag{5.16}$$

where, $\tau_i = R_iC_i$

In the second step of RVM test, the insulation is kept short circuited for a time period of T_d. Then the voltage U_c^i is given by

$$U_c^i = U_c\left(1 - e^{-\frac{t}{\tau_i}}\right)e^{-\frac{t}{\tau_i}},\ i = 1,\ldots,n;\ 0 \le t \le T_d \tag{5.17}$$

In the third and final step, the insulation is open circuited and the recovery voltage U_r is caused by the residual polarization. Then the voltage U_c^i satisfy the following equations

$$C_0 \frac{dU_r(t)}{dt} + \frac{U_r(t)}{R_0} + \sum_{i=1}^{n} C_i \frac{dU_c^i(t)}{dt} = 0 \tag{5.18}$$

$$\frac{U_r(t) - U_c^i(t)}{R_i} = C_i \frac{dU_c^i(t)}{dt}; \quad i = 1, \ldots, n \tag{5.19}$$

So the equivalent circuit parameters can be estimated from the measured RVM data with the help of the above equations [28].

One recovery voltage measurement does not provide enough information about the different polarization processes within an insulation system, which may have widely varying time constants. The RVM data for short charging time periods give information about the fast polarization processes, while the RVM data for long charging times provide information about the slow polarization processes. As a result, a polarization spectrum is required for proper diagnosis of insulation condition using RVM data.

5.6.2 Computation of Recovery Voltage Data from Equivalent Circuit Parameters

Once the equivalent circuit parameters are determined, then the recovery voltage data can be computed by using the following state-space equations, which are deduced from Eqs. (5.16) and (5.17).

$$\begin{bmatrix} \dot{U}_c^1 \\ \dot{U}_c^2 \\ \cdot \\ \cdot \\ \dot{U}_c^n \\ \dot{U}_r \end{bmatrix} = \begin{bmatrix} -\frac{1}{R_1C_1} & 0 & \cdots & 0 & \frac{1}{R_1C_1} \\ 0 & -\frac{1}{R_2C_2} & \cdots & 0 & \frac{1}{R_2C_2} \\ \cdot & \cdot & \cdots & \cdot & \cdot \\ \cdot & \cdot & \cdots & \cdot & \cdot \\ 0 & 0 & \cdots & -\frac{1}{R_nC_n} & \frac{1}{R_nC_n} \\ \frac{1}{R_1C_0} & \frac{1}{R_2C_0} & \cdots & \frac{1}{R_nC_0} & \frac{1}{C_0}\left(\frac{1}{R_0} - \sum_{i=1}^{n} \frac{1}{R_1}\right) \end{bmatrix} \cdot \begin{bmatrix} U_c^1 \\ U_c^2 \\ \cdot \\ \cdot \\ U_c^n \\ U_r \end{bmatrix} \tag{5.20}$$

The initial conditions of the state variables in Eq. (5.20) are given by

$$U_c^i(0) = U_c(1 - e^{-\frac{T_c}{\tau_i}})e^{-\frac{T_d}{\tau_i}}, \; i = 1, \ldots, n \; ; \quad U_r(0) = 0 \tag{5.21}$$

This can be a way of model validation from the measured RVM data. This is important in synergizing both PDC and RVM data for an integrated approach toward insulation condition assessment.

5.7 Interpretation of Time Domain Data Using Dielectric Response Model

The most important part of time domain measurement and dielectric response modeling is to interpret the condition of the insulation from such a model. From utility point of view, unless and until they know the condition of the insulation both qualitatively and quantitatively (more desirable), the whole procedure of dielectric response measurement goes in vain. Therefore, a mapping function is required to map the obtained insulation model (more precisely the values of different model parameters) into some quantitative measure of insulation condition (e.g. moisture content of paper or remaining life for the insulation). Despite having many papers published in this direction and addressing this problem, no *standard* methodology for such mapping exists yet. Hence, unique, certain, and unambiguous interpretation about insulation condition from such dielectric measurements is still a challenge for the researchers.

One possible way of relating the values of model parameters (obtained from Debye model of the insulation mentioned above) with the insulation condition is discussed in [26]. Before we proceed into the steps of such interpretation, let us discuss the effect of model parameters on the PDC and RV values [22].

Recall the following observations as discussed in the previous sections:

- Initial parts of the PDC are affected by mainly oil condition.
- Final or tail portions of PDC are governed by paper or solid insulation condition
- Dominant time constant of RV spectra corresponds to the oil condition and any subdominant maximum represents the condition of solid insulation.
- With increase in conductivity of the insulation the CTC of RV spectra decreases, i.e., peak of RV spectra occurs earlier.

Let us consider now, an example of insulation model of a 45 MVA power transformer [22]. The Debye model parameters as obtained from PDC measurement are shown in Table 5.1.

Now, by varying the R_i & C_i values for different branches keeping the time constant of that particular branch same as before, the effects on PDC and RV are observed. They are shown in the Figs. 5.31, 5.32, 5.33, 5.34, 5.35, and 5.36.

Table 5.1 Debye model parameters of a 45 MVA power transformer

R–C branch number of Debye model	Time constant of the R–C branch, τ_i (s)	Resistance of the branch, R_i (GΩ)	Capacitance of the branch, C_i (nF)
1	6850	7.2	900
2	890	6.1	145
3	150	4.6	35
4	35	3.6	8.6
5	5.6	6.2	0.9
6	0.9	2.4	0.25

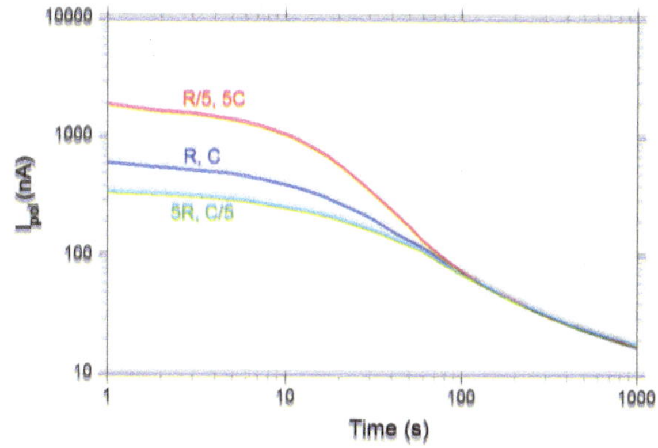

Fig. 5.31 Effect of the values of R_i & C_i on polarization current for the smaller time-constant branches of the insulation model

It is seen in Figs. 5.31, 5.32, and 5.33 that the variations in the R_i & C_i values of the smaller time-constant branches (i.e., for the branches where $\tau_i < 10$ s), affects the magnitude of the initial parts of the PDC, which are dependent on the condition of the oil insulation. Moreover, the variations in the R_i & C_i values of the smaller time-constant branches shift the dominant central time constants of the RV spectra without influencing the position of the secondary peaks which occur at much higher times. Hence, it can be concluded that, the values of R_i & C_i for the smaller time-constant branches correspond to the condition of the oil as dominant time constant of RV spectra corresponds to the oil condition. It, therefore, affects the initial portion of the PDC and only the first dominant peak of the RV spectra.

On the other hand, it is evident from Figs. 5.34, 5.35, and 5.36, the magnitude of the final or, tail parts of the PDC depend only on the values of R_i & C_i of the larger time-constant (i.e., τ_i around 1000 s or more) branches, which are dependent on the condition of the paper insulation. Moreover, it is also clear from the figures that the effects of changing the values of R_i & C_i of larger time-constant branches have almost no effect on the first CTC of the RV spectra. Contrary to this, the subsidiary peaks (second peak) of the RV spectra vary significantly with the

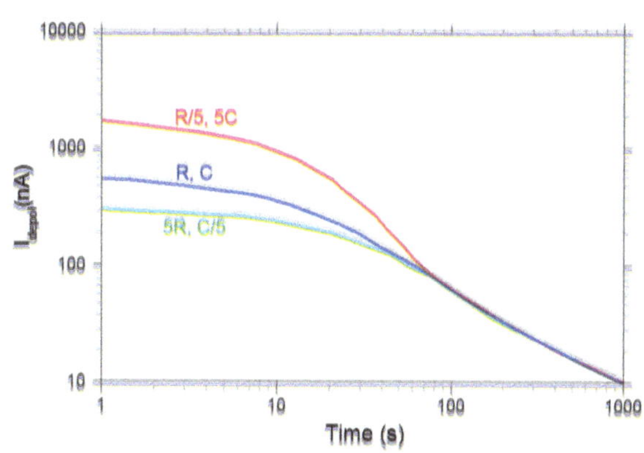

Fig. 5.32 Effect of the values of R_i & C_i on depolarization current for the smaller time-constant branches of the insulation model

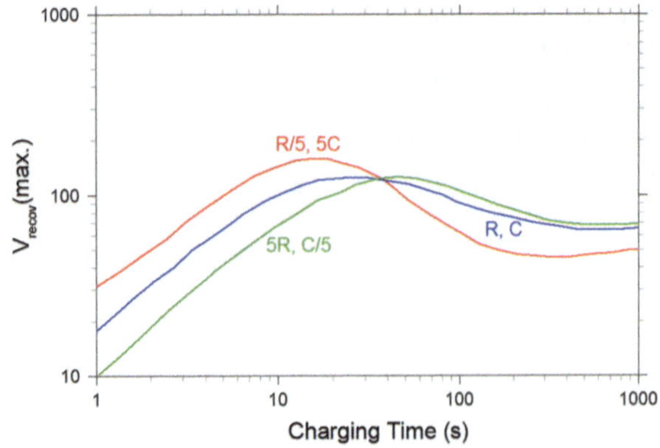

Fig. 5.33 Effect of the values of R_i & C_i on RV spectra for the smaller time-constant branches of the insulation model

variation of larger time-constant branches. This is an indication that the larger time-constant branches of the insulation model correspond to the condition of the paper.

In a similar manner, by varying the values of R_O and C_O their effects on the dielectric response have also been observed. A higher value of R_O indicates better condition of insulation whereas lower R_O corresponds to aged and moist insulation. Higher values of R_O tend to reduce the polarization current magnitudes but its effect on first dominant peak of the RV spectra is found to be insignificant. However, a noticeable shift in the subsidiary peak can be observed of the RV spectra with the change of the value of R_O.

The value of the capacitance C_O is dependent on the geometry of the insulation arrangements of the transformer. Hence, change of C_O alters the magnitude of the RV spectra by shifting the central time constants. However, on PDC the variation of C_O does not have a significant effect.

Thus by obtaining the model parameters for different branches from PDC and RV measurements, one can have a qualitative idea about insulation condition (for oil and paper separately). However, to have a quantitative idea, an expert system or

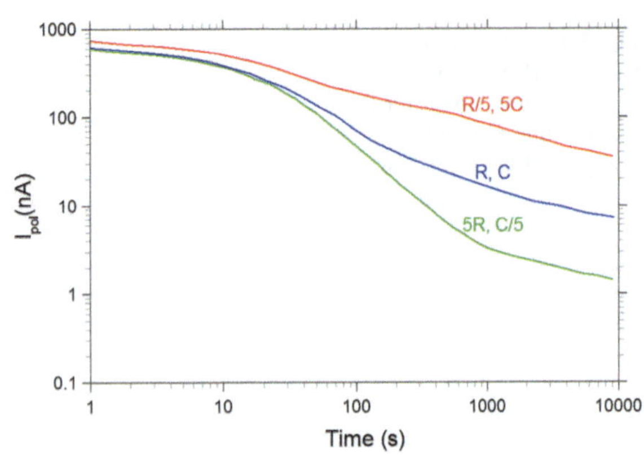

Fig. 5.34 Effect of the values of R_i & C_i on polarization current for the larger time-constant branches of the insulation model

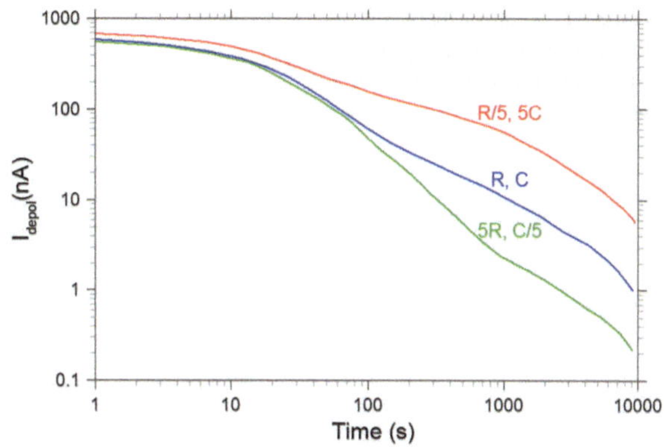

Fig. 5.35 Effect of the values of R_i & C_i on depolarization current for the larger time-constant branches of the insulation model

similar approach may be used. It is worth mentioning here that, there is no universally accepted method for such quantitative judgment about insulation condition using time domain dielectric response data, till date. The following section explains some of practiced approaches for this purpose.

5.7.1 Quantitative Measure of Insulation Condition from RV Measurements

As discussed before, many researchers found out that there exists a relation between CTC or dominant time constant and moisture content of paper. Csepes et al. [29] presented results on the variation of CTC with paper moisture content at different temperatures as shown in the figure below. By measuring the RV following the procedure stated before, one can get the RV spectra. Noting the temperature at which the experimentation is performed and using the curves given in Fig. 5.37 the paper moisture can be obtained. As suggested in [30], "*The actual insulation temperature should be used for evaluation of the dielectric*

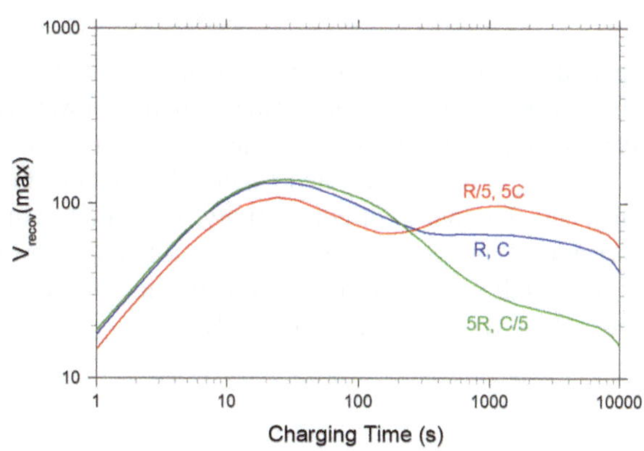

Fig. 5.36 Effect of the values of R_i & C_i on RV spectra for the larger time-constant branches of the insulation model

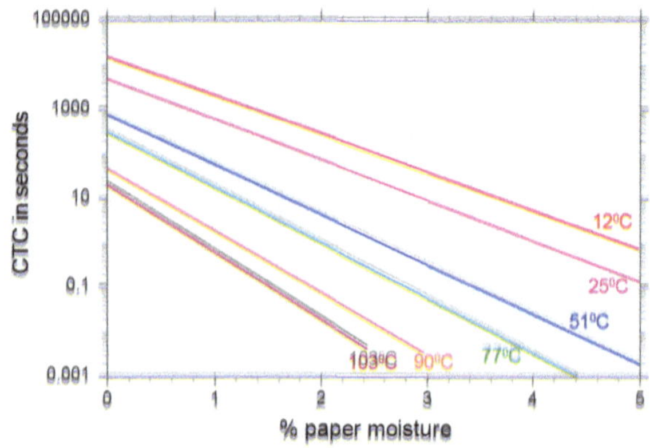

Fig. 5.37 Relation between dominant time constants or, CTC and moisture content of paper

measurements, while the in-service temperature, corresponding to moisture equilibrium, should be used for evaluations based on analyses of oil samples." However, CIGRE Task Force 15.01.09 suggests that, "*For the RVM technique, the old interpretation based only on simple relationship between the dominant time constant of the polarization spectrum and the water content in cellulose is not correct. An improved interpretation requires taking into account the complete measured curve*" [14].

5.7.2 Quantitative Measure Of Insulation Condition from PDC Measurements

Another way of getting a quantitative measure about both oil and paper insulation is to use Polarization–depolarization curves. From PDC, oil and paper conductivity can be measured using the following expression [7, 14, 31–36],

$$\sigma = \frac{\varepsilon_0}{U_0 \frac{C}{\varepsilon_r}}(i_{pol} - i_{depol}) \tag{5.22}$$

here, the U_0 is the charging voltage and ε_0 is the effective relative permittivity of the composite oil-paper insulation system at power frequency. C_0 is the geometric capacitance of the insulation system, as discussed before. If ε_0 is in F/m, C_0 in F, i_{pol} and i_{depol} in amperes, and U_0 in volts then conductivity of oil or paper, σ is in Siemens. To obtain oil conductivity, the initial values of polarization and depolarization currents are considered as i_{pol} and i_{depol} respectively. For paper moisture content, tail or final values of polarization and depolarization currents are considered as i_{pol} and i_{depol} respectively.

For proper estimation of this effective permittivity, it is required to have correct and complete knowledge about the design and composition of the oil-paper insulation system of the transformer. It is easy to understand that, most of the cases this information about the design and composition of the transformer insulation system neither is available from the utilities nor it is possible. As a rule of thumb, it is sufficient to assume a series arrangement of paper/pressboard and oil duct of the transformer insulation system with around 20 % paper and 80 % oil [7, 26].

Once the conductivity values are known, the oil moisture content (in ppm) or DDF and paper moisture (in %) can be found out using moisture content versus conductivity curves for both oil and paper insulations. It is better for the user to get the oil and paper moisture data from the curves generated at the own laboratory using known samples (i.e., developed on the basis of controlled aging of the samples). The preparation of such samples has been discussed in [8, 9]. Some of the representative data for this purpose are shown in Figs. 5.38, 5.39, and 5.40 [37].

5.8 Future Directions and Challenges

There are many issues in time domain dielectric response measurements yet to be sorted out. It is required to find out a standard methodology for quantitative computation of insulation condition from such measurements. Moreover, separate interpretations from PDC and RV differ from each other. Hence, many researchers feel that it is also required to synergize the PDC and RV measurements to converge into a unique solution about insulation condition. As stated earlier, there exists no such unanimously accepted method over the globe. However, works are going on so that, different methods converge into an internationally accepted scheme.

In many cases, certain oscillatory nature in the PDC curve is observed. However, the reasons behind such oscillations and up to what extent the insulation condition has a role to play in this are yet to be investigated in detail.

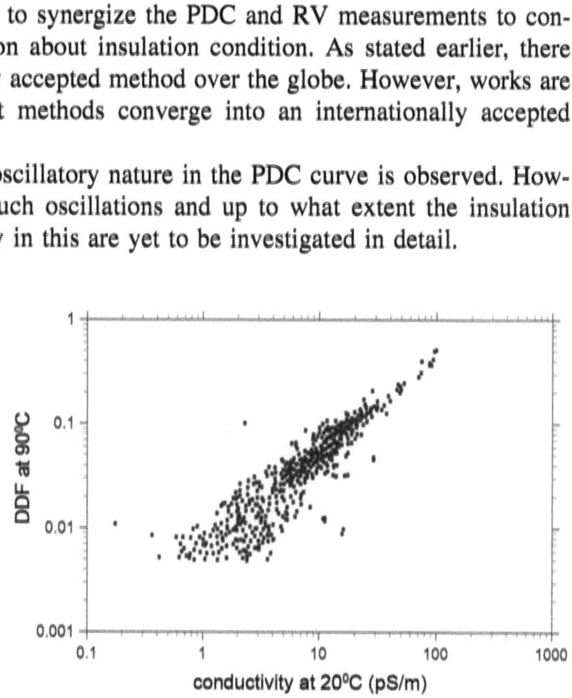

Fig. 5.38 DDF (IEC60247) at 90 °C versus conductivity (IEC61620) at 20 °C

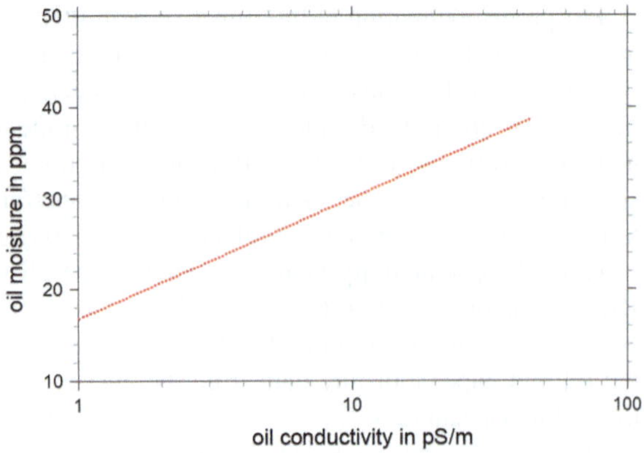

Fig. 5.39 Oil moisture (ppm) versus oil conductivity at ambient temperature around 28 °C

The insulation system of a transformer is very complex. Several dielectrics with a number of different interfaces exist in it. Therefore, is it possible for the linear Debye model approach for proper analysis of the insulation, is another question that the researchers are interested into. How can the conventional Debye model be modified to tackle the complexity and non-linearity of the insulation system of transformer, is another challenge.

Analysis of time domain dielectric response measurement may be viewed as the separation and determination of different polarization mechanisms that are superposed on each other. In another development in recent times [15, 38], the differential time domain spectroscopy (DTDS) function is defined as the product of the discharging time t and the discharging current $i(t)$. In DTDS, $i(t)$ is the differential coefficient of the time domain response function, and can be expressed as,

$$t \cdot i(t) = \sum_{n=1}^{N} q_n f(t, \tau_n, \alpha_n) \tag{5.23}$$

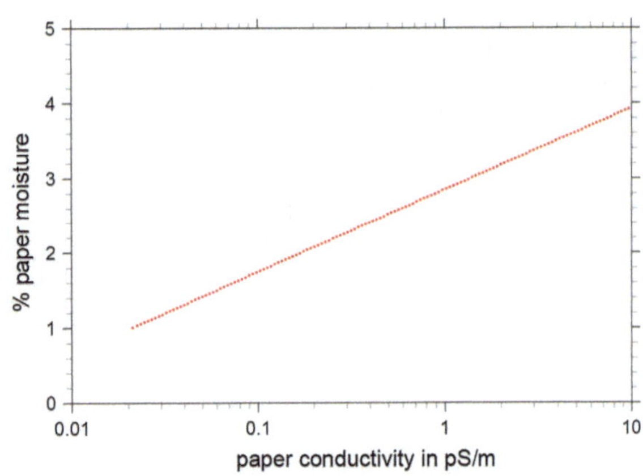

Fig. 5.40 Paper moisture (%) versus paper conductivity at ambient temperature around 28 °C

where,

$$f(t, \tau_n, \alpha_n) = \alpha_n \left(\frac{t}{\tau_n}\right)^{\alpha_n} \exp\left[-\left(\frac{t}{\tau_n}\right)^{\alpha_n}\right] \tag{5.24}$$

and N is the polarization mechanism number, and q_n, τ_n, α_n are the parameters in DTDS of the n-th polarization mechanism. Qn is called polarization component, indicating the polarization intensity and the contribution of the n-th mechanism in the total charge in equilibrium stored only by dielectric polarization; $\tau_n$polarization response time; and $\alpha_n$line-shape parameter. These parameters can be investigated for a better insight of the whole polarization process, which may be another interesting way of interpreting the results of dielectric response measurements in time domain in the near future.

It is well known that, during normal operating conditions, the temperature inside a transformer is higher than the ambient, depending upon the operating condition. As field measurements of time domain dielectric response last for hours, the final temperature may be much lower than the initial. Hence, during on-site measurements the water migration is commonly occurring inside the transformer making it in a non-equilibrium state. This transient can lead to wrong interpretation of insulation condition. Researchers are proposing methods to tackle these problems [39]. However, there exists ample scope of research in his direction.

References

1. Zaengl WS (2003) Dielectric spectroscopy in time and frequency domain for HV power equipment, part i: theoritical considerations. IEEE Electr Insul Mag 19(5):6–19
2. Yao ZT, Saha TK, Darveniza M (1999) Effects of moisture on the recovery voltage measurement for aged transformer. In Proceedings of International Conference on Power Engineering, vol 1. pp. 349–353
3. Gafvert U, Frimpong G, Fuhr J (1998) Modeling of dielectric measurements on power transformers. In Proceedings of CIGRE, Paris. Paper no. 15/103
4. Griffin PJ, SSokolov V, Vanin B (1999) Moisture equilibrium and moisture migration within transformer insulation system, CIGRE SC12 transformer colloquium, Budapeste. (Julho)
5. Saha TK (2003) Review of time-domain polarization measurements for assessing insulation condition in aged transformers. IEEE Trans Power Delivery 18(4):1293–1301
6. Saha TK, Purkait P (2008) Investigations of temperature effects on the dielectric response measurements of transformer oil-paper insulation system. IEEE Trans Power Delivery 23(1):252–260
7. Frimpong G, Gafvert U, Fuhr J (1997) Measurement and modeling of dielectric response of composite oil/paper insulation. In proceedings of 5th international conference of properties and applications of dielectric materials. Seoul, Korea, pp 86–89
8. Koch M (2008) Reliable moisture determination in power transformers. Ph.D. thesis, Institute of energy transmission and high voltage engineering. University of Stuttgart, Sierke Verlag Göttingen, Germany
9. Kheng TC (2000) Impact of moisture on the strength of paper-insulated conductor. B.E Thesis, University of Queensland, Australia

10. Gfiert U, Ildstad E (1994) Modelling return voltage measurements of multi-layer insulation systems, ICPADM 94. Brisbane
11. Leibfried T, Kachler AJ (2002) Insulation diagnostics on power transformers using the polarization and depolarization current (PDC) analysis. IEEE international symposium on electrical insulation, Boston, pp 170–173
12. Saha TK, Purkait P (2004) Investigation of polarization and depolarization current measurements for the assessment of oil-paper insulation of aged transformers. IEEE Trans Dielectr Electr Insul 11:144–154
13. Osvath P, Csepes G, Webb M (1992) Polarisation spectrum analysis for diagnosis of insulation systems. Sixth international conference on dielectric materials, measurements and applications, Manchester, pp 270–273
14. Dielectric response methods for diagnosis of power transformers. In Electra no. 202, 2002, CIGRE Task Force 15.01.09
15. Wei JL et al (2011) Novel characteristic parameters for oil-paper insulation assessment from differential time-domain spectroscopy based on polarization and depolarization current measurement. IEEE Trans Dielectr Electr Insul 18(6):1918–1928
16. Dey D, Chatterjee B, Chakravorti S, Munshi S (2007) A hybrid filtering scheme for proper denoising of real-time data in dielectric spectroscopy. IEEE Trans Dielectr Electr Insul 14(5):1323–1331
17. Yin L, Yang R, Gabbouj M, Neuvo Y (1996) Weighted Median Filters: A Tutorial. IEEE Trans Circuits Syst II Analog Digital Signal Process 43:157–192
18. Arce GR (1998) A general weighted median filter structure admitting negative weights. IEEE Trans Signal Processing 46:3195–3205
19. Arce GR, Paredes JL (2000) Recursive weighted median filters admitting negative weights and their optimization. IEEE Trans Signal Process 48:768–779
20. Proakis JG, Manolakis DG (1998) Digital signal processing: principles, algorithms, and applications. Prentice-Hall, New Jersey
21. Bognar A, Kalocsai L, Csepes G, Nemeth E, Schmidt J (1990) Dielectric tests of high voltage oil paper insulating systems (in particular transformer Insulation) using DC dielectrometrics. In Proceedings of CIGRE, 33rd Session, Paris vol 2. pp 15/33–08
22. Saha TK, Purkait P, Muller F (2005) Deriving an equivalent circuit of transformers insulation for understanding the dielectric response measurements. IEEE Trans Power Del 20(1):149–157
23. Mohamed G, Németh E (1991) Computer simulation of dielectric processes. In Proceedings of 7th international symposium on high voltage engineering, Dresden. pp 309–331
24. Jota PRS, Islam SM, Jota FG (1999) Modelling the polarisation in composite oil/paper insulation systems. IEEE Trans Dielectr Electr Insul 6(2):145–151
25. Urbani GM, Brooks RS (1998) Using the recovery voltage method to evaluate aging in oil-paper insulation. In Proceedings of IEEE international conference conduction breakdown in solid dielectrics, Vasteras. pp 93–97
26. Saha TK, Purkait P (2004) Investigation of an expert system for the condition assessment of transformer insulation based on dielectric response measurements. IEEE Trans Power Delivery 19(3):1127–1134
27. Zaengl WS (2003) Dielectric spectroscopy in time and frequency domain for HV power equipment part I: theoretical considerations. IEEE Electr Insul Mag 19(5):5–19
28. Xu S, Middleton R, Fetherston F, Pantalone D (2003) A comparison of return voltage measurement and frequency domain spectroscopy test on high voltage insulation. In Proceedings of the 7th international conference on properties and applications of dielectric materials, Nagoya. 1-5 June 2003
29. Csepes G, Hamos I, Brooks I, Karius V (1998) Practical foundations of the RVM (Recovery voltage method for oil/paper insulation diagnosis). In proceedings of IEEE conference on electrical insulation dielectric phenomena. pp 345–355

30. Gubanski SM et al. (2006) Reliable diagnostics of hv transformer insulation for safety assurance of power transmission system, REDIATOOL—a european research project CIGRE D1-207. Paris
31. der Houhanessian V, Zaengl WS (1998) On-site diagnosis of power transformers by means of relaxation current measurementsin Proc. IEEE Int Symp Electr Insul 1:28–34
32. Gafvert U, Adeen L, Tapper M, Ghasemi P, Jonsson B (2000) Dielectric spectroscopy in time and frequency domain applied to diagnostics of power transformers. In proceedings of 6th international conference properties and applications of dielectric materials. vol 2, Xian. pp 825–830
33. Kachler AJ (2000) On-site diagnosis of power and special transformers. In Proceedings of IEEE International Symposium on Electrical Insulation. pp 362–367
34. Saha TK, Purkait P, Yao ZT (2002) Condition monitoring of transformer insulation by polarization and depolarization current measurements. In proceedings of australasian universities power engineering conference, Melbourne
35. Hassig M, Braunlich R, Gysi R, Alff JJ, Der Houhanessian V, Zaengl WS (2001) On-site applications of advanced diagnosis methods for quality assessment of insulation of power transformers. In proceedings of conference electrical insulation and dielectric Phenomena, pp 441–447
36. Yao ZT, Saha TK (2002) Analysis and modeling of dielectric responses of power transformer insulation. In proceedings of IEEE power engineering society summer meeting, vol 1, Chicago. pp 417–421
37. Bhumiwat S, Lowe S, Nething Ph, Wickramasuriya P, Kuansatit P (2010) Performance of oil and paper in transformers based on IEC 61620 and dielectric response techniques. IEEE Elec Insul Mag 26(3):16–23
38. Li J, Chen M, Zheng F, Zhou Z (1997) Diffusion theory of slow responses. Science in China (Series A) 40(3):290–296
39. Fofana I, Hemmatjou H, Meghnefi F (2011) Effect of thermal transient on the polarization and depolarization current measurements. IEEE Trans Dielectr Electr Insul 18(2):513–520

Chapter 6
Frequency Domain Spectroscopy

6.1 Introduction

Dielectric spectroscopy tests on transformer insulation could be performed in time domain as well as in frequency domain. Polarization and Depolarization Current (PDC) measurement and Recovery Voltage Measurement (RVM), as discussed in Chap. 5, are examples of time domain measurements. On the other hand, the principle of frequency domain spectroscopy is to measure the dielectric dissipation factor ($tan\delta$) and complex capacitance/permittivity as a function of frequency. In other words, Frequency Domain Spectroscopy (FDS) is a generalization of the conventional measurement of $tan\delta$ and complex capacitance/permittivity at different frequencies, which are performed traditionally at power frequency. The importance of FDS technique lies in the fact that it allows characterization of electrical properties of the insulating material at different frequencies, which in turn provide indication of the general ageing status of the oil-paper insulation of transformer.

It is well known that the dielectric response of the transformer insulation system depends on the properties of insulating oil, properties of solid insulation like paper and pressboard and the geometrical arrangement of the insulation components. FDS is gaining in popularity in practice because of the fact that the measurement of dielectric dissipation factor is stated to be independent of the geometry of transformer insulation system. However, to be precise FDS measurements are independent of geometry when insulation structure is homogeneous and isotropic [1]. Moreover, FDS measurements are severely influenced by temperature. Hence, proper analysis of the FDS results requires a clear understanding of the physical behaviour of the insulation system.

S. Chakravorti et al., *Recent Trends in the Condition Monitoring of Transformers*,
Power Systems, DOI: 10.1007/978-1-4471-5550-8_6, © Springer-Verlag London 2013

6.2 Basic Theory of Frequency Domain Spectroscopy

The impedance of a test object is determined by applying a sinusoidal voltage across the object and accurately measuring the voltage across and current through the object. Considering a phase difference between the voltage $U(\omega)$ and the current $I(\omega)$ at a specific frequency, the impedance of the test object can be described as follows:

$$\overline{Z}(\omega) = Z'(\omega) + jZ''(\omega) = \overline{U}(\omega)/\overline{I}(\omega) \tag{6.1}$$

Assuming the test sample to be a complex capacitance, the relationship between $U(\omega)$ and $I(\omega)$ can be rewritten as

$$\overline{I}(\omega) = j\omega\overline{C}(\omega)\overline{U}(\omega) \tag{6.2}$$

Such that $\overline{Z}(\omega) = \frac{1}{j\omega\overline{C}(\omega)}$,

where

$$\overline{C}(\omega) = C'(\omega) - jC''(\omega) \tag{6.3}$$

Since the complex capacitance of the test object is correlated with the complex permittivity ($\overline{\varepsilon}$), $\overline{\varepsilon}$ can be determined as follows

$$\overline{\varepsilon}(\omega) = \varepsilon'(\omega) - j\varepsilon''(\omega) = \frac{\overline{C}(\omega)}{C_0} \tag{6.4}$$

where, C_0 is the geometrical capacitance of the test object.

Then the dielectric dissipation factor ($tan\delta$) can be written as

$$\tan\delta = \frac{\varepsilon''(\omega)}{\varepsilon'(\omega)} = \frac{C''(\omega)}{C'(\omega)} \tag{6.5}$$

A rigorous treatment of this methodology has been given by Zaengl in [2]. In this paper it has been shown that an analytical transition from time to frequency domain can be made by using Laplace or Fourier transform as follows. In time domain, the total current density, $J(t)$, within an insulation system generated by electric field intensity, $E(t)$, is given by

$$J(t) = \sigma_0 E(t) + \varepsilon_0 \frac{\partial E(t)}{\partial t} + \frac{\partial P(t)}{\partial t} \tag{6.6}$$

where, on the RHS the first term is conduction current, the second and third terms are vacuum and polarization displacement currents, respectively. σ_0 is the dc conductivity, which represents the movement of free charges within the insulation and is not involved in polarization processes.

An ideal step response for the total current density of a dielectric response function is given as

$$J(t) = \sigma_0 E(t) + \varepsilon_0 \frac{dE(t)}{dt} + \varepsilon_0 \frac{d}{dt} \int_0^t f(t-\tau) E(\tau) d\tau \tag{6.7}$$

Taking Laplace transform of Eq. 6.7 and substituting the Laplace operator by the complex frequency $j\omega$, one can write

$$\overline{J}(\omega) = \sigma_0 \overline{E}(\omega) + j\omega\varepsilon_0 \overline{E}(\omega) + j\omega\varepsilon_0 \overline{F}(\omega)\overline{E}(\omega) \tag{6.8}$$

where, $\overline{F}(\omega)$ is the Fourier transform of the dielectric response function *f(t)* and is also known as complex susceptibility $\overline{\chi}(\omega)$.

So,

$$\overline{\chi}(\omega) = \overline{F}(\omega) = \chi'(\omega) - j\chi''(\omega) = \int_0^\infty f(t)\, e^{-j\omega t}\, dt \tag{6.9}$$

Hence, from Eqs. 6.8 and 6.9,

$$\overline{J}(\omega) = [\{\sigma_0 + \varepsilon_0 \omega \chi''(\omega)\} + j\,\omega\varepsilon_0\{1 + \chi'(\omega)\}]\overline{E}(\omega) \tag{6.10}$$

Equation 6.10 can be rewritten in terms of complex relative dielectric permittivity, $\overline{\varepsilon_r}(\omega)$, as follows:

$$\overline{J}(\omega) = j\omega\, \varepsilon_0\, \overline{\varepsilon}_r(\omega)\, \overline{E}(\omega) \tag{6.11}$$

where,

$$\overline{\varepsilon}_r(\omega) = \varepsilon_r'(\omega) - j(\omega)\, \varepsilon_r'' = [1 + \chi'(\omega)] - j\left[\chi''(\omega) + \frac{\sigma_0}{\omega\varepsilon_0}\right] \tag{6.12}$$

The real part of complex permittivity represents the capacitance of a test object whereas the imaginary part represents the losses within the insulation.

Hence, the dielectric dissipation factor ($tan\delta$) in frequency domain is given by [2]

$$\tan\delta = \frac{\varepsilon_r''(\omega)}{\varepsilon_r'(\omega)} = \frac{\chi''(\omega) + \frac{\sigma_0}{\omega\varepsilon_0}}{1 + \chi'(\omega)} \tag{6.13}$$

Both complex permittivity and $tan\delta$ depend on frequency. Figure 6.1 shows the typical sensitivity of the different sections of the frequency domain response on properties of oil and solid insulation. Dielectric response in the low frequency range $f < 0.01$ Hz and also in the high frequency range $f > 10$ Hz is mainly influenced by the properties of solid insulation. In frequency range $0.01 \text{ Hz} \leq f \leq 10$ Hz, the response is influenced by the properties of insulating oil [3]. Consequently, in FDS method these quantities are monitored over a wide frequency range to ascertain the insulation condition appropriately.

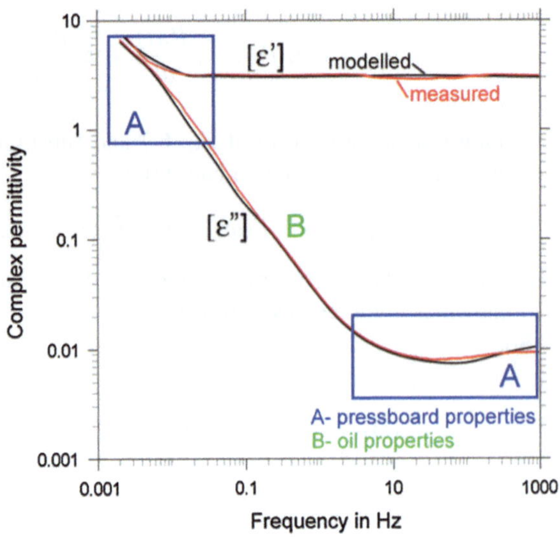

Fig. 6.1 Frequency domain response of a real life transformer

6.3 Frequency Domain Measurements

With reference to Fig. 6.2 the measurement principle can be described as follows: A sinusoidal test voltage of desired frequency is generated with the help of a digital signal processing unit. This voltage after being raised to the appropriate level is applied to the test object. The voltage across and the current through the test object are measured with high accuracy. Unlike time-domain measurements, in FDS measurements it is not necessary to discharge the test object in between the measurements at different frequencies.

A frequency range of 1 mHz–1 kHz is commonly used in practice. However, frequencies in the range of 0.1 mHz as well as in the range higher than 1 kHz have

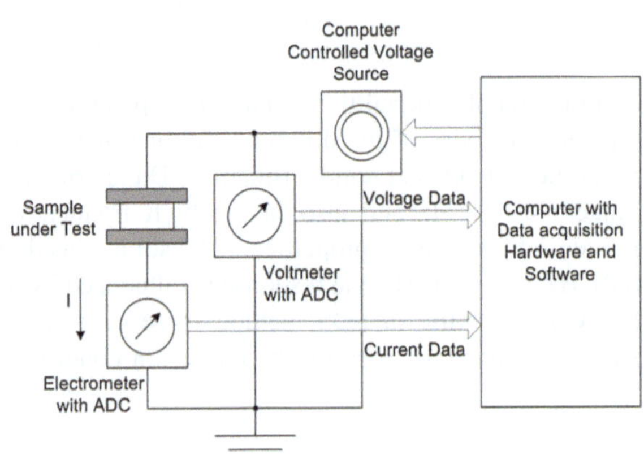

Fig. 6.2 Basic circuit arrangement for FDS measurement

been used by researchers. FDS measurements are not necessarily faster than time-domain measurements, because at low frequencies such measurements become very time consuming. As a standard practice, at least two cycles of the sinusoidal test voltage are recorded to accurately ascertain the amplitude and phase differences between the voltage and current signals. Therefore, for a frequency of 1 mHz, $\approx$2000 s is required to complete a single measurement [2]. In order to reduce the effects of polarization memory, the measurements are started at the highest practicable frequency and continued to the lowest practicable frequency in steps of discrete frequencies. Temperature has significant effect on FDS results. Hence, the measurements for all the frequencies are to be taken at the same ambient temperature.

The test voltage may vary from 5 V to 200 V(rms). By using proper electromagnetic shielding of the instrumentation and the test cells, a test voltage of only 3 V(rms) was also found to be adequate in laboratory for testing samples [2]. Experimental results have revealed that the losses decrease in the frequency range of less than 10 Hz with increasing test voltage [5]. Such non-uniformity in the behaviour of the dielectric arises due to the field dependent ionic oscillations, which are also known as Garton effect. This non-linearity becomes negligible when the test voltage is small of the order of 5 V(rms). However, it has also been found that the results obtained for test voltages below 5 V(rms) are not reproducible, primarily because of very small currents that limit the measurement accuracy of the instruments. However, for on-site testing of power transformers the test voltages are generally kept more than one hundred volts.

FDS measurements always give a pair of magnitudes for each test frequency, e.g. real and imaginary parts of complex permittivity or capacitance, as well as frequency dependence of *tanδ*. Figures 6.3 and 6.4 show typical FDS measurement results for *tanδ* and complex permittivity, respectively.

FDS measurements are non-invasive terminal measurements and in most of the cases are carried out by applying a sinusoidal test voltage between the HV and LV terminals (so called CHL measurements, where H and L indicate HV and LV windings, respectively) as shown in Fig. 6.5a. In such cases, FDS results typically characterize the average condition of the composite insulation between the HV and

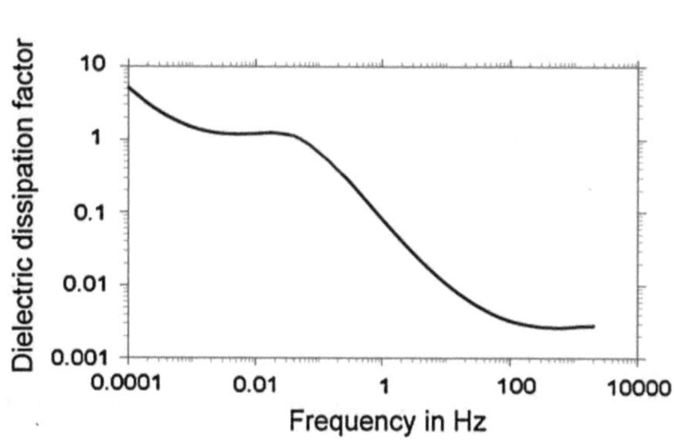

Fig. 6.3 Typical FDS results showing frequency dependency of *tanδ*

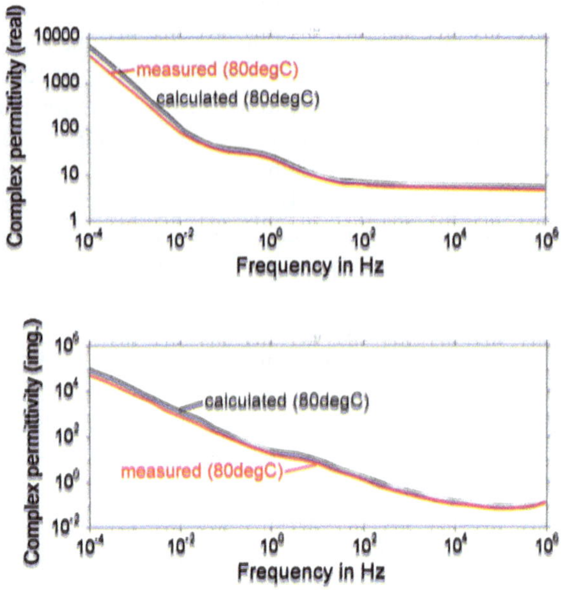

Fig. 6.4 Typical FDS results obtained for pressboard samples

LV windings. For three-phase transformers, the three HV or LV terminals are connected together so that the insulation of the three phases is tested in parallel in one go.

However, FDS measurements could also be taken between windings and ground (CH or CL), in which cases it becomes possible to evaluate different parts of the transformer insulation system. In CH measurements, as shown in Fig. 6.5b, the capacitance and the losses of the insulation between HV winding and the tank are measured. On the other hand, in CL measurements the insulation between the LV winding and the core is mainly measured.

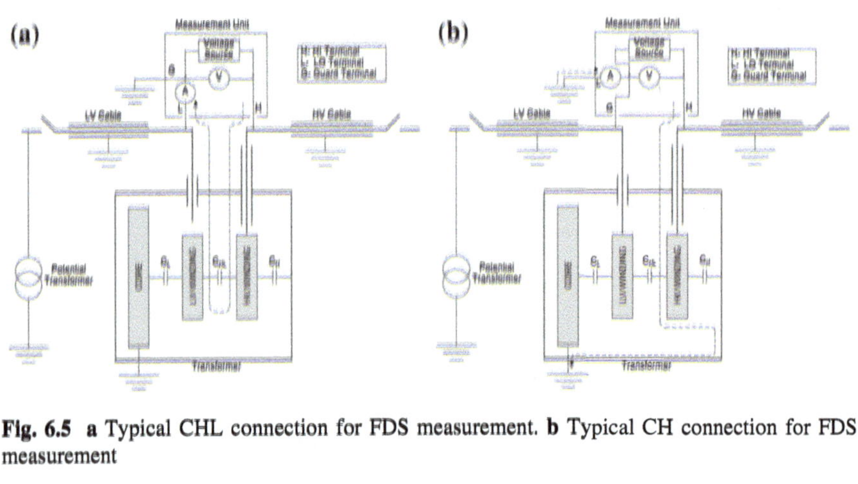

Fig. 6.5 **a** Typical CHL connection for FDS measurement. **b** Typical CH connection for FDS measurement

In three electrode measurements, there are two measuring electrodes and one guard electrode. An "ungrounded specimen test configuration" is obtained when none of the transformer windings, which act as measuring electrodes, is grounded. A "grounded specimen test configuration" is obtained when one of the two transformer windings is grounded. In both the cases, the parts of the transformer which are not associated with the measuring electrodes should be connected to the guard [3].

Ideally FDS measurements are to be taken after the transformer has been taken off the service and has been kept deenergized for a long time. But due to practical constraints, field measurements are often taken not long after deenergization of the transformer. Consequently large temperature variation may come into play and the moisture distribution within the transformer insulation may not be in a state of equilibrium. Further, large temperature instabilities may occur as the low frequency measurements require very long time. FDS results being highly dependent on operating conditions, it is of utmost importance to take proper care in the interpretation of FDS results [7].

6.3.1 Frequency Response of Transformer Insulation

As discussed earlier, frequency changes the dielectric properties significantly as it changes the polarization processes within the insulation system. If depicted in the frequency domain, this influence manifests itself as a measured spectral function along the frequency axis. Figure 6.6 shows the plot of real and imaginary capacitance values against frequency for the scaled down laboratory model of a transformer [8].

Typical frequency response of insulating oil used in transformers is shown in Fig. 6.7. It is relevant to mention here that transformer oil is a weakly polar liquid and hence its losses under low electric field stresses are primarily caused by ionic

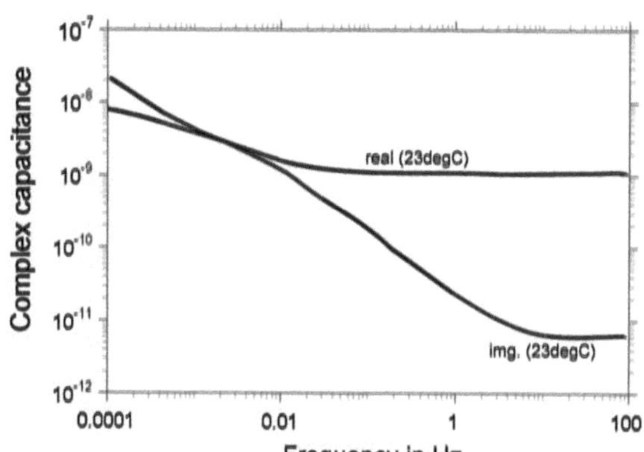

Fig. 6.6 Frequency response of laboratory model of transformer insulation

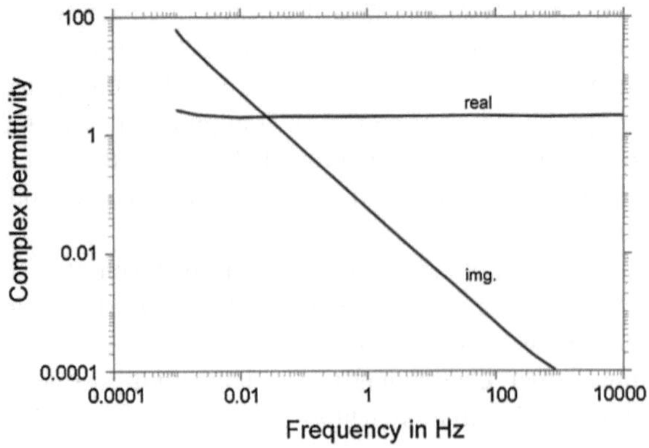

Fig. 6.7 Frequency response of oil

impurity conduction [9]. The frequency response of pure transformer oil, as depicted in Fig. 6.7, can thus be written as follows [5]:

$$\bar{\varepsilon}_{\omega,T} = 2.2 - j\frac{\sigma(T)}{\omega\varepsilon_0} \tag{6.14}$$

Figure 6.8 shows the frequency response of oil impregnated paper and pressboard measured at a temperature of 50 °C. As the pressboard has a denser structure, it has a higher permittivity. From Fig. 6.8 it may be seen that the loss in impregnated paper is slightly higher than the loss in impregnated pressboard at low frequencies in the range of 0.1–10 mHz. On other hand, at high frequencies in the range of 1 Hz–1 MHz this relationship is reversed.

An important parameter of insulation diagnostics, known as Capacitance Ratio (CR), is often computed from the FDS results, in the same way Polarization Index (PI) is calculated from time domain results. CR is defined as the ratio of C at 1 mHz to the C at power frequency (50 Hz or 60 Hz). In the case of high voltage transformers, the insulation system comprises oil ducts in series with the pressboard. In such cases, a CR value in the range of 3–5 may be taken as an indicator of good insulation condition [10].

6.3.2 *Effect of Conductivity*

As may be seen from Eq. 6.12, the real and imaginary components of complex permittivity and hence the complex capacitance are interdependent. If the conductivity is dominant, then imaginary component of complex permittivity will have a slope of −1 in log–log scale, particularly at low frequencies. Figure 6.9 shows the effect of dc conductivity of oil on the imaginary component of complex permittivity for an oil impregnated paper insulation model [11]. Figure 6.9 depicts that the frequency dependent losses, represented by ε'', are significantly influenced by oil conductivity in the frequency range of 0.01–10 Hz.

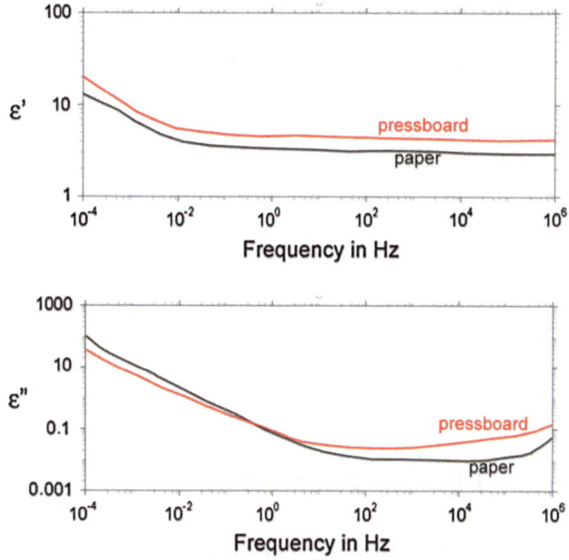

Fig. 6.8 Frequency response of impregnated paper and pressboard

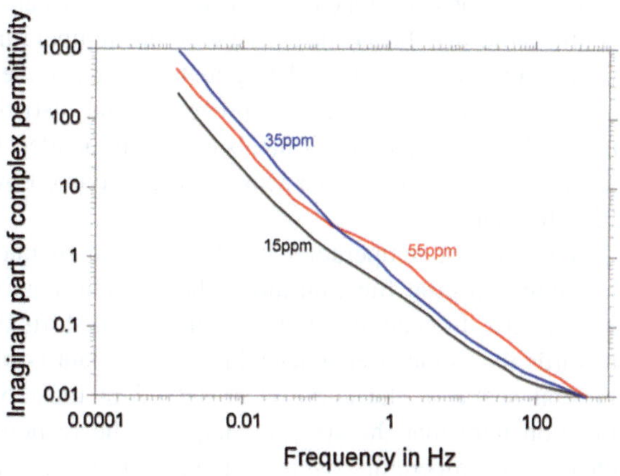

Fig. 6.9 Effect of oil conductivity on imaginary component of complex permittivity

In Fig. 6.10 the measured values of *tanδ* for four different transformers are shown [12]. In this study the transformers T1 and T2 were similar and T3 was similar to T4. Fig. 6.10 shows that the spectra of the two pairs of transformers are shifted notably along the frequency axis. Investigations revealed that the oil conductivity of the transformers T1 and T2 was higher than that for T3 and T4, which caused such shift in frequency response.

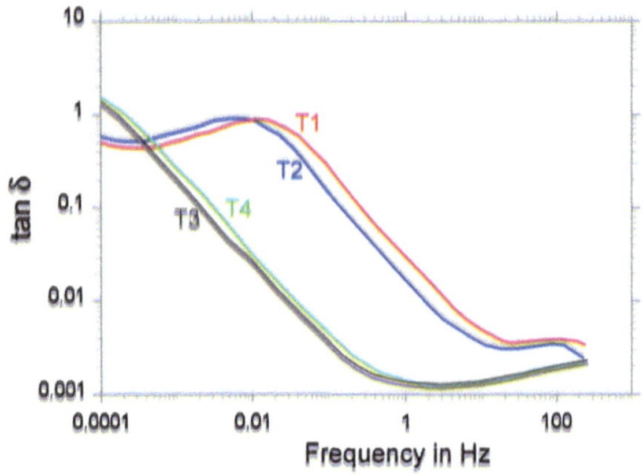

Fig. 6.10 Effect of oil conductivity on *tanδ* for four different transformers

6.3.3 Effect of Temperature

CIGRE task force has identified that FDS results are severely influenced by several environmental factors, the most important of which is temperature [13]. In outdoor substations the external environmental conditions are not predictable as well as not controllable and hence in such cases the temperature effect is more prominent. Dielectric properties of insulating materials are immensely affected by temperature as it changes the polarization processes and also the conductivity of insulating materials. Consequently, it has been recommended that sufficient time should be allowed for the transformer temperature to come down to ambient before starting FDS measurements.

Variation of temperature has been found to manifest itself in a shift of the measured spectral function along the frequency axis, commonly termed as horizontal shift. In general, a higher temperature shifts the frequency response curves towards the higher frequencies [14]. If the data is plotted on a log–log scale, the general shape of the curve is preserved in most of the cases. It is relevant to mention here that the spectral shape of the response does not change with the change in temperature over the temperature range in which the structure of the insulating material does not get altered [5]. Figure 6.11 shows the *tanδ* vs frequency curve of an 110 kV power transformer at two different temperatures [15]. The transformer was taken off the service for a long time so that moisture in oil and solid insulation within the transformer could reach equilibrium. It is clear from Fig. 6.11 that the curves shift to the right with the increase of temperature keeping the shape of the curve more or less unchanged.

With the increase in temperature, the intermolecular interaction force increases, which in turn increases the degree of polarization and causes FDS curves towards higher frequency. However, it has been shown that the horizontal shift in FDS curves, due to temperature change without a change of the shape of the curve,

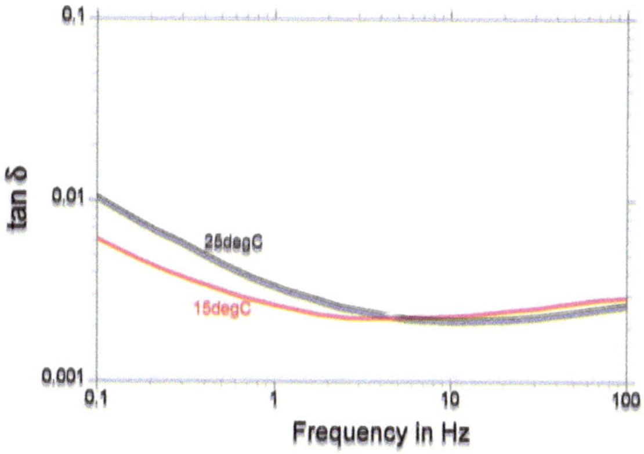

Fig. 6.11 Frequency dependence of *tanδ* of a transformer at two different temperatures

occurs only when a single process is responsible for the losses [16]. In such cases, dielectric susceptibility can be written as follows:

$$\chi(\omega, t) = A_y(t)\, F\left(\frac{\omega}{A_\omega(t)}\right) \tag{6.15}$$

Each A can most often be expressed with the help a single Arrhenius type activation energy W as follows [16]:

$$A = \exp(-W/kT) \tag{6.16}$$

where, k is Boltzmann constant and T is absolute temperature. It has also been reported that dc conductivity also has Arrhenius type temperature dependence [16]. However, the activation energies (W) are not necessarily the same in y and ω directions or for χ and σ. Moreover, the shifts in the y-direction are in general negligible [16].

From the above discussion it follows that a "Master Curve" (MC) can be created by shifting the FDS curves measured at different temperatures along the frequency axis until they form a continuous curve at a particular reference temperature [14]. Figure 6.12 shows a set of FDS curves measured at different temperatures and the master curve formed with the help of these curves. The advantages of master curve technique are two fold: Firstly the range of frequency is greatly enhanced compared to the original measured curves; secondly the spurious data points could be efficiently filtered out using the regularity of the master curves. MC technique is applicable for both real and imaginary components of the complex permittivity.

If it is possible to use a single Arrhenius process to describe the temperature dependence over the entire frequency range, the master curve fit will be perfect, as shown in Fig. 6.13. In that case the activation energy can be obtained from the plot of log of frequency shift against the inverse of the absolute temperature, as shown in Fig. 6.14, and the activation energy will be the main parameter for describing temperature dependence. For reasonably good fit of master curves, the activation

Fig. 6.12 A set of FDS curves measured at different temperatures and the corresponding master curve

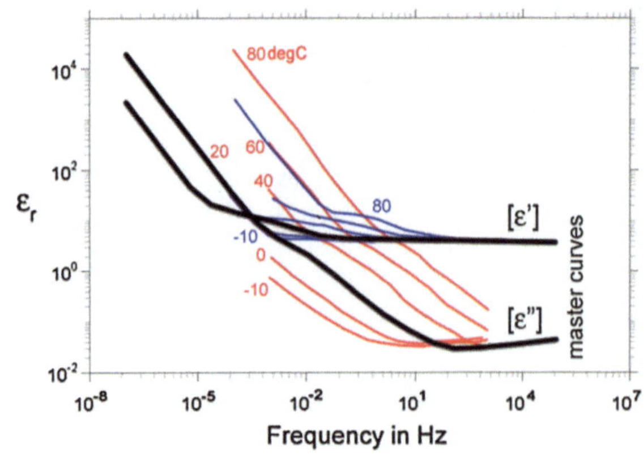

energy is found to be in the range of 0.98–10.08 eV [14]. But in many cases, multiple mechanisms are responsible for the dispersion, which can have different temperature dependence. This phenomenon is reflected in that part of the spectrum which deviates from the master curve, as shown in Fig. 6.15.

In real life, it is common practice that winding temperature is estimated by measuring the temperature of oil samples taken from the top and bottom oil taps of transformer tank. But, temperature along the length of the winding within the tank of a transformer is not constant. In general, temperature is higher at the top of the winding compared to temperature at the bottom. This difference in some cases could be in the range of 20–30 °C. Because of such temperature difference between the longitudinal parts of the insulation, there exists a moisture distribution along the length of the winding. Figure 6.16 shows such longitudinal moisture distribution in pressboard barrier along a real life transformer [3]. This makes the analysis of FDS results complex, as the question that arises is which temperature,

Fig. 6.13 An example of good fit of master curve

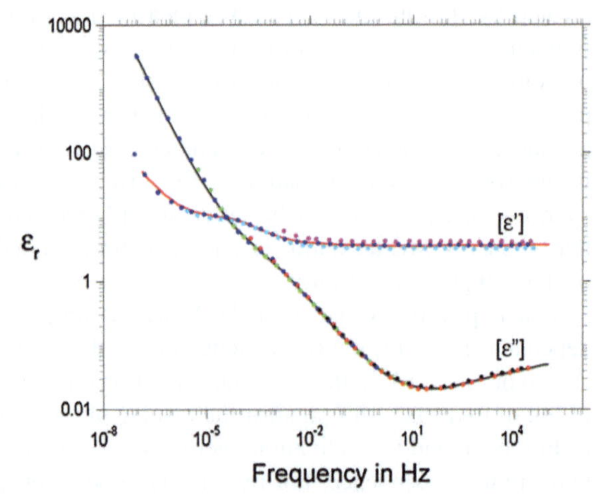

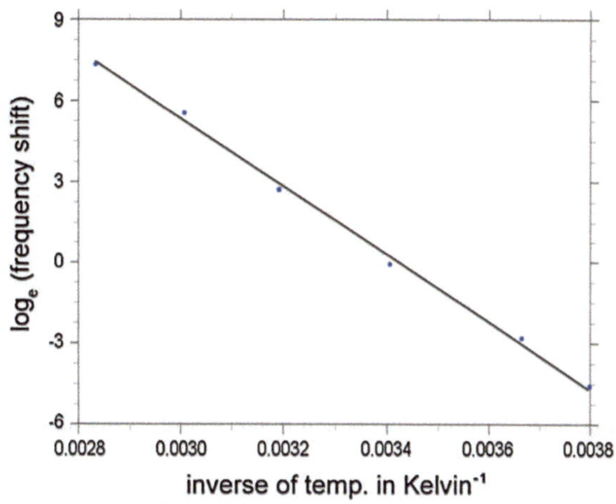

Fig. 6.14 Activation energy as a function of temperature

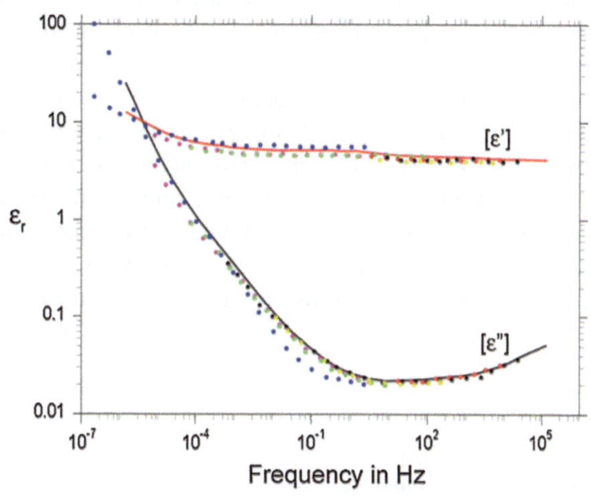

Fig. 6.15 An example of imperfect fit of master curve

i.e. top or bottom, should be used for analysis. Researchers [3] have shown that the best results are obtained if the temperature value, taken from the middle of the transformer winding, is used for analysis. In this context, it is to be kept in mind that thermometers mounted on transformer tanks usually have limited accuracy. On the other hand it is difficult to physically locate temperature sensors inside the transformer tank. These calls for an appropriate calculation of temperature at the middle of the winding from the measured values of top and bottom oil temperatures.

It is a common practice that sufficient time is given for the transformer to cool down to ambient after it is taken off the service before performing the FDS test. Because of the large thermal time constant of the transformer, it takes very long, sometimes exceeding 10 h, for the transformer to cool down to ambient. In many

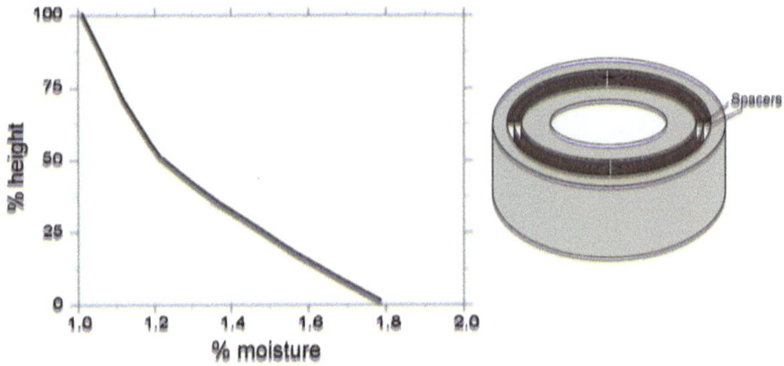

Fig. 6.16 Moisture distribution in pressboard barrier along transformer winding

cases, due to operational constraints, the testing personnel are forced to start the measurements while the temperature of the transformer is still decreasing. In such cases, the temperature decreases up to about 10 °C during the low frequency measurements. Figure 6.17 shows the *tanδ* results obtained for a transformer measured at a constant temperature of 85 °C and also when the temperature was allowed to cool from 85 °C to ambient temperature during the measurements. It is clear from Fig. 6.17 that the *tanδ* curves for both constant and transient temperature conditions are similar in nature until the temperature started to decrease after some time, which is reflected in the low frequency range [17]. Similar results were also obtained by other researchers for FDS measurements carried out during cooling period of the transformer, as shown in Fig. 6.18 [18].

Fig. 6.17 *tanδ* under constant and also transient temperature conditions

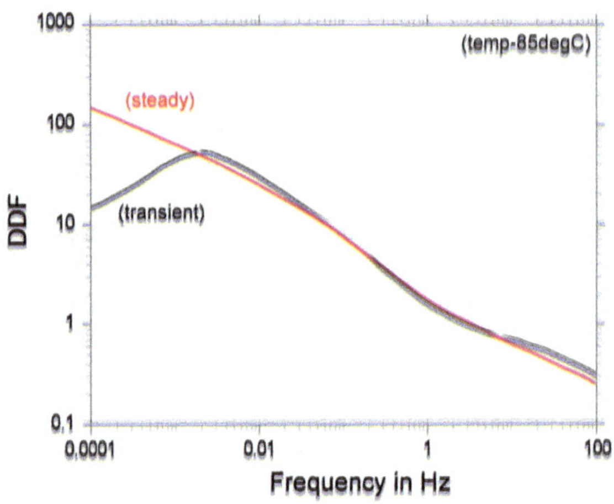

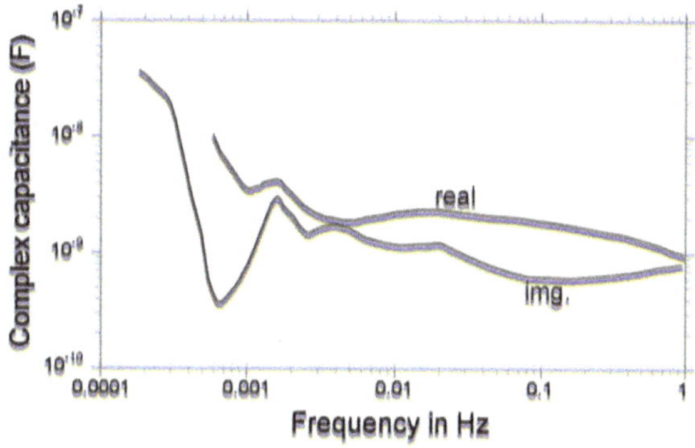

Fig. 6.18 Complex capacitance components under transient temperature conditions

6.3.4 *Effect of Moisture in Insulation*

FDS results are significantly influenced by the presence of moisture in insulation. In general, a higher dielectric dissipation factor is obtained with the increase of moisture content in cellulose insulation. Figure 6.19a shows that at frequencies lower than 0.01 Hz a sharp rise of real part of complex permittivity occurs in the case of pressboard samples having moisture content ≥2 % [5]. From Fig. 6.19b it may be seen that the dielectric losses in impregnated pressboard increases considerably with the increase in moisture content particularly for frequencies lower than 100 Hz [5]. For frequencies higher than 10 kHz, the difference between the loss curves of pressboards having different moisture content becomes insignificant.

With increasing moisture content in paper insulation, the value of the minimum of imaginary part of complex permittivity increases and the curves for both real and imaginary parts of complex permittivity are shifted towards higher frequencies, as shown in Fig. 6.20 [14].

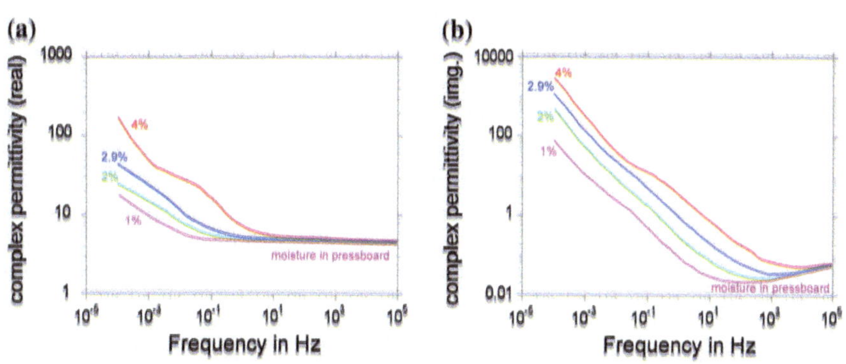

Fig. 6.19 **a** Real part of dielectric permittivity of pressboard having different moisture content. **b** Imaginary part of dielectric permittivity of pressboard having different moisture content

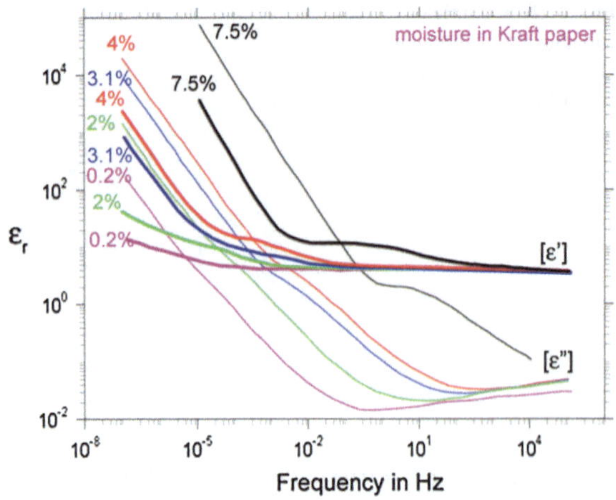

Fig. 6.20 Variation of complex permittivity components with moisture content in Kraft paper

Figure 6.20 shows a small relaxation peak in the curves at frequencies below the frequency corresponding to the minimum in ε''. This peak is not noticeable for nearly dry paper samples, while it becomes gradually prominent in both the real and imaginary parts of complex permittivity with increasing moisture content. As a consequence of the relaxation peak, the part of the curve on the high frequency side of the peak is shifted more in frequency than the part of the curve on the low frequency side of the peak [14]. Figure 6.20 also shows that the difference between the highest values of the relaxation peaks is small for different moisture contents.

Figure 6.21 presents a comparison of FDS results obtained for impregnated and unimpregnated Kraft paper [14]. The effect of moisture content has been found to be similar in both the cases. Figure 6.22 shows that the effect of moisture on the FDS results obtained for pressboard is different from the FDS results obtained for paper insulation [14]. For both pressboard and paper insulation the general shift towards higher frequency with moisture content could be noted. Similarly the increase in the value of the minimum of imaginary part of complex permittivity could also be seen for both paper and pressboard. But the moisture dependent relaxation peak is not seen in the case of pressboard, while such peaks are clearly present for moist paper insulation. However, the FDS results are still inconclusive in respect to relaxation peak. In [5], it was found that the peak is present in pressboard and not so significant in paper. Another study in this respect [19] did not find any such relaxation peak in FDS results. A possible cause of such discrepancy may be the type of pressboard used in the respective experiments, particularly the difference in the density of the pressboards under experimentation.

Several researchers have tried to develop expression for moisture content from the FDS results. Equations 6.17 and 6.18 were proposed in [20] so that the moisture content can be calculated by determining the real and imaginary part of complex permittivity of the insulation.

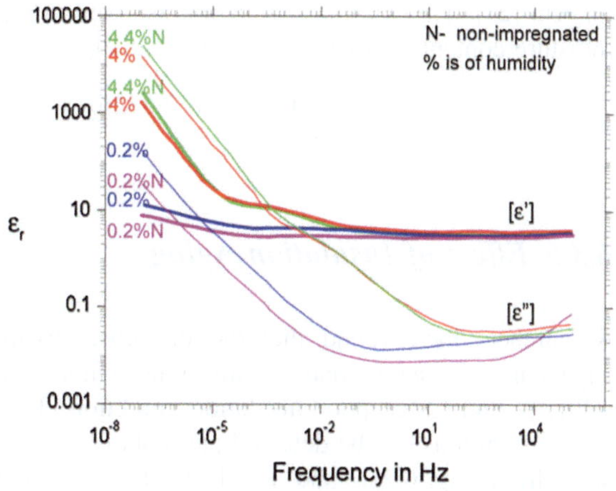

Fig. 6.21 FDS results for impregnated and unimpregnated paper insulation

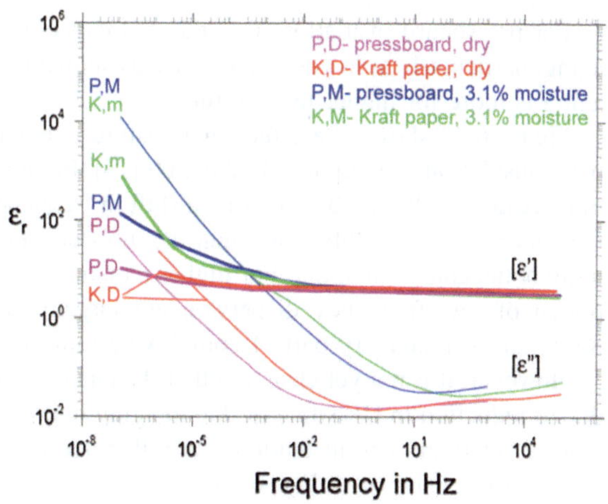

Fig. 6.22 FDS results for impregnated pressboard and paper insulation

$$m = \left(f^{0.0011T+0.233}\left(\frac{\varepsilon'}{\varepsilon_0} - 3\right)10^{0.6499-0.0081T}\right)^{1/-0.0851\log(f)+1.778} \tag{6.17}$$

$$m = \left(f^{0.0032T+0.2265}\left(\frac{\varepsilon''}{\varepsilon_0}\right)10^{-0.0177T+1.2571}\right)^{1/-0.27971\log(f)+2.1328} \tag{6.18}$$

where, m is the moisture content, f is the frequency and T is Temperature in Kelvin.

Attempt has also been made to quantify moisture content in oil-paper insulation system of cables from FDS results with the help of the minimum value of loss tangent [21]. The non-linear function as given in Eq. 6.19 is proposed in [21] for moisture content evaluation in cable insulation

$$m = 15.297 + 2.53267 \times \ln(\tan \delta_{min}) \tag{6.19}$$

6.3.5 Effect of Insulation Aging

Researchers have found that the dielectric dissipation factor as well as the capacitance values increases with aging. Higher capacitance means higher permittivity and hence poor insulation condition. Thus from FDS results different aging conditions can be detected particularly at low frequencies. This fact can be seen from Fig. 6.23a and b where the *tanδ* and capacitance values of oil impregnated paper insulation have been found to increase with aging at frequencies below 0.1 Hz [22].

For pressboard insulation, too, the real capacitance curves shift upwards with aging, particularly at lower frequencies, as shown in Fig. 6.24 [23]. The impact of aging is more for higher temperatures.

Figure 6.25 shows the effect of moisture content on complex permittivity of paper insulation aged up to DP 280–360 [14]. Figure 6.26a and b show such effect in more details. It may be seen that a shift in frequency occurs for the minimum of the curves. Figure 6.26b shows that the moisture related relaxation peak becomes more noticeable with aging. For higher moisture content the frequency shift in the region of low frequency dispersion are larger than the horizontal shift at the minimum of imaginary part of complex permittivity [14].

However, it is not yet clear whether the changes in FDS results due to aging are due to reduction in DP value or due to aging byproducts like acids. Further, in the case of relatively dry insulation, the difference in dielectric response is comparatively low as shown in Fig. 6.26a.

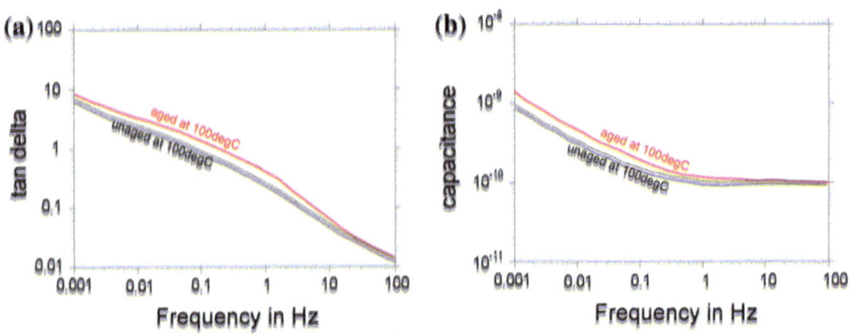

Fig. 6.23 **a** Variation of *tanδ* with frequency for aged and unaged paper. **b** Variation of capacitance with frequency for aged and unaged paper

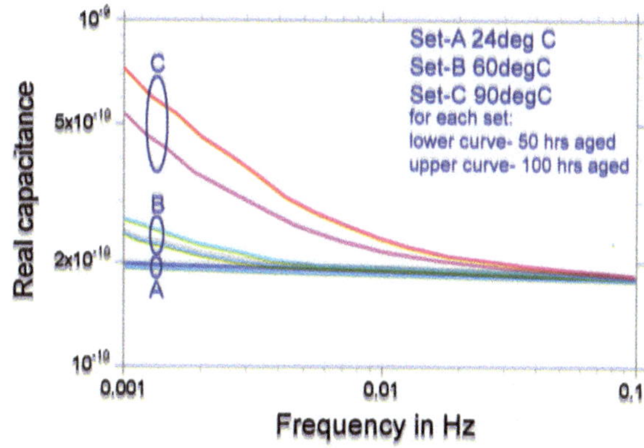

Fig. 6.24 Effect of temperature and aging on real part of complex capacitance of pressboard insulation

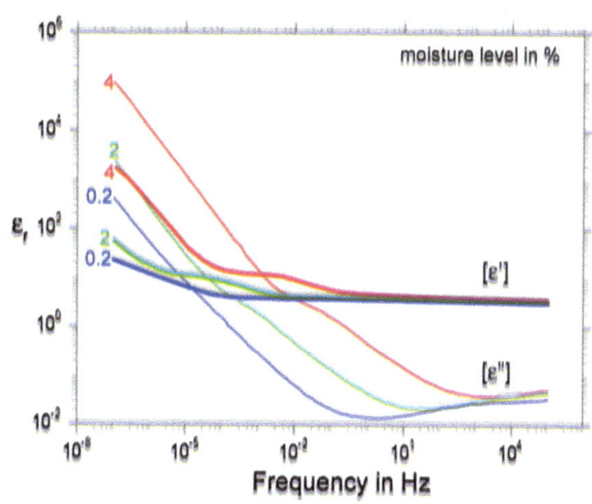

Fig. 6.25 Effect of moisture content on complex permittivity for aged paper

Figure 6.27 shows the variation of *tanδ* with frequency for four different transformers which were in service for different period of time. It could be seen from this figure that the minimum value increases and shifts towards higher frequency with the increase in service duration [15].

6.3.6 *Effect of Acids*

It has been reported in literature that there are two types of acids found in the insulation of aged transformers [24]. Low molecular acids consisting of formic, acetic and levulinic acids are known as "light acids". These are hydrophilic acids, which are water-soluble and prefer to stay in the paper insulation and contribute to paper aging. Light acids have a partitioning ratio between paper and oil of about 100:1 at 20 °C. The high molecular acids such as naphthenic and stearic acids are

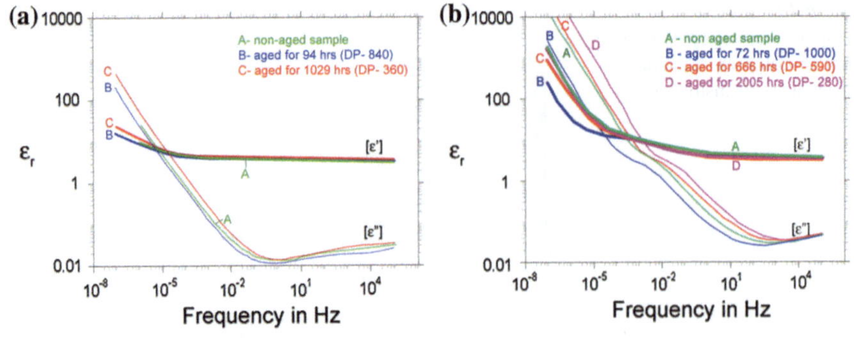

Fig. 6.26 **a** Effect of aging of impregnated Kraft paper with 0.2 % moisture. **b** Effect of aging of impregnated Kraft paper with 4 % moisture

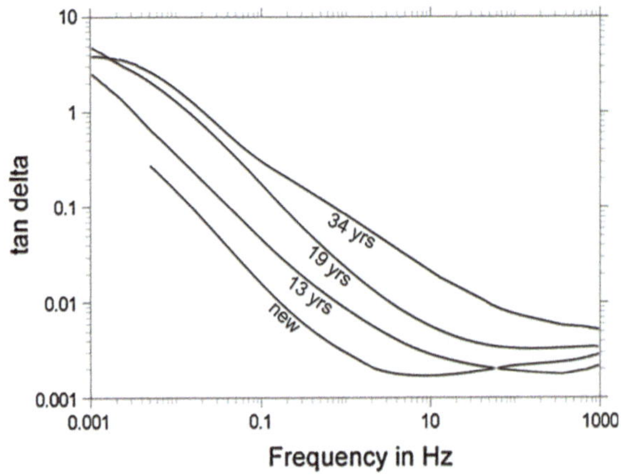

Fig. 6.27 Variation of *tanδ* for four transformers having different service times

known as "heavy acids". These are hydrophobic acids, which are fat-soluble and prefer to stay in oil insulation. In real-life transformers, degradation of oil gives rise to various long-chained acids, while degradation of paper causes mainly short-chained acids [24].

The effect of acids on both dry and wet paper is shown in Fig. 6.28, which shows that the light acids cause the curves to shift towards higher frequencies. For paper having higher moisture content, the minimum of ε" is higher with acids than without acids. Comparing these results with those for the effect of moisture content, it may be stated that the effect is somewhat similar to the effect of moisture in insulation. However, unlike moisture, acids on their own do not cause any relaxation peak [14]. It may also be seen that the difference between the curves with and without acids is minimal for heavy acids.

The effect of acids on pressboard, as shown in Fig. 6.29, is a little bit different from the effect on paper. In general, light acids shift the minimum of ε" to a higher value for pressboard, which is similar to the effect of light acids on paper. But the

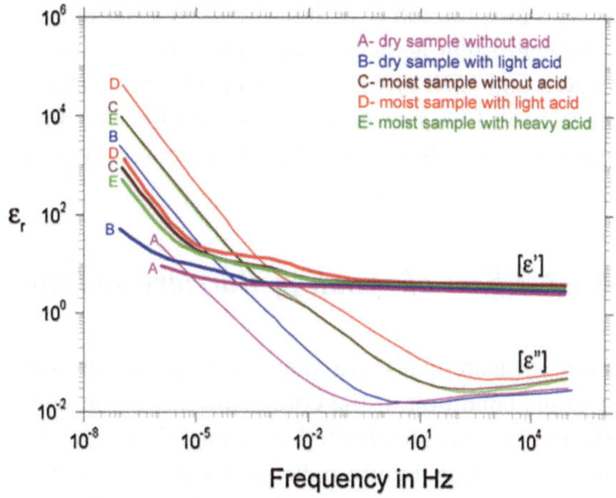

Fig. 6.28 Effect of acids on Kraft paper

difference is that from the frequency corresponding to the minimum of ε" downwards, ε" is shifted upwards in a way similar to the effect of moisture in insulation without however the relaxation peak [14].

Moisture and light acids are both important indicators of aging of solid insulation. But large concentration of acid is required for a contribution on FDS results comparable to that of moisture. Researchers have found that an acidity of paper of 2–10 mg of KOH/g of paper is needed to give the same effect as 1 % moisture content [14]. Aging experiments on the other hand showed that the acidity of paper becomes about 2 mg of KOH/g of paper when the DP of paper is about 300, i.e. in the final stages of aging. In other words, in a transformer acids will contribute at a level equal to that of moisture only when the transformer is very aged or otherwise contaminated.

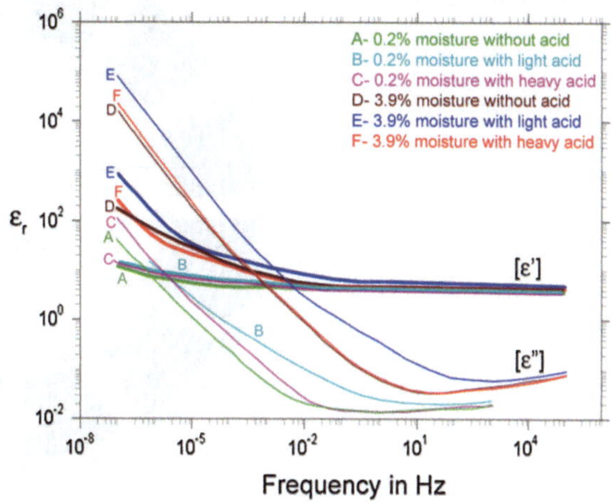

Fig. 6.29 Effect of acids on pressboard

In short, the effect of light acids and the effect of moisture are similar but not identical. Light acids and moisture affect different parts of the dielectric response curves and such effects are again different for pressboard and paper. Light acids have been found to have very little effect on the activation energy for the temperature dependence of dielectric response [14] of paper and pressboard.

6.3.7 *Effect of Winding Barriers and Spacers*

FDS measurements are done at transformer terminals and hence indicate the average condition of the composite insulation within the transformer. As a result, the FDS results are affected by the axial cylindrical barriers and radial spacers used in the winding. Figure 6.30 shows the typical arrangement of cylindrical pressboard barriers in series with oil ducts and spacers in a transformer. Such an insulation structure can be represented by a lumped model, known as X–Y model, that uses the relative dimensions of the spacers, barriers and the oil duct, as shown in Fig. 6.31 [25]. In this X-Y model, the parameter X is the sum of the radial thickness of all the barriers divided by the radial width of the duct, while the parameter Y is the sum of the width of all the spacers along periphery of the duct divided by the total peripheral length of the duct.

The effect of barrier on the real part of complex capacitance is shown in Fig. 6.32 [1]. It can be seen from Fig. 6.32 that the real capacitance at 0.1 mHz increases as the value of X decreases. Figure 6.33 shows the effect spacers on the real part of complex capacitance [1]. In Fig. 6.33 the real capacitance curves are shifted upwards as the value of Y decreases.

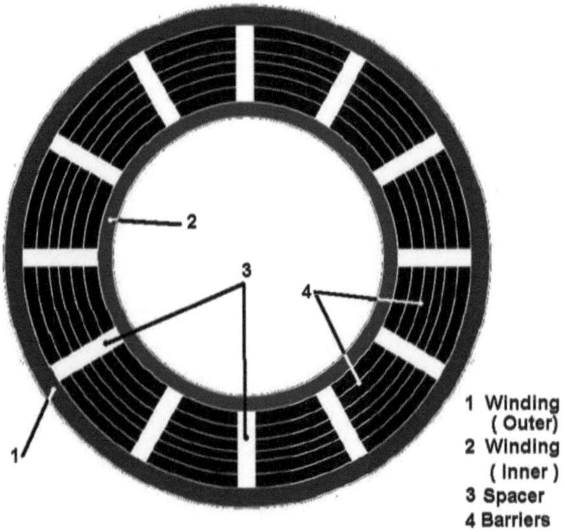

Fig. 6.30 Typical transformer winding arrangement with oil ducts

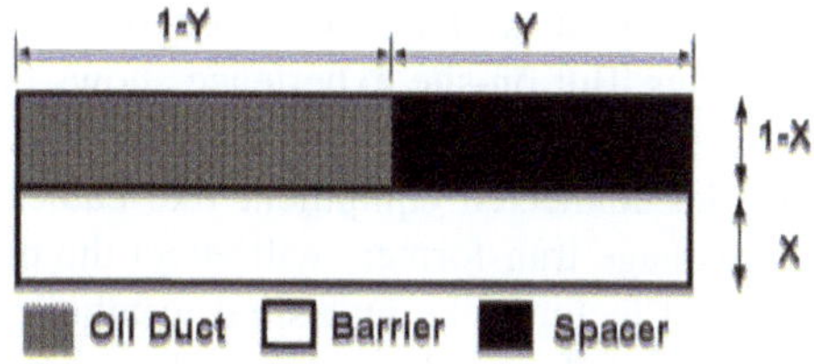

Fig. 6.31 X-Y model of transformer insulation structure

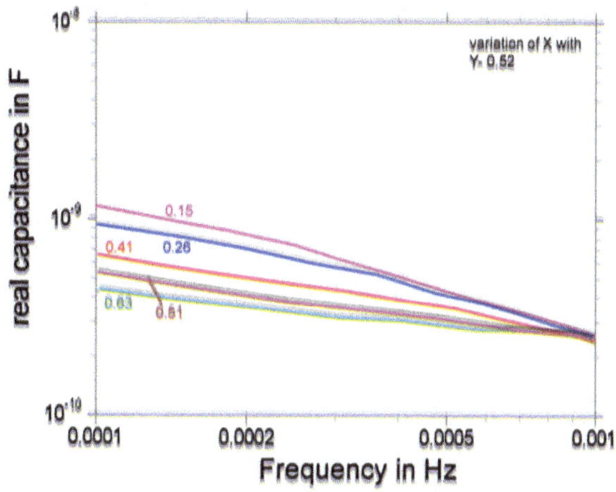

Fig. 6.32 Effect of barrier on real part of complex capacitance

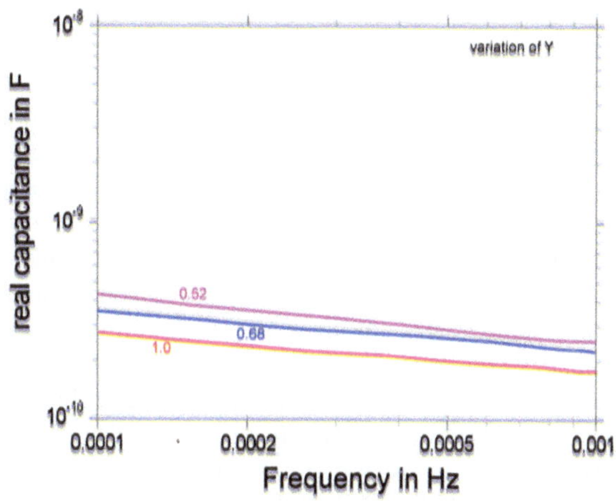

Fig. 6.33 Effect of spacer on real part of complex capacitance

6.3.8 Various Factors Affecting Dielectric Response Measurements in Field Conditions

Dielectric response measurements are carried out off-line. Ideally the transformer should be completely isolated from the network. In such ideal case, the current that is being measured is influenced only by the internal leakage and creep from bushings. But on-site experience shows that in many cases the measurements are to be taken by opening the disconnectors only. In such cases the leakage due to parallel connected equipment like cables, lightning arrestors, support insulators and voltage transformers will affect the measurements. In addition there are other external factors, like undesired weather conditions like rain and electromagnetic interference from other energized equipment located near the test site, which also affect dielectric response measurements [3].

6.3.8.1 Effect of External Leakage Paths on CHL Measurements

As shown in Fig. 6.5a, in CHL measurements, the test voltage is applied to the HV terminal of the transformer and the dielectric response current is measured on the LV side. The current path shown in Fig. 6.5a is the path through which current flows when the guard cable is connected to the grounded transformer tank and no other network element is connected to the bushings. Such connection is necessary for proper evaluation of the capacitance and losses of insulation between the HV and LV windings.

On the other hand, when the equipment like cables, insulators etc. remain connected to the transformer bushings, then the current drawn by the capacitances of these equipments load the measuring instrument. Sometimes the capacitances of cables could be high so that the current through the measuring instrument may cross its limit. Significant error in measurement arises when a voltage transformer remain connected to the transformer terminal at the LV side. The impedance of the voltage transformer is frequency dependent and at low frequency could be comparable to the impedance of the transformer insulation system. In the case of PDC measurement, which is performed with dc test voltage, the voltage transformer acts like a short-circuit that makes measurement impracticable.

Figure 6.34 shows the comparison between a reference CHL measurement, which was taken with all shunt paths disconnected, with the measurement taken with all shunt paths connected including voltage transformers [3]. It is quite clear from Fig. 6.34 that the measurements were severely affected by the presence of voltage transformer. The deviation from the reference measurement increases with decreasing frequency and due to major disturbances it was not possible to take measurements below 1 Hz. It has been reported in [3] that the interpretation of CHL measurements, without the voltage transformer but with other shunt paths connected, is similar to the reference case. But the estimation of moisture content using the FDS results obtained with the voltage transformer connected is nearly twice as high as that obtained using the results of reference case.

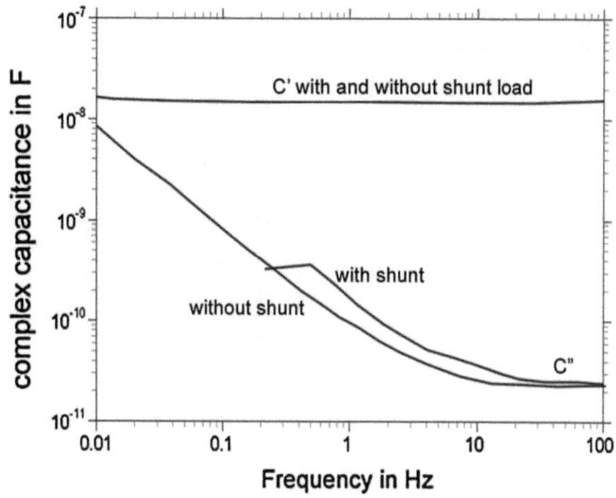

Fig. 6.34 Effect of connected shunt paths on CHL measurement of FDS

6.3.8.2 Effect of External Leakage Paths on CH and CL Measurements

In CH measurement the test voltage is applied to the HV terminals and the response current returning through the ground is measured. The circuit arrangement is shown in Fig. 6.5b, in which the guard is connected to LV winding so that inter-winding capacitance does not come into the measurement. Thus in CH measurement only the capacitance between the HV winding and the tank is investigated. Similarly in CL measurement the capacitance between the LV winding and the tank is tested. In both CH and CL measurements, even when all the shunt elements are disconnected from the transformer terminals, bushing capacitance and internal creeps are measured in parallel to the HV-to-tank or LV-to-tank capacitance, as the case may be. If cables are left connected to the transformer terminal, then the capacitance of the cable, which could be of length 100 m in some cases, come in parallel to the winding to tank capacitance. Many cables have oil impregnated paper insulation and have properties similar to that of transformer insulation. Cable capacitances being high, the measurement results in such cases represent the cable insulation more and transformer insulation less [3]. Moreover, in both CH and CL measurements the effect of voltage transformer needs to be considered if one is connected to the transformer during testing.

Figure 6.35 shows the results of a CH measurement in which all the shunt elements were disconnected along with the results of a CH measurement in which the cables were left connected to the transformer [3]. As expected, the measured capacitances with the cables connected were found to be much higher than that without the cables.

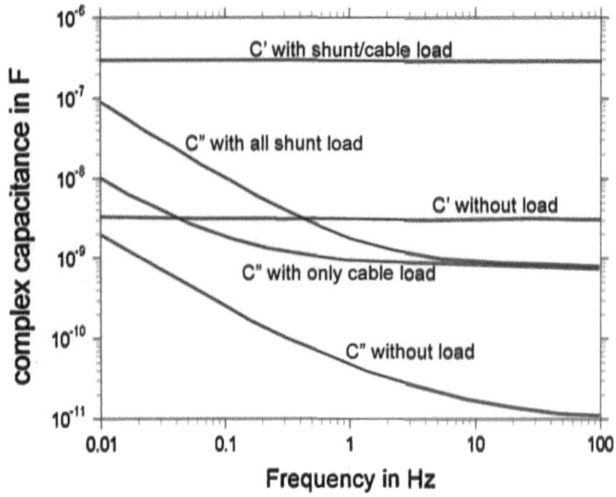

Fig. 6.35 Effect of connected shunt paths on CH measurement of FDS

6.3.8.3 Effect of Weather Conditions

The leakage current that flows along the surface of the transformer bushing affects the FDS measurement depending on the test configuration. Such surface leakage current increases when the bushing surface is contaminated, notably in the presence of high humidity or rain. CHL measurement is not affected by bushing creep as the surface leakage current flows to the tank which is guarded. But CH or CL measurements are affected significantly by bushing creep, as the tank in such test configurations is not guarded. Hence, the surface leakage currents will add to the winding to tank current, and consequently influence the recorded response. Figure 6.36 shows the FDS test results for CH connection taken on a sunny as well as on a rainy day without any shunt element connected to the transformer [3]. It is apparent from Fig. 6.36 that the results under rainy condition are different from the results under dry condition. In [3] a guard ring was fitted on the bushing surface, which was connected to the guard of the FDS measurement system. In this way the surface leakage current were eliminated from the currents measured by the FDS system. These results, as shown in Fig. 6.36, show that the results obtained with guard ring fitted on the bushing surface are quite similar to the reference results. It indicates that such surface guard rings are necessary to take accurate FDS results under undesired weather condition during field testing.

6.3.8.4 Effect of Electromagnetic Interference

Although the transformer is deenergized while performing dielectric response measurements, other installations in substations remain energized that are potential sources of electromagnetic disturbances. Segments of lines, which remain connected to transformer bushings even when the transformer is isolated from the

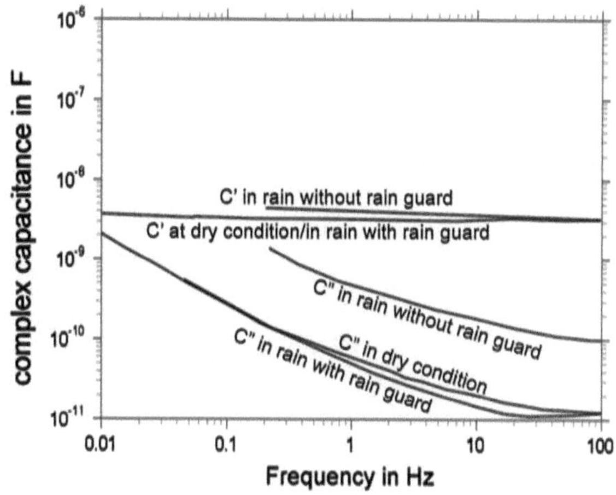

Fig. 6.36 Effect of bushing creep on CH measurements in dry and wet weather conditions

network, act as antennas which pick up electromagnetic disturbances through air. Conductively coupled disturbances may appear if conductive paths exist to the ground system of substation, e.g. through the voltage transformer. Parasitic voltage could appear at the terminal of the transformer, where the measurements are being taken, due to a neighboring transformer which may still be operating or from other high voltage lines passing close by. In literature [3] it has been mentioned that in some cases such parasitic voltages are too high making the dielectric measurements impossible.

6.4 Modelling Dielectric Response in Frequency Domain

From the equivalent circuit model of a linear dielectric, as shown in Fig. 6.37, the complex capacitance can be calculated from the complex admittance as follows:

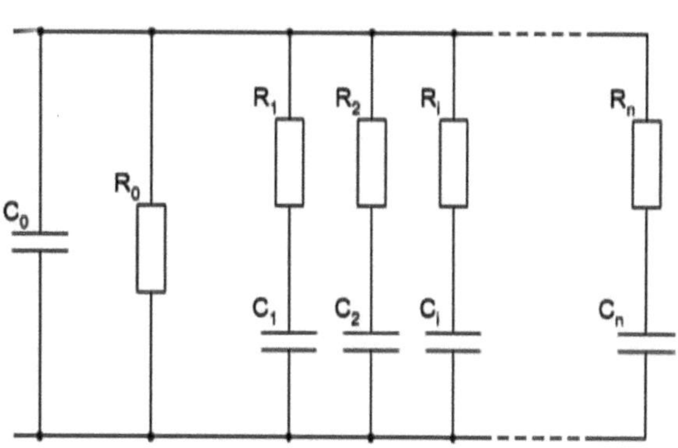

Fig. 6.37 Equivalent circuit model of a linear dielectric

$$\bar{C}(\omega) = \frac{\bar{Y}(\omega)}{j\omega} = C_0 + \frac{1}{j\omega R_0} + \sum_{i=1}^{n} \frac{C_i}{1 + j\omega R_i C_i} \tag{6.20}$$

From Eq. 6.20, the real and imaginary parts of complex capacitance as well as *tanδ* can be written as follows:

$$C'(\omega) = C_0 + \sum_{i=1}^{n} \frac{C_i}{1 + (\omega R_i C_i)^2} \tag{6.21}$$

$$C''(\omega) = \frac{1}{\omega R_0} + \sum_{i=1}^{n} \frac{\omega R_i C_i^2}{1 + (\omega R_i C_i)^2} \tag{6.22}$$

and

$$\tan\delta = \frac{C''(\omega)}{C'(\omega)} = \frac{\frac{1}{\omega R_0} + \sum_{i=1}^{n} \frac{\omega R_i C_i^2}{1+(\omega R_i C_i)^2}}{C_0 + \sum_{i=1}^{n} \frac{C_i}{1+(\omega R_i C_i)^2}} \tag{6.23}$$

Equations 6.21 through 6.23 show that if the equivalent circuit parameters are known, then the FDS parameters could be simulated. Researchers have simulated FDS data by using equivalent circuit parameters obtained from time-domain measurement, e.g. PDC or RVM data. Figure 6.38 shows simulated FDS data obtained from PDC data [7], while Fig. 6.39 shows FDS data obtained from RVM data [26].

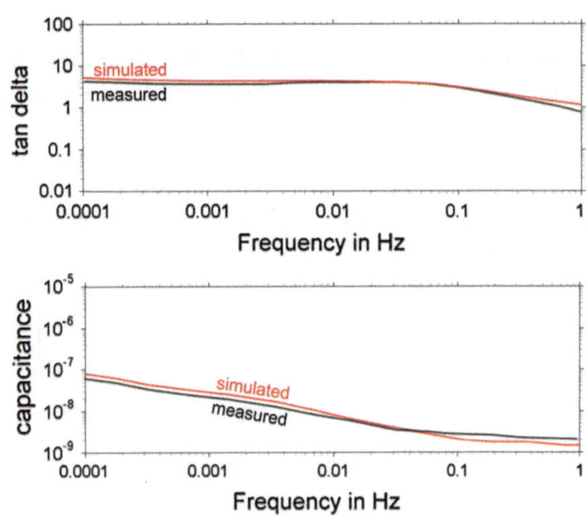

Fig. 6.38 Comparison of measured FDS data with that calculated from PDC data

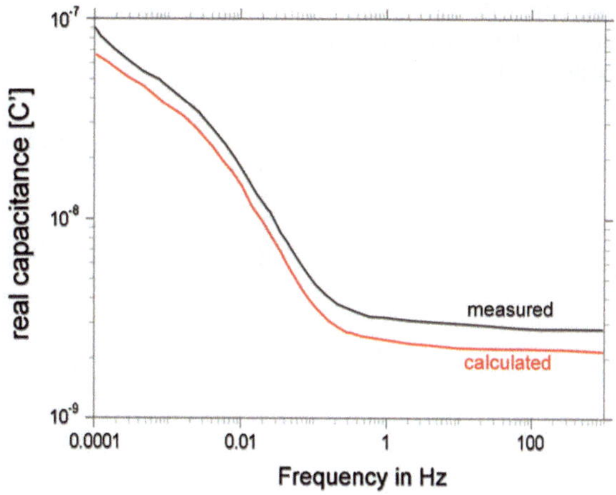

Fig. 6.39 Comparison of measured FDS data with that calculated from RVM data

It has been reported in literature [7] that the error is less than 5 % of the measured value in simulated FDS results obtained from PDC data, when PDC measurements are taken at low electric field below 10 V/mm.

6.4.1 Determination of Equivalent Circuit from FDS Data

With the measured values of complex capacitance and *tanδ*, as obtained from the FDS test results, the complex admittance of the tested insulation can be computed using Eq. 6.24

$$\bar{Y}(\omega) = j\omega\, C'(\omega)\,(1 + j\,\tan\delta(\omega)) = j\omega C_0 + \frac{1}{R_0} + \sum_{i=1}^{n}\frac{j\omega\, C_i}{1 + j\omega R_i C_i} \tag{6.24}$$

So the equivalent circuit parameters can be determined from the measured FDS data using Eq. 6.24. Suitable mathematical technique like non-linear least square optimization with the constraint that all the parameters are positive could be used for equivalent circuit parameter estimation. In such optimization, the objective function could be [26]

$$J = \min\left[\sum_{k=1}^{L}\left\{\bar{Y}(\omega_k) - j\omega_k C_0 - \frac{1}{R_0} - \sum_{i=1}^{n}\frac{j\omega_k\, C_i}{1 + j\omega_k\, R_i C_i}\right\}^2\right] \tag{6.25}$$

where, $\bar{Y}(\omega_k)$is calculated from the measured FDS data at *k*th recorded frequency out of total *L* no. of frequencies recorded.

After determining the equivalent circuit parameters from FDS data, time domain results can be simulated. A comparison of measured RVM data with RVM

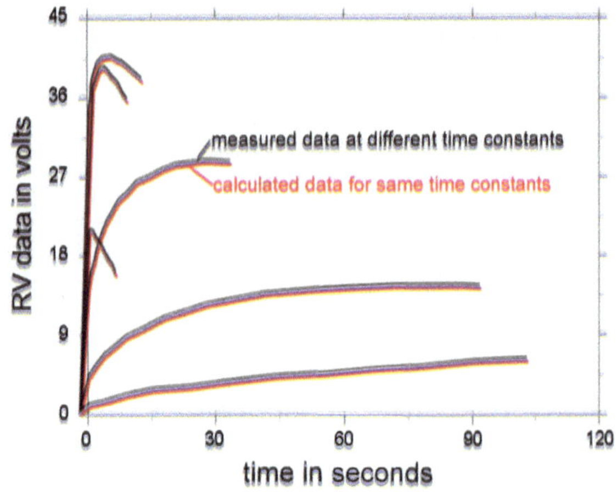

Fig. 6.40 Comparison of measured RVM data with that calculated from FDS data

data computed from FDS data is shown in Fig. 6.40 [26]. Figure 6.40 indicates that time domain data can be recovered from frequency domain data with good accuracy.

However, it has been reported [26] that simulating time domain data from frequency domain and vice versa fits well for low moisture content of transformer insulation. For relatively higher moisture content of transformer insulation, difference between experimental and simulated data becomes more.

6.5 Advantages and Disadvantages of Dielectric Response Measurements in Time-Domain and Frequency-Domain

Dielectric response measurement techniques in time domain, viz. PDC and RVM, as well as that in frequency domain provide useful information on the condition of transformer insulation. Practical experience shows that all the three techniques rank the transformers in the same order [12]. Salient features of the three measurement techniques are discussed below.

RVM being a high input impedance method, leakage current, e.g. surface leakage over the bushings, affect the measurement. RVM has been found to be more sensitive to systematic errors and the interpretation of RVM results is not always unambiguous, especially in the presence of multiple peaks in RVM spectrum for old transformers. Moreover, RVM considers insulation geometry and oil conductivity only to a limited extent. It takes very long time to record the complete polarization spectrum. A more practical approach has been suggested in [12] in which three recovery voltage measurements are recommended to be taken with charging times 10, 100 and 1000 s.

Interpretation of PDC results is straightforward and it provides moisture content in solid insulation and both oil and paper conductivities. Other diagnostic parameters like *tanδ* and polarization spectra can be calculated from PDC results. But, drift and jitter free voltage source is required for accurate analysis. Since the currents are low, the measurements are affected by offset currents and interference in the field. But it is possible to implement guarded measurement to eliminate the surface leakage currents from measurements. Another important aspect of PDC measurement is the switching transients at the start of polarization and depolarization which means that the first few seconds are difficult to measure, which in turn makes the lower limit of the equivalent frequency about 1 Hz.

Interpretation of FDS results is also straightforward. It could separate real and imaginary parts of complex capacitance and complex permittivity. It also provides the moisture content in solid insulation and diagnostic quantity like C-ratio. FDS method is relatively insensitive to interference due to its inherent small bandwidth. There is no need of any high voltage power supply. It is possible to exclude creep currents on bushings by implementing three terminal measurements. But, low frequency measurements are very time consuming.

6.6 Future Works in FDS

FDS measurement has been found to be rugged for on-site application. However, the effects of electromagnetic interference, especially their limits and the steps to be taken to make FDS measurements more immune to such disturbances need further investigations.

PDC test lasts about 1.2 h to get time domain information up to 2000 s. On the other hand 2000 s is necessary to take only one FDS measurement for two cycles of 1 mHz. For two cycles of 0.1 mHz, it will take about 12 h for one set of data, which is practically not feasible for on-site measurement. Because of abnormally high time required to take very low frequency FDS measurements, researchers are working on the transfer of time domain measurements (PDC) to frequency domain as an alternative.

As FDS results are significantly affected by temperature and the measuring times for low frequency are very high, temperature changes within the transformer during FDS measurement cycles may lead to inaccurate moisture analysis. Instead of taking oil temperature and then using it as insulation temperature, novel temperature determination methods are under investigation. One such upcoming method is measurement of winding resistance as a function of time during FDS measurement [18]. The measured resistance values are then used to determine winding temperature, which could provide more accurate moisture estimation from FDS results.

It has been noted that FDS can identify the effects of moisture and acids, both of which are significant parameters in relation to the aging of transformer insulation. However, more work is needed for better understanding and quantification of the effects of acids on FDS results. Better differentiation of the effects of moisture and acids will definitely improve the accuracy of condition assessment of transformer insulation in the light of insulation aging. As a generalization of the above mentioned fact, it is to be noted that all the parameters that affect FDS results are simultaneously present in most of the cases of real-life transformers at site. Present knowledge available in literature does not make it possible to estimate the contribution of the individual parameters. The separation of individual contribution of different parameters is an issue that needs detailed investigation.

The dielectric parameter such as resistances and capacitances are geometry dependent. Hence, to make FDS analysis geometry independent, poles computed from resistances and capacitances may be used. Researchers have already reported that a lower value of poles indicate better condition of insulation while a higher value corresponds to higher moisture content [10]. Further studies are required to determine the capability of poles in assessing the overall condition of oil-paper insulation system of transformers.

References

1. Pradhan MK, Yew JH, Saha TK (2008) Influence of the geometrical parameters of power transformer insulation on the frequency domain spectroscopy measurement. In: Proceedings of IEEE power and energy society general meeting, Pittsburg, USA
2. Zaengl WS (2003) Dielectric spectroscopy in time and frequency domain for hv power equipment, part I: theoretical considerations. IEEE Electr Insul Mag 19(5):5–19
3. Blennow J, Ekanayake C, Walczak K, Garcia B, Gubanski SM (2006) Field experiences with measurements of dielectric response in frequency domain for power transformer diagnostics IEEE Trans PWRD 21:681–688
4. G.E. Services IDA 200: Insulation diagnostic system user's manual
5. Ekanayake C, Gubanski SM, Graczkowski A, Walczak K (2006) Frequency response of oil impregnated pressboard and paper samples for estimating moisture in transformer insulation. IEEE Trans PWRD 21:1309–1317
6. Jaya M, Koch M (2010) Information within the dielectric response of power transformers for wide frequency ranges. In: Proceedings of ISEI-2010, San Diego, June 2010
7. Setayeshmehr A, Fofana I, Eichler C, Akbari A, Borsi H, Gockenbach E (2008) Dielectric spectroscopic measurements on transformer oil-paper insulation under controlled laboratory conditions. IEEE Trans Dielectr Electr Insul 15:1100–1111
8. Yew JH, Saha TK, Thomas AJ (2006) Impact of temperature on the frequency domain dielectric spectroscopy for the diagnosis of power transformer insulation. In: IEEE power engineering society general meeting, Montreal
9. Zaky AA, Hawley R (1973) Conduction and breakdown in mineral oil. Peregrinus, London
10. Fofana I, Hemmatjou H, Meghnefi F, Farzaneh M, Setayeshmehr A, Borsi H, Gockenbach E (2010) On the frequency domain dielectric response of oil-paper insulation at low temperatures. IEEE Trans Dielectr Electr Insul 17:799–807
11. Supramaniam GK, Hussien ZF, Aizam M (2008) Application of frequency domain spectroscopy (fds) in assessing dryness and ageing state of transformer insulation systems. In: Proceedings of the 2nd IEEE international conference on power and energy (PECon-08), Johor Baharu, Malaysia, Dec 1–3 2008
12. Gäfvert U, Adeen L, Tapper M, Ghasemi P, Joensson B (2000) Dielectric spectroscopy in time and frequency domain applied to diagnostics of power transformers. In: Proceedings of

the 6th international conference on properties and applications of dielectric materials, Xi'an, China, June 21–26 2000
13. CIGRE Task Force 15.01.09 (2002) Dielectric response methods for diagnostics of power transformers. Electra 202:25–36
14. Linhjell D, Lundgaard L, Gäfvert U (2007) Dielectric response of mineral oil impregnated cellulose and the impact of aging. IEEE Trans Dielectr Electr Insul 14:156–169
15. Yang S, Dong M, Zhang G, Zhang Z (2009) Condition evaluation of oil-paper insulation based on dielectric spectroscopy. Annual report of ieee conference on electrical insulation and dielectric phenomena, Virginia Beach, USA, pp 77–80
16. Linhjell D, Hestad ØL, Gäfvert U, Lundgaard LE (2005) Dielectric response of oil-impregnated cellulose from 0.1 mHz to 3 MHz. In: Proceedings of IEEE international conference on dielectric liquids, June 26–July 01 2005, pp 277–280
17. Yew JH, Pradhan MK, Saha TK (2008) Effects of moisture and temperature on the frequency domain spectroscopy analysis of power transformer insulation. In: Procedings of IEEE power and energy society general meeting, Pittsburgh, USA, 20–24 July 2008
18. Jaya M, Helmlinger O, Leibfried T (2010) Novel method to determine insulation temperature during dielectric measurements on power transformers. In: Proceedings of 20th Australasian universities power engineering conference (AUPEC), Christchurch, 5–8 Dec 2010
19. der Houhanessian V (1998) Measurement and analysis of dielectric response in oil-paper insulation systems, PhD Thesis, Swiss Federal Institute of Technology, Zürich, Switzerland
20. Wei G, Saha TK (2011) Study on moisture in oil-paper insulation by frequency domain spectroscopy. In: Asia-Pacific power and energy engineering conference (APPEEC), Wuhan, China, 25–28 March 2011
21. Zaengl WS (2003) Applications of dielectric spectroscopy in time and frequency domain for HV power equipment. IEEE Electr Insul Mag 19(6):9–22
22. Supramaniam GK, Hussien ZF, Aizam T M (2008) Application of frequency domain spectroscopy (FDS) in assessing dryness and ageing state of transformer insulation systems. Proceedings of 2nd IEEE international conference on power and energy (PECon-08), Johor Baharu, Malaysia, 1–3 Dec 2008, pp 55–61
23. Wang S, Wei J, Yang S, Dong M, Zhang G (2009) Temperature and thermal aging effects on the frequency domain spectroscopy measurement of oil-paper insulation. Proceedings of the 9th international conference on properties and applications of dielectric materials, Harbin, China, 19–23 July 2009, paper no. C-15
24. Lundgaard LE, Hansen W, Linhjell D, Painter TJ (2004) Aging of oil impregnated paper in power transformers. IEEE Trans Power Delivery 19:230–239
25. Gaefvert U (2005) Influence of geometric structure and material properties on dielectric frequency response of composite oil cellulose insulation. In: Proceedings of international symposium on electrical insulating materials (ISEIM 2005), Kitakyushu, Japan, 5–9 June 2005
26. Xu S, Middleton R, Fetherston F, Pantalone D (2003) A comparison of return voltage measurement and frequency domain spectroscopy test on high voltage insulation. In: Proceedings of the 7th international conference on properties and applications of dielectric materials, Nagoya, 1–5 June 2003

Chapter 7
Frequency Response Analysis

7.1 Introduction

In the middle of 1980s, successful introduction of data acquisition and analysis with the help of digital measurement technique was reported in high voltage impulse testing [1]. This resulted in vigorous research activities in this field in subsequent years and a major development was the introduction of transfer function method in fault diagnosis during impulse testing [2]. In this method, the transfer function was estimated using the time-domain voltage and current data recorded during testing. Over the years, further developments took place and Frequency Response Analysis (FRA) gained prominence in which the transfer function is measured in frequency domain. In FRA, problems associated with time-domain measurements such as signal-to-noise ratio, frequency resolution, and aliasing were greatly reduced or completely eliminated. One significant reason for increasing prominence of FRA is its application in monitoring the mechanical integrity of transformer windings either at manufacturing end or in service. It is known that mechanical forces generated due to short circuits, unskilled handling during manufacturing or rough transportation could disturb the geometry of transformer windings. The cumulative effect of several such effects could lead to winding displacement or deformation which could eventually cause major fault in transformers. Very often small mechanical damages go unnoticed as the transformer continues to perform its primary duty of power transfer unaltered. But it is crucial to detect such small winding damages at the earliest opportunity so that such small damages do not develop into major transformer failure. FRA is capable of detecting mechanical damages in transformer winding as information related to winding deformation is embedded in the transfer function. Following detection of any mechanical fault, a major step forward is to identify the location of the fault, preferably by means of terminal measurements without detanking the transformer winding as detanking is expensive and time consuming. In this context studies have now proved that FRA, which is a non-invasive terminal measurement method, can locate mechanical fault of transformer winding with reasonably good accuracy [3, 4].

S. Chakravorti et al., *Recent Trends in the Condition Monitoring of Transformers*, Power Systems, DOI: 10.1007/978-1-4471-5550-8_7, © Springer-Verlag London 2013

7.2 Methods of Frequency Response Analysis

Stated in simple words, FRA consists of measuring the impedance/admittance of transformer winding over a wide range of frequencies. The magnitude of the impedance/admittance of the transformer winding as function of frequency gives a so-called fingerprint of the winding. Such fingerprints from two or more tests are compared to detect winding deformation. There are two commonly used methods to carry out FRA, viz. Impulse Frequency Response Analysis (IFRA) and Swept Frequency Response Analysis (SFRA).

In IFRA the required wide range of frequencies is injected as a voltage impulse into one terminal. The response in the form of voltage or current is measured at another winding or terminal. The measured signals are filtered, sampled, and stored in time-domain. Subsequently, the time-domain data is transformed into the frequency domain and the transfer function is calculated. Thus, transfer functions at several frequencies can be measured simultaneously. IFRA is similar to a time-domain test method known as Low-Voltage Impulse (LVI) measurement, which was first reported in 1966 [5]. Further developments on LVI could be found in [6, 7]. But, LVI requires special instrumentation to get repeatable results, needs extensive calibration procedures and also is subject to interference effects. Hence, LVI is not used often now-a-days.

In SFRA, which gives more repeatable results, sinusoidal voltage of frequency varying over a wide range is applied to one terminal and the response is measured at another terminal or winding. Then the transfer function is computed for the entire test frequency range. However, measurement for only one frequency can be made at a time and hence simultaneous determination of transfer function at more than one frequency is not possible.

The most significant advantage of IFRA over SFRA is its shorter measurement time. The time required for each measurement in IFRA is typically 1 min, while that for each measurement in SFRA typically varies from 1 to 10 min. In fact, the time taken to sweep one single frequency in SFRA depends on how much signal processing is involved. The advantages of SFRA over IFRA are manifold, e.g., (i) better signal-to-noise ratio, (ii) nearly same accuracy over the entire frequency range, (iii) wider range of injected frequencies, (iv) finer frequency resolution at lower frequencies, and (v) lesser requirement of measuring equipment [8]. Relative merits and demerits of these two methods are discussed in details in [9].

In recent years, SFRA is being increasingly used in practice for detection and localization of mechanical faults in transformer windings. Consequently, this chapter focuses on SFRA hereafter.

7.3 Swept Frequency Response Analysis

Swept frequency response analysis, as it is currently used, was first introduced by Dick and Erven [10]. Subsequent major advancement was reported in [11]. SFRA requires the use of suitable equipment to generate the sinusoidal signal of variable frequency over a wide range, make the measurements and analyze the results. Commercially available network/impedance analyzers are now extensively used for performing SFRA tests. In SFRA a series of narrow band measurements are made which increases the dynamic range as well as the signal-to-noise ratio. Moreover, measurements are made only at the exciting frequency because of which measurements at different frequencies are not affected by non-linearity of the test object.

An IEEE working group is working actively for the preparation of a guide for the application of FRA and consequent interpretation of results for oil-immersed transformers [12]. The main aim of this guide is to provide necessary guidelines for making FRA measurements, test connections and procedures, preparation of test object, etc.

7.3.1 Basic Measurement

A basic measurement circuit is shown in Fig. 7.1 [8]. Such an instrumentation system is easily portable and can be readily set-up and operated. The impedance under test, Z_T, is the impedance of the transformer winding. The standard impedance is Z_S, which could be the impedance of the measuring cables. In Fig. 7.1, S, R, T are injected signal, excitation measurement, and response measurement, respectively. According to Dick and Erven [10] the modulus is determined as

$$k = 20\log_{10}\left(\frac{T}{R}\right) = 20\ \log_{10}\left(\frac{Z_S}{Z_S + Z_T}\right). \tag{7.1}$$

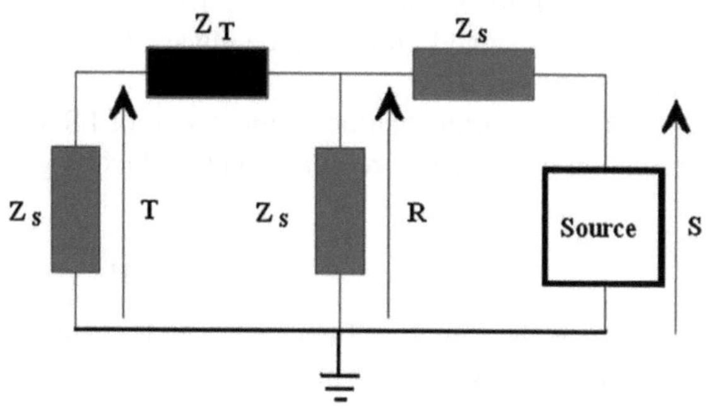

Fig. 7.1 Basic measurement circuit

The modulus is often called the amplitude, the voltage ratio, the voltage gain, the impedance, and the admittance. The modulus is presented in decibels on a 50 Ω basis, when 50 Ω co-axial cables are used for measurements.

The argument is defined as

$$\phi = \angle(T/R). \tag{7.2}$$

The argument is commonly known as phase.

The range of measured frequency is very wide, typically from 5/10 Hz to 10/20 MHz. The results are usually depicted as amplitude or phase plots against frequency. Commonly, the *x*-axis representing frequency is plotted in logarithmic scale, while the *y*-axis representing the amplitude or phase is plotted in linear scale. The logarithmic scale for frequency is commonly used to show the entire frequency range, while the linear scale for frequency is useful for looking at specific frequency bands to detect small differences at particular frequencies.

All connections are made by screened co-axial cables, with the screens being connected to ground at the source end as well as at the transformer tank. All the cables are terminated by their characteristic impedance to minimize any problem arising out of reflections. The low-voltage power source needed to drive the entire instrumentation is supplied through a shielded isolating transformer. The chassis of all the instruments, screens of the co-axial cables, and the shield of the isolating transformer are connected to the same ground as that of the tank of the transformer under test.

Certain good practices need to be followed to make repeatable and accurate FRA measurements: (i) the resolution bandwidth is to be set to about 10 % of the measured frequency, (ii) for testing transformers of different sizes, cables of different length are to be used with the goal that shortest possible length of cable is used, (iii) cables are to be kept more or less straight and not knotted together, (iv) the transformer is to be completely disconnected at every terminal, because even a small length of external connection at any one terminal will affect FRA measurements significantly, and (v) measuring cables are to be placed away from power cables [13].

Commonly, frequency sweeps are made with logarithmically spaced points to reduce the number of required sweeps. However, due to the limitation of the source, if linear sweeps are to be made then the frequency spacing between sweep points needs to be varied. But the frequency difference between two adjacent sweep points should preferably be always less than 2 % of the overall frequency range.

In digital instrumentation, computation of FRA results is affected by sampling resolution of frequency responses [11], which is given by

$$\Delta f = \frac{f_s}{M_s} \tag{7.3}$$

where, f_s is the sampling frequency and M_s is the maximum number of samples. There are two ways by which sampling resolution can be improved, either by decreasing f_s or by increasing M_s. But, f_s should be more than the Nyquist frequency by at least two times to avoid the problem of aliasing. On the other hand, due to increased variance of power spectrum signals, the frequency responses contain more noise when M_s is increased. Due to the above reasons there is a lower limit of sampling resolution, because of which meaningful information may be missed in the low frequency range of the frequency responses [14].

7.3.2 Basic Features of SFRA Response

In SFRA measurement at every sweep frequency the input impedance can be computed with the help of the input voltage and current. Typical amplitude and phase plots of the input impedance as measured for a continuous disk type winding is shown in Fig. 7.2 [3]. Figure 7.2 shows clear peaks and valleys in the plot, which can be explained with the help of the simplified *n*-stage lumped element ladder network of a transformer winding, as shown in Fig. 7.3 [15]. As may be seen from Fig. 7.3, the leakage inductance and series capacitance of the winding forms a parallel L-C combination, which produces parallel resonance at a certain frequency in the SFRA response. Such parallel resonance blocks the signal at that frequency and manifests itself as a local minimum or the so-called 'valley', which has a large negative decibel value. On the other hand, when the shunt capacitance is connected in between two leakage inductances, it forms a series L-C combination, which produces series resonance at a certain frequency. Such series resonance magnifies the signal at that frequency and appears as the local maximum or the so-called 'peak', which has a small negative decibel value. Conductive and dielectric losses, which are frequency dependent, attenuate the sharpness of the peaks and valleys, as shown in Fig. 7.4 [15].

However, the equivalent circuit of a two-winding transformer involve multiple inductive and capacitive elements as well as mutual couplings between the winding elements, which produces multiple series and parallel resonances in the SFRA response for a wide range of frequencies, as shown in Fig. 7.2. Figure 7.5 shows the equivalent circuit of a two-winding transformer with secondary short-circuited or open-circuited as the case may be [16].

With reference to the equivalent circuit shown in Fig. 7.5, the amplitude of driving point impedance becomes maximum and the phase becomes zero at certain frequencies. These frequencies are known as open-circuit natural frequencies (*ocnf*). At some other frequencies, the amplitude of driving point impedance becomes minimum and the phase again becomes zero, which are known as short circuit natural frequencies (*scnf*). The *ocnf*'s and *scnf*'s can be determined by proper inspection of amplitude and phase plots as shown in Fig. 7.2 [3], as these are related to the poles and zeros of the transfer function.

Meaningful assessment of mechanical integrity of transformer winding is possible by analyzing the transformer equivalent circuit as shown in Fig. 7.5, i.e., the poles and zeros of the transfer function, taking into consideration all the

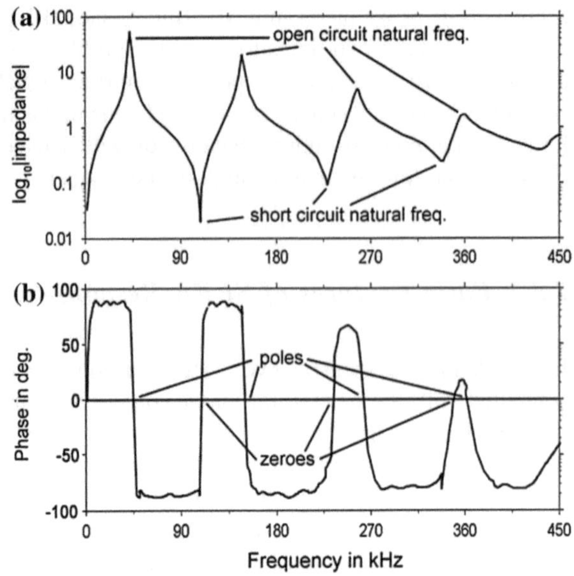

Fig. 7.2 Input impedance of a transformer winding. **a** amplitude and **b** phase

windings and the mutual interactions. It is because of the fact that dynamics of any electrical network is fully known if all the poles and zeros are known. However, it is to be noted here that the lumped parameter model of transformer winding is prepared using quasi-static approximations. Hence, such models are not valid above 2 MHz [16].

As a transformer winding is equivalent to a network comprising multiple distributed parameters, the frequencies of series and parallel resonances depend upon the leakage series inductance (L_s), series capacitance (C_s) and shunt capacitance (C_g). This gives rise to two extreme cases of FRA response: (i) when $C_s = 0$ and (ii) when C_s is very large such that in comparison C_g could be taken as zero. Figure 7.6a shows the FRA response for negligible series capacitance in which there are series of peaks and valleys with steady amplitudes, which in literature has been termed as 'U'-shaped FRA feature. Figure 7.6b shows the FRA response for high C_s in which there are very few resonances, and which shows a clearly increasing capacitive trend [15].

The poles and zeros of the transfer function and the winding structure are related to each other as the natural frequencies are determined by the electrical parameters of the equivalent circuit. As a result all the factors that influence the equivalent circuit parameters, e.g., type and number of windings, spatial position of windings, terminal connections and interaction between neighboring windings, affect the natural frequencies, i.e., the poles of the transfer function. Hence, sensitivity of detection of mechanical faults will be dependent on the number of poles that are known. The higher the number of known poles, the better the sensitivity. However, the zeros also significantly influence the shape of the transfer function [17]. For example, if a pole lies very close to a zero, then pole-zero cancellation

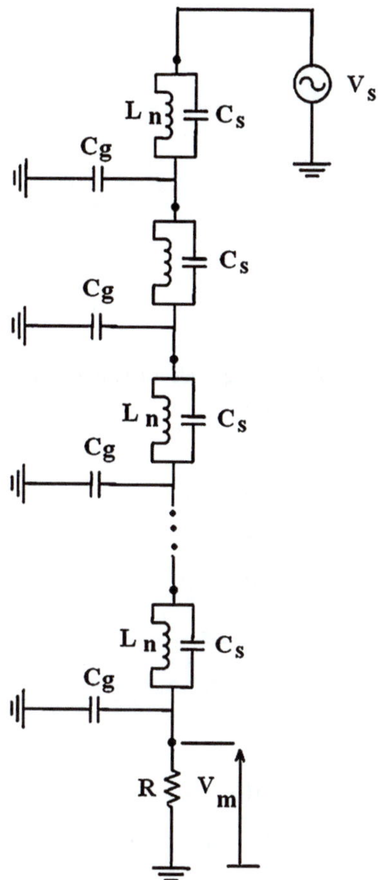

Fig. 7.3 Simplified *n*-stage lumped element ladder network of a transformer winding

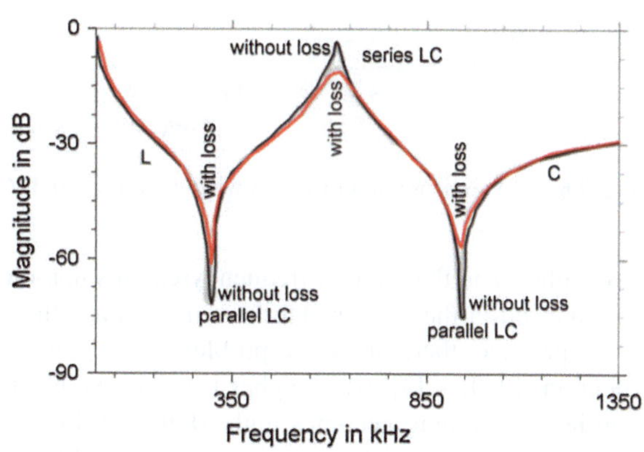

Fig. 7.4 Peaks and valleys in FRA response

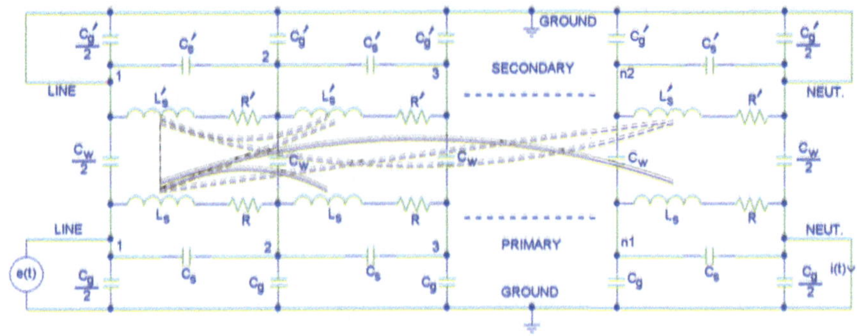

Fig. 7.5 The equivalent circuit of a two-winding transformer with secondary short-circuited or open-circuited

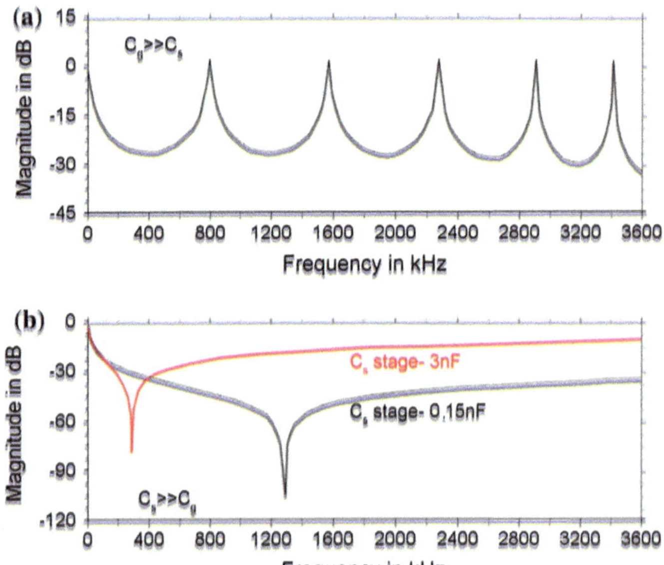

Fig. 7.6 FRA response of an isolated winding. **a** $C_s = 0$, **b** $C_g = 0$

takes place and that natural frequencyremains non-observable. Such effect is more pronounced in the case of interleaved windings than the disk windings.

In practice, there are some problems in identifying the natural frequencies as mentioned below [3]: (i) at higher frequencies the skin effect becomes prominent and hence damping becomes high. Hence, in FRA plots *ocnf*s and *scnf*s are less pronounced at higher frequencies, (ii) in real-life measurements a few peak-like kinks may appear in the FRA plots due to noise and limited sensitivity of the measuring instruments, (iii) sometimes peak/valley appear in the amplitude plot which is not accompanied by a corresponding phase zero, as shown in Fig. 7.2. Such peak/valley should not be considered as *ocnf*/*scnf* and (iv) The current drawn from the source becomes nearly zero at frequencies corresponding to the peaks, i.e., *ocnf*s, at which the driving point impedance becomes maximum. At those frequencies zero crossing of phase angle becomes difficult to identify accurately, which is very important for differentiating the actual *ocnf*s from stray kinks.

7.3.3 FRA Measurement Connection Methods

There are three connection methods which are commonly used in practice for FRA measurements [18]:

(a) End-to-End voltage ratio measurement (FRA_{EE})
(b) Input impedance/admittance measurement (FRA_{IM}/FRA_{AD})
(c) Transfer voltage ratio measurement (FRA_{TR}).

For a conventional two-winding transformer, the above-mentioned connection methods are shown in Fig. 7.7 [18], where R_C is the characteristic impedance of the cables used in measurement. The value of R_C is usually 50 Ω. For measuring the current, a current shunt r is used, where typically $r = R_C$. While performing FRA tests for three-phase transformers, all the terminals including the neutral, which are not being tested, are usually kept floating, unless there is specific requirement of neural grounding.

The transfer functions in the above-mentioned three connection methods are given by

$$TF_{EE} = \left|\frac{V_2(f)}{V_1(f)}\right| \tag{7.4}$$

$$TF_{IM} = \left|\frac{V_1(f)}{I_1(f)}\right|, \quad TF_{AD} = \left|\frac{I_1(f)}{V_1(f)}\right| \tag{7.5}$$

$$TF_{TR} = \left|\frac{V_2(f)}{V_1(f)}\right|. \tag{7.6}$$

Figure 7.8 shows the FRA results in the form of the plot of transfer function against frequency for all the connections shown in Fig. 7.7 [18]. Studies have revealed that the sensitivity of detection of resonance peaks is maximum over the entire frequency range for End-to-End connection and is minimum in higher frequency range for Input Impedance/Admittance measurement connection. Figure 7.8 also shows that the frequencies for resonance peaks in FRA_{IM}/FRA_{AD} and FRA_{TR} are same as those for FRA_{EE}, when the neutral is grounded. But, when the neutral is left floating as is the common practice, then the resonance frequencies are different in FRA_{IM}/FRA_{AD} and FRA_{TR} from those in FRA_{EE}.

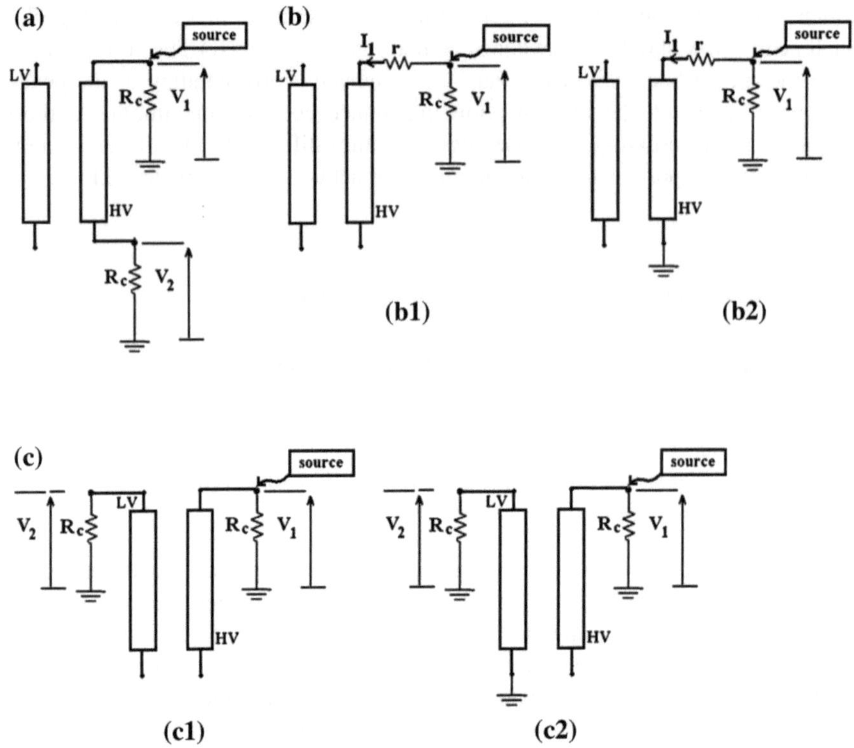

Fig. 7.7 FRA measurement connection methods

7.3.4 Transfer Function Features with Reference to Windings

It has been identified two decades earlier that a transformer has major resonances governed by winding structure [11]. It was also reported that a transformer having single-layer winding without any tap-changer has lesser resonances compared to a transformer having disk winding or a transformer having large number of taps. Depending on the size of the transformer, the major resonances are in the range of a few kHz to a few hundreds of kHz. Smaller transformers generally have higher resonance frequencies than larger transformers [11].

In a transformer depending upon the type and physical arrangement of the windings, series, and shunt parameters of the equivalent circuit shown in Fig. 7.5 gets altered. Moreover, due to the presence of two windings per phase in a real-life transformer, mutual inductive and capacitive couplings between the sections of the same winding as well as that of different windings significantly change the transfer function from transformer to transformer.

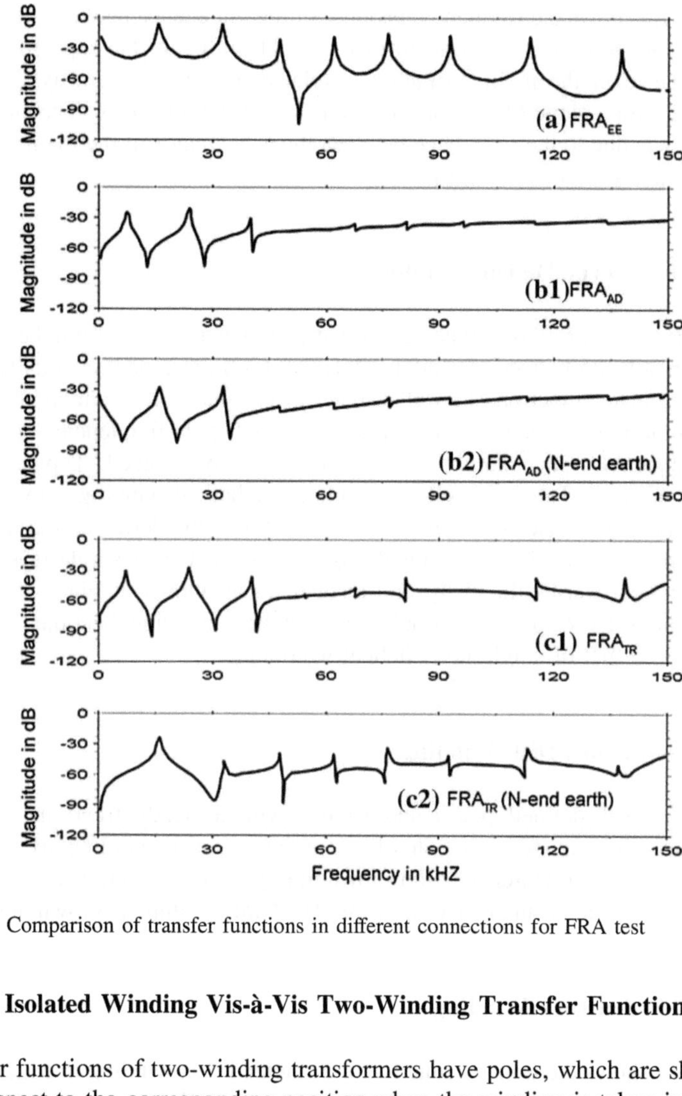

Fig. 7.8 Comparison of transfer functions in different connections for FRA test

7.3.4.1 Isolated Winding Vis-à-Vis Two-Winding Transfer Function

Transfer functions of two-winding transformers have poles, which are shifted left with respect to the corresponding position when the winding is taken in isolation [19]. The shift of the secondary poles has been found to be greater than the shift of the primary poles, particularly in high frequency range. It has also been reported [19] that an important feature, which is useful for identification of poles and the corresponding winding to which such poles belong, is the so-called 'distance rule' In this 'distance rule' the distance between any two adjacent poles in frequency terms is computed. Then the ratio of such distances in the case of isolated winding and two-winding transformers could be studied. Existing literature [19] reports that such ratios remain more or less the same for two winding transformers compared to isolated winding.

Figure 7.9 shows the frequency response of the impedance between the primary and secondary terminals of a transformer [20]. The effect of the capacitive part of the impedance in the main insulation is related to the continuously decreasing magnitude with a slope of almost minus one in logarithmic scale. The subsequent deep valley is due to the resonance between the capacitance of the main insulation and the winding inductance [20].

7.3.4.2 Layer-Type Helical Winding

Transfer function of a layer-type helical winding generally has a number of poles which typically lie between 0 and 1.0 MHz and are more or less equally spaced [17]. The transfer function has well-defined peaks and valleys which are directly related to the poles. Transfer function amplitudes at pole frequencies vary and at higher frequencies such amplitudes get progressively reduced. Typical transfer functions of single-layer as well as double-layer helical windings are shown in Fig. 7.10 [15]. Because of the inter-coil capacitance the series capacitance of a double-layer helical winding is much higher than that of a single-layer helical winding, which causes the sharp anti-resonances as shown in Fig. 7.10. Consequently, the FRA response of double-layer helical winding becomes distinctly different from that of single-layer helical winding.

7.3.4.3 Continuous Disk Winding

A series of well-defined peaks and valleys with a steady trend in amplitude characterize the transfer function of a continuous disk winding as shown in Fig. 7.11 [15]. The transfer function shown in Fig. 7.11 is for a transformer having low series capacitance and hence show the 'U' shaped feature as shown in Fig. 7.6.

7.3.4.4 Shielded Disk Winding

Figure 7.12 shows the transfer function of a shielded disk winding in which half of the winding was shielded while rest half was like normal continuous disk winding [15]. The transfer function initially has a series of 'U'-shaped resonances and anti-resonances followed by a so-called 'camel-bump' with damped resonances and anti-resonances. The damped responses are caused by the increased series capacitance of the shielded part of the winding. In the high frequency range after the 'camel-bump' the transfer function looks like an inverted 'U' [15].

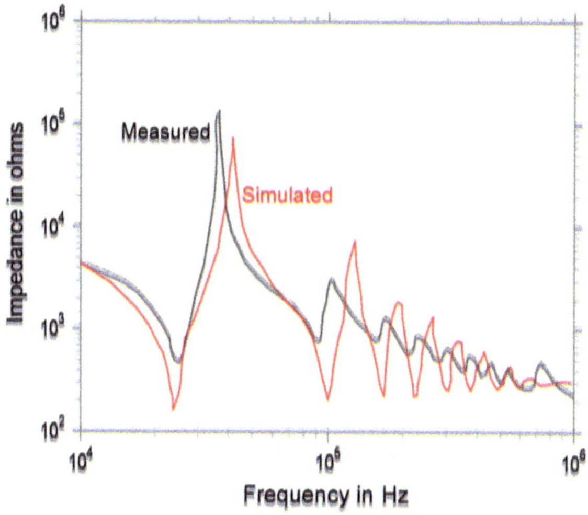

Fig. 7.9 FRA response of impedance between the primary and secondary terminals

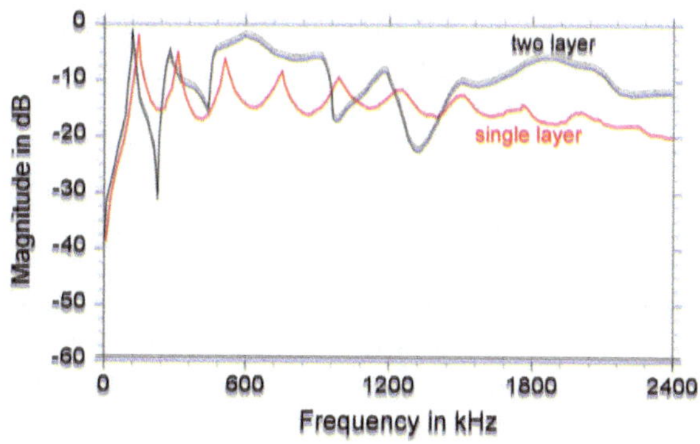

Fig. 7.10 FRA response of single-layer and two-layer helical windings

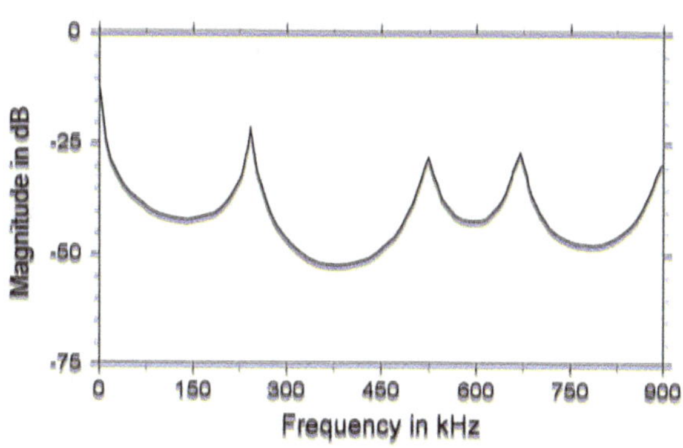

Fig. 7.11 FRA response of continuous disk winding

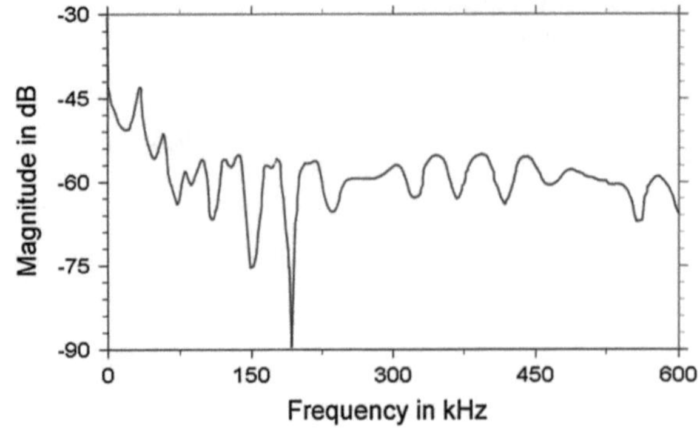

Fig. 7.12 FRA response of shielded disk winding

7.3.4.5 Interleaved Disk Winding

Due to the high series capacitance of interleaved winding, the transfer function of an isolated fully interleaved winding shows very few resonances, most of which occur well within 300 kHz, as shown in Fig. 7.13 [15]. The amplitude at higher frequency resonance peaks gradually decreases and after the last peak, the transfer function becomes almost a straight line.

7.3.4.6 Vertically Interleaved Tap Winding

Figure 7.14 shows the FRA response of the 132 kV winding of a 275/132 kV power transformer in which the main winding was shielded disk type and the tap winding was vertically interleaved type. Both the main and the tap windings were electrically connected and concentrically arranged [15]. From Fig. 7.14 it may be seen that the transfer function of the 132 kV winding along with the tap winding has the features of alternating resonances and anti-resonances up to 300 kHz similar to that for a shielded disk winding as shown in Fig. 7.12. But above

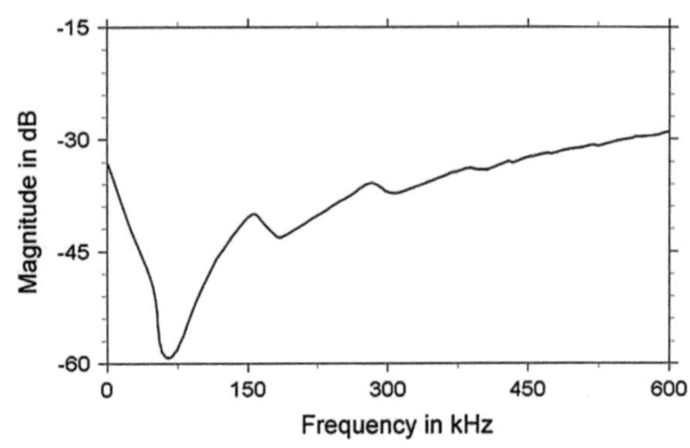

Fig. 7.13 FRA response of interleaved disk winding

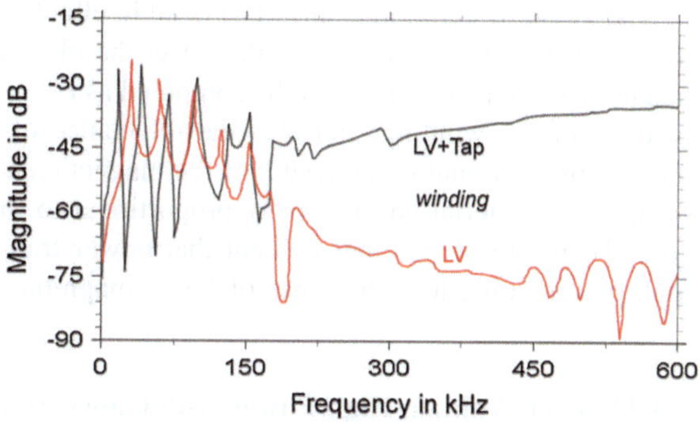

Fig. 7.14 FRA response of shielded disk type main LV winding with vertically interleaved tap winding

500 kHz the high series capacitance of the vertically interleaved tap winding smoothes the response into a capacitive trend of increasing magnitude similar to that for an interleaved disk winding as shown in Fig. 7.13 [15].

7.3.4.7 Windings Placed on Outer and Middle Limbs

Figure 7.15 shows the open-circuit impedance of a three-phase transformer as seen from the LV side [20]. The transformer had a continuous disk type HV winding and a layer-type helical LV winding. The FRA response shows major resonances at low frequencies close to 1 kHz. There are two peaks in the response for the outer limb whereas there is only one peak in the response for the middle limb. The reason for such difference is that the outer limb has two magnetic paths of different reluctances through the middle and the other outer limbs, but the middle limb has identical magnetic paths of same reluctance through the two outer limbs on both the sides.

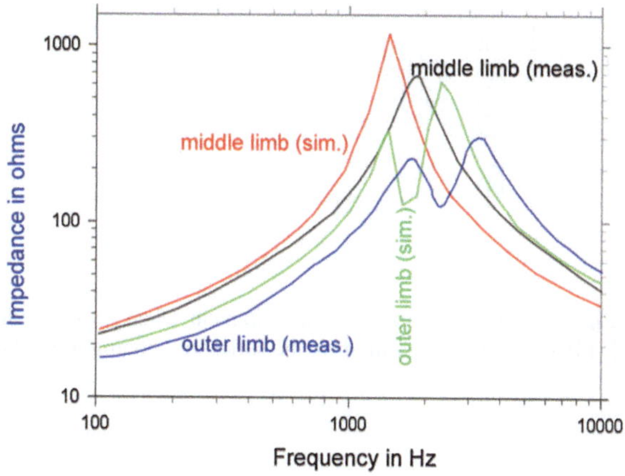

Fig. 7.15 FRA response of LV winding placed on outer and middle limbs

The value of core inductance has notable effect on the FRA response, because the larger the core inductance is, the lower the FRA magnitude becomes [15]. The magnetizing inductance L_m can be computed as $L_m = N^2/R$, where N is the number of turns of the winding, and R is the reluctance of the magnetic path, which is directly proportional to the length of the magnetic path and magnetic permeability of the core material and inversely proportional to the cross-sectional area of the core. Therefore, it becomes evident that power transformers of high voltage and high power rating have response of lower magnitude at lower frequencies.

7.3.4.8 High Voltage High Power Autotransformers

Presently, high voltage high power transformers used in power transmission network are mostly autotransformers constructed with windings designed to have high series capacitance. In such transformers the resonant frequencies are limited in number and must be investigated carefully for detection of mechanical faults. Experiences from performing FRA tests in UK transmission system reveal that the high voltage high power autotransformers display a feature of two distinct resonances with an anti-resonance situated in between the resonances within 2 and 20 kHz in the end-to-end voltage ratio measurement taken on series and common windings. This feature is now known as 'double-peak' feature for autotransformer FRA responses [21]. Figure 7.16a shows the FRA measurement set-up for an autotransformer whose series winding is interleaved disk type and the common winding is shielded disk type. Both windings of the autotransformer have high or moderately high series capacitance. Corresponding FRA responses are shown in Fig. 7.16b [21].

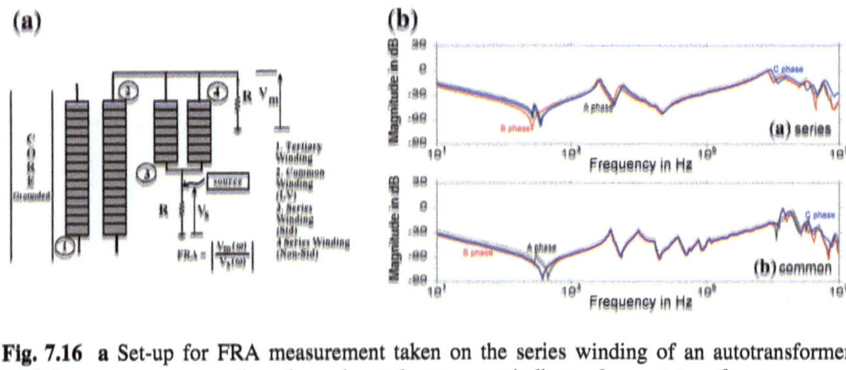

Fig. 7.16 **a** Set-up for FRA measurement taken on the series winding of an autotransformer. **b** FRA responses measured on the series and common windings of an autotransformer

7.3.5 Assessment of Mechanical Integrity of Transformer Winding

Deviations of inductances and capacitances, which are in turn dependent upon the physical dimensions and materials used in a transformer, cause changes in resonant frequencies and/or magnitudes. Consequently, mechanical faults such as winding displacements and deformations could be diagnosed by proper analysis of changes in the resonances. For a fault to be detectable, it must cause significant change in either the inductance and/or capacitance of a significant part of the winding. If a fault does not cause such change, e.g., partial discharge, then that fault will not be detectable by FRA.

7.3.5.1 Sub-Band Division of Frequency Response

There are many effects that change the low frequency response, the significant ones of which are as follows: (i) Short-circuited turns that change the magnetization characteristics of a transformer, (ii) Circulating current loops that are capable of redirecting the leakage flux into the core, and (iii) ungrounded core that changes the shunt capacitance of the winding closest to the core [8].

The medium frequency response is affected by those faults which cause a change in the parameters of the entire winding. Axial movement of transformer winding is indicated by a significant increase in medium frequency response. On the other hand, a significant decrease in medium frequency response usually indicates radial movement of the inner winding, often known as 'hoop buckling' [8]. Slight differences are caused by what is known as 'winding settling' FRA response of a transformer winding having 'winding settling' after short circuit test is shown in Fig. 7.17 [11].

Those faults which cause changes in the parameters of only part of a winding affect the high frequency response. Some such faults are mentioned hereafter:

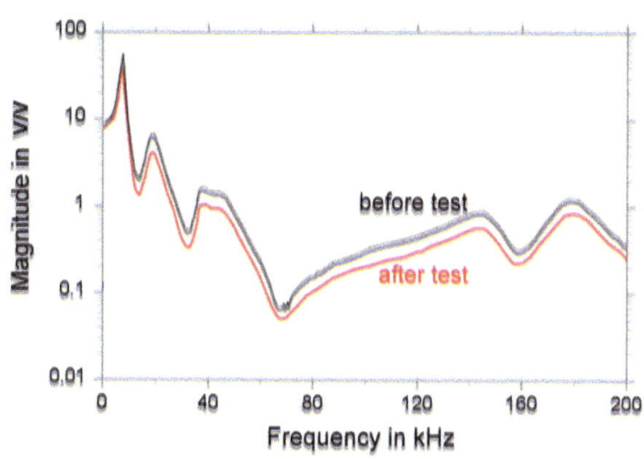

Fig. 7.17 Transfer function before and after short circuit testing showing mechanical settling

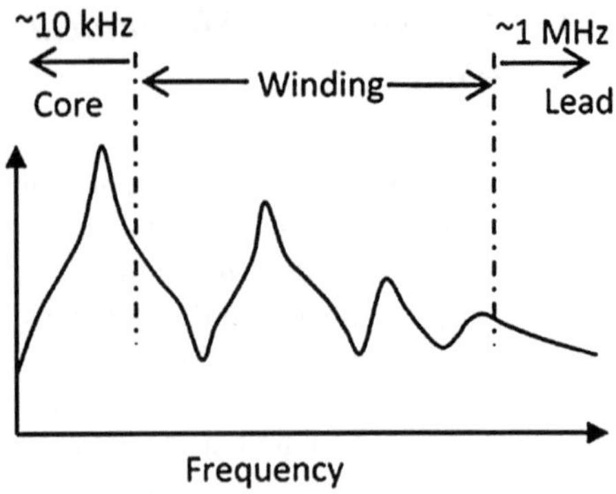

Fig. 7.18 Frequency band sub-division of frequency response [23]. Reproduced with kind permission © IET 2012

(i) Localized winding damage which causes random changes in the high frequency response often by creating new resonant frequencies, (ii) Poor tank or cable grounding, out of which poor tank grounding is easy to detect as it affects all the windings. Poor cable grounding is difficult to detect as it does not usually create a new resonant frequency [8].

Below 10 kHz, transformer winding response is dominated by inductive components. Combination of inductances and capacitances cause multiple resonances over the frequency range 10 kHz to 1 MHz. In the frequency range 2–10 MHz, distributed capacitances of the winding shunt the winding inductances [22]. Figure 7.18 shows such sub-band frequency division of frequency response [23].

Based on experimental case studies [24, 25], another division of frequency range of FRA response has been proposed, in which the entire frequency range has been sub divided into four sub bands each of which is specifically sensitive to different winding faults as given in Table 7.1 [22].

7.3.5.2 Major Types of Winding Deformation

Three types of deformations are very common in transformers, viz. axial displacement which mostly occur in HV windings, radial or hoop buckling which mostly occur in LV winding and axial bending which mostly occur in HV winding. FRA features of these three types of winding deformations are reported in details in [18].

Table 7.1 Sub bands of Frequency range of FRA response

Sub bands	Sensitivity of failure
Less than 2 kHz	Core deformation, open circuits, shorted turns and residual magnetization
>2 kHz, <20 kHz	Movement of bulk winding (between clamping and windings)
>20 kHz, <400 kHz	Main or tap windings deformation
>400 kHz up to 1 MHz	Main and tap windings movement, variations of ground impedance

Axial Displacement of HV Winding

HV winding often shifts axially with respect to the LV winding, as shown schematically in Fig. 7.19a, under the influence of the axial electromagnetic forces [18]. The electrical parameters affected most by this type of deformation are inter-winding capacitance and mutual inductance between the windings. Figure 7.19b [18] shows typical FRA results for axial displacement of HV winding obtained by means of all the connection methods as shown in Fig. 7.7. It may be noted here that this type of deformation could also be detected by measurement of electrical parameters such as capacitance and leakage inductance.

Hoop Buckling of LV Winding

Hoop buckling typically occurs locally on the inner winding, i.e., the LV winding, when the conductors bend between the axial supports under the influence of compressive stresses due to the radial electromagnetic forces, as shown schematically in Fig. 7.20a [18]. The electrical parameters affected most by this type of

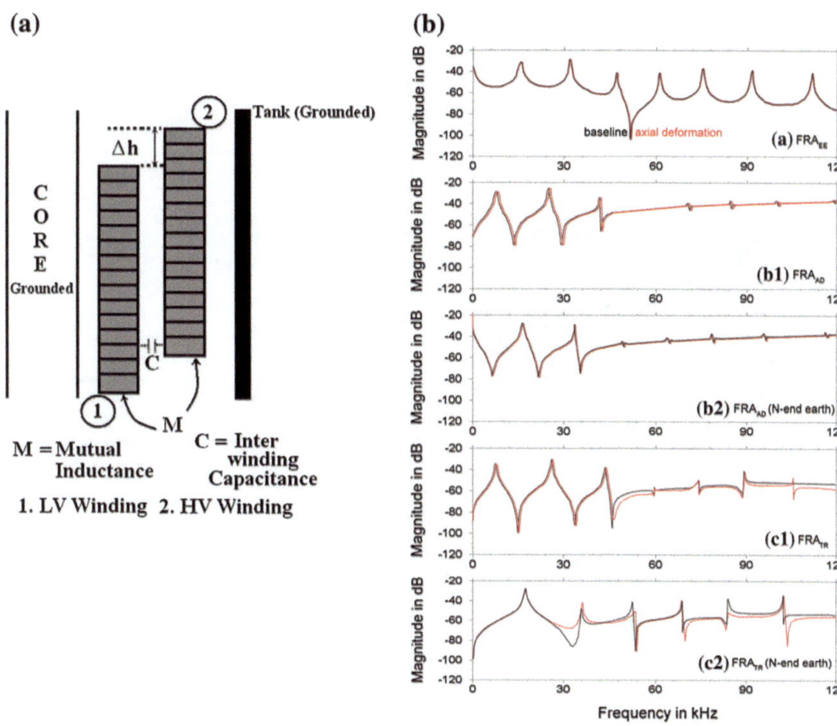

Fig. 7.19 **a** Axial displacement between HV and LV windings. **b** FRA results for axial displacement of HV winding

deformation are also inter-winding capacitance and the mutual inductance between the windings, which are reduced due to hoop buckling. Figure 7.20b [18] shows typical FRA results for hoop buckling of LV winding obtained by means of all the connection methods as shown in Fig. 7.7. Radial deformation could also be detected by capacitance and leakage inductance measurements.

Axial Bending of Conductors of HV Winding

Typical deformation of transformer winding conductors, mostly for HV winding, as shown schematically in Fig. 7.21a, occurs due to axial bending of conductors of HV winding between radial spacers [18]. Such deformation is also caused by axial electromagnetic forces. The electrical parameters affected most by this type of deformation are inter-disk capacitance and mutual inductance between disks. Figure 7.21b [18] shows typical FRA results for axial bending of HV winding conductors obtained by means of all the connection methods as shown in Fig. 7.7. It may be noted here that this type of deformation is very difficult to detect by measurement of conventional electrical parameters such as capacitance and leakage inductance.

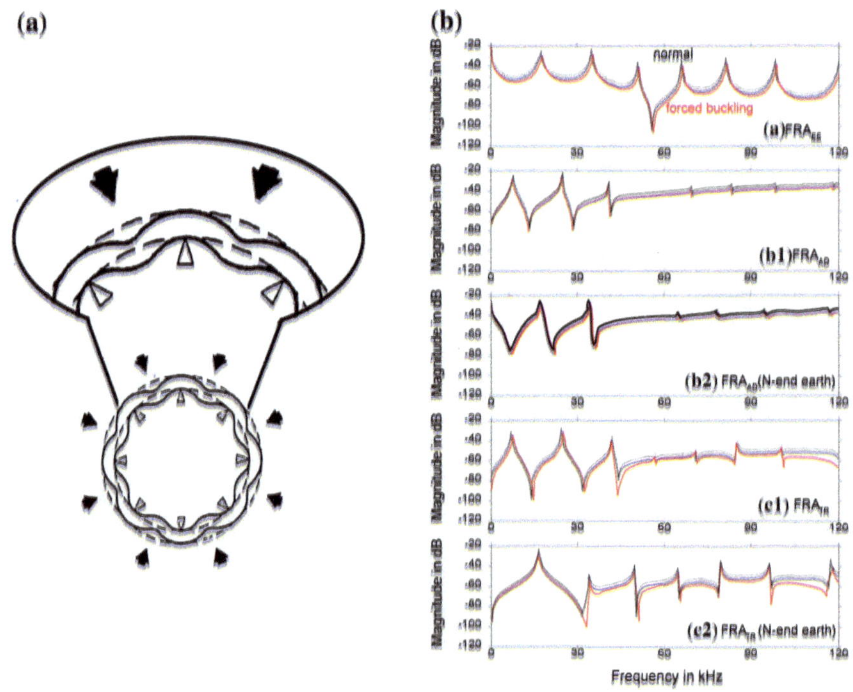

Fig. 7.20 **a** Hoop buckling of LV winding. **b**. FRA results for Hoop buckling of LV winding

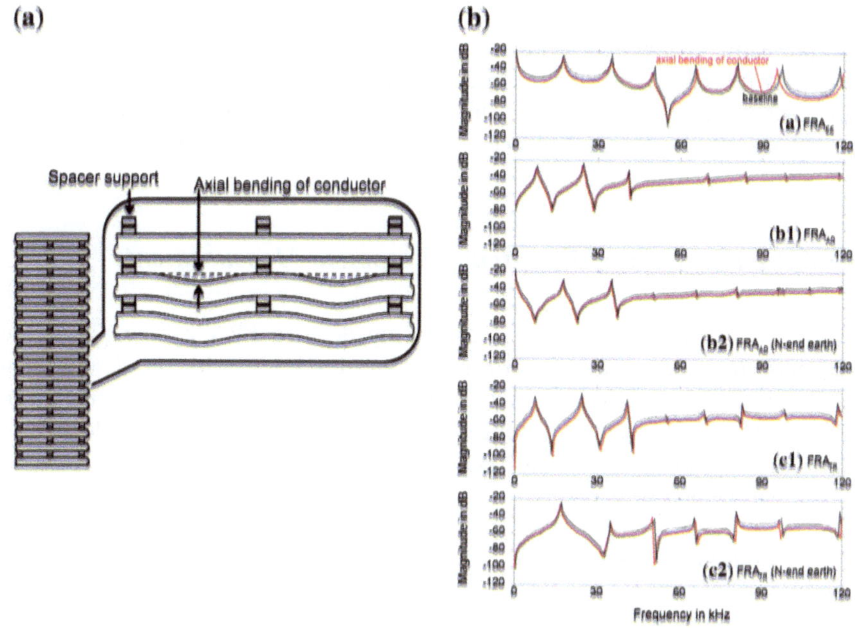

Fig. 7.21 **a** Axial bending of conductors of HV winding. **b** FRA results for axial bending of conductors of HV winding

Sensitivity of FRA Connection Methods in Detecting Winding Deformation

A detailed study has been reported in [18] on the sensitivity of different FRA connection methods as shown in Fig. 7.7 in detecting major types of mechanical deformation of transformer windings, as discussed in the previous sub-sections. Researchers have found the following [18]: (i) for detection of axial displacement *transfer voltage ratio method* (FRA_{TR}) is most sensitive, (ii) for detection of radial buckling *transfer voltage ratio method* (FRA_{TR}) is most sensitive, and (iii) for detection of axial bending of conductors *end-to-end voltage ratio method* (FRA_{EE}) is most sensitive. It has also been reported in [18] that *input impedance/admittance method* (FRA_{IM}/FRA_{AD}) is least sensitive for detection of winding deformation. It has, therefore, been recommended that both *end-to-end voltage ratio* (FRA_{EE}) and *transfer voltage ratio* (FRA_{TR}) measurements be carried out in order to detect all the major types of mechanical deformation of transformer winding.

Detection of Mechanical Faults by Conventional Measurements

There are conventional measurements by which several faults that cause changes in the low frequency response can be detected, e.g., measurement of magnetizing current or turns ratio to detect short-circuited turns, dissolved gas analysis or thermal scan to detect circulating currents and capacitance measurement to detect no core earth. However, the faults that cause changes in the medium and high frequency responses are difficult to detect using conventional techniques like turns ratio, capacitance, and leakage inductance measurements. Measurement of capacitance and leakage inductance can be effective in some cases but not all [8].

7.3.6 Comparison of Frequency Response Measurements

Diagnosis of mechanical faults is commonly done in FRA by comparing the frequency response of a transformer winding under test with the results of a reference measurement over a wide range of frequencies. The main comparison procedures that are used in practice for assessment of mechanical integrity of transformer windings are detailed below.

7.3.6.1 Reference Comparison

Reference Comparison (RC), which is also known as *time-based comparison*, is carried out by comparing the present FRA response of the winding under test with the reference response of the same winding taken earlier. It is considered to be the most accurate method of comparison of FRA responses. In the case of real-life transformers, the two measurements should be taken on the same tap-position and with the same accessories like bushings, etc. For oil-filled transformers, the relative permittivity of the oil should preferably be the same [8].

Reference comparison made on a test transformer over a gap of 12 months is shown in Fig. 7.22 [11]. Since the responses are in good agreement, hence it may be concluded that the transformer has not developed any mechanical fault in the span of 12 months from the date of baseline measurement.

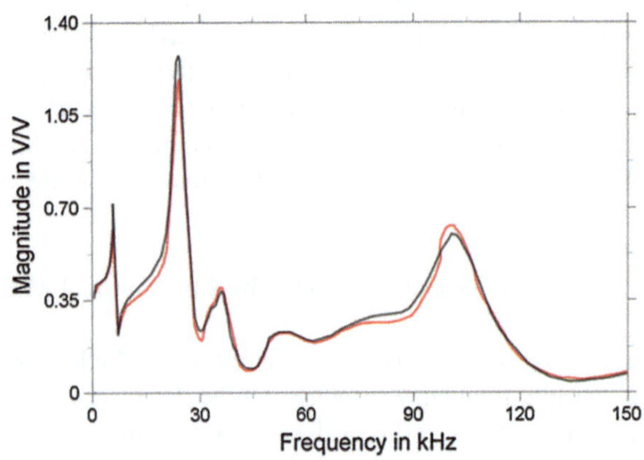

Fig. 7.22 Reference comparison after 12 months

RC is often carried out to detect mechanical faults that may develop in a transformer winding due to short circuit testing. Figure 7.23 shows the RC of a power transformer before and after a short circuit test. Since there is no significant deviation in the FRA responses before and after the testing, the transformer may be considered to have withstood the mechanical stresses developed in the short circuit test [8].

7.3.6.2 Inter-Phase Comparison

Inter-Phase Comparison (IPC), which is also known as *construction-based comparison*, is possible for three-phase transformers. In IPC, the comparison of FRA response of a particular winding of a transformer is made with the FRA response of the winding of another phase of the same transformer. Commonly, such comparison always shows some deviations in the low frequency range due to differences in the flux paths and hence magnetizing inductances of the different phases of a three-phase transformer. In the medium and high frequency ranges the FRA responses of two different phases of the same transformer generally show good agreement. But in some cases, depending upon the differences in winding external clearances or lead configurations, the agreement is not as good as expected [8].

7.3.6.3 Sister-Unit Comparison

FRA responses taken from transformers of the same batch from the same manufacturer are compared in Sister-Unit Comparison (SUC), which is also known as *type-based comparison*. It is very useful for three-phase transformer banks comprising three single-phase transformers of same design. While differences appear in the low frequency range in SUC, the responses agree reasonably well at higher frequencies. Major changes in resonance frequencies in the FRA response of the

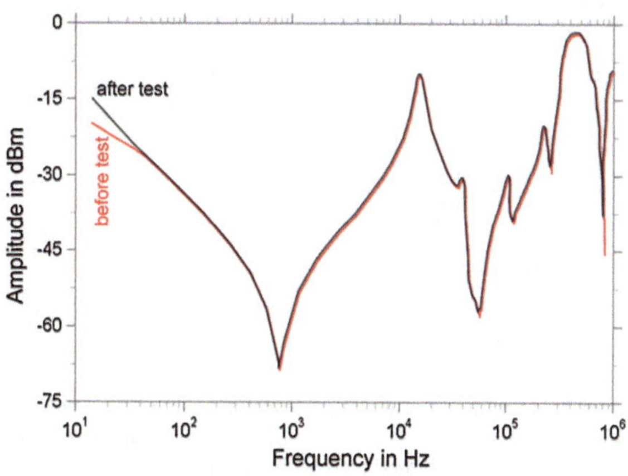

Fig. 7.23 Reference comparison before and after short circuit testing

same phase windings of the same type of transformers indicate mechanical faults in the winding under test, if the winding of the sister transformer unit is known to be in normal condition [22].

7.3.6.4 Significance of Multiple Comparisons

For RC in the case of oil-filled transformers, if it is known that oil has been changed since the reference measurement was taken or there has been a significant change in the moisture content of the oil, then adequate precautions are necessary while making the comparison. It is advisable in such cases to make IPC to supplement the RC [8].

For IPC the FRA responses of the three phases of a power transformer are shown in Fig. 7.24a, which depicts that the responses are in good agreement up to 600–700 kHz [22]. Above 700 kHz, phase C response shows some deviations at a number of resonance peaks with respect to A and B phases. The concern about the mechanical integrity of phase C winding was eliminated by making SUC carried out with respect to the FRA response of the phase C winding of a sister transformer of the same type, which is known to be in good condition. The graphical representation of the SUC is shown in Fig. 7.24b [22], which shows good resemblance of the sister-unit responses. Small variations in Fig. 7.24b are possibly related to design features of the transformers.

7.3.6.5 Comparison Parameters

Comparison of FRA responses is done to study significant differences, which are considered to be: (i) changes in the shape of the response curve, (ii) creation of new resonant frequencies, (iii) elimination of existing resonant frequencies, and

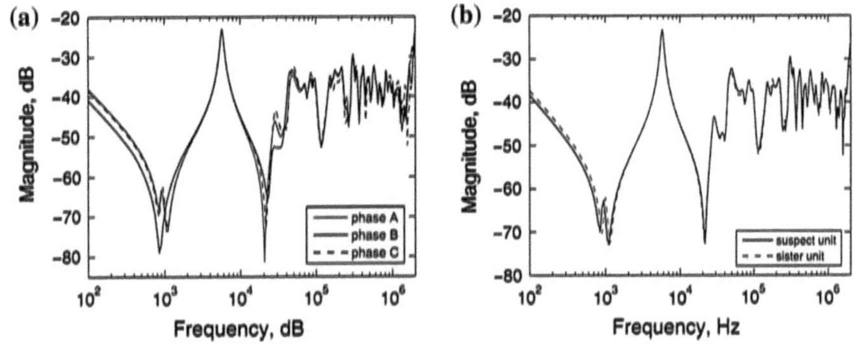

Fig. 7.24 **a** Inter-phase comparison for a three-phase power transformer [22]. Reproduced with kind permission ©IET 2012. **b** Sister-unit comparison for phase C of a three-phase power transformer [22]. Reproduced with kind permission ©IET 2012

(iv) major shifts in the existing resonant frequencies [8]. When there are sufficient resonant frequencies in the FRA response, then useful information can be extracted out of such comparison. On the other hand, too many resonant frequencies cause confusion in decision making.

7.3.7 Key Factors Affecting Frequency Response Measurements

There are several key factors that affect FRA measurements, e.g., current measuring shunt impedance, lead arrangement, temperature, moisture content, and aging of insulation. It is also very important that the layout of FRA measurement should remain the same for the set of measurements which are being compared for fault diagnosis.

7.3.7.1 Effect of Shunt Impedance

A current measuring shunt impedance, mostly resistive in nature, is used in FRA to measure the response current. Typically the value of shunt impedance is taken as 50 Ω, which is equal to the characteristics impedance of the measuring cables. At frequencies less than 1 MHz, this impedance of the current shunt is not significant compared to the impedance of the transformer winding. But at higher frequencies, in which range the small winding movements usually show up, the shunt impedance becomes comparable to transformer winding impedance. It is also to be noted that in the case of real-life FRA measurements the transfer function includes transformer, shunt as well as lead impedances. Higher values of shunt impedance dampen the resonant peaks at higher frequencies, which could be significant in identification of small winding movements. On the other hand, a lower value of shunt impedance increases the quality factor of the whole network and a higher quality factor in turn will increase the sensitivity of winding movement detection. Figure 7.25 shows that the higher the shunt impedance is, the lower is the sensitivity of FRA measurement [26].

7.3.7.2 Effect of the Length of Measuring Leads

The effects of the length of measuring leads on FRA measurements have been studied in details in [26]. Typical results are shown in Fig. 7.26, which shows that the differences between the responses for shorter and longer leads are very high at frequencies above 2 MHz. This is due to the fact that a shorter lead has much higher admittance, i.e., lower impedance, at higher frequencies than the longer leads. In fact at higher frequencies the lead admittance becomes the main

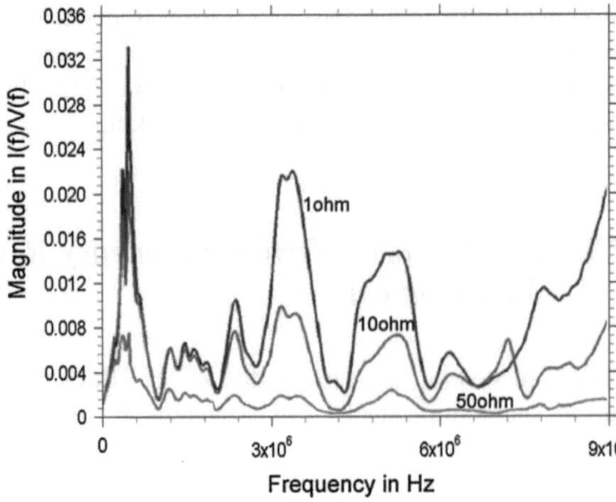

Fig. 7.25 Effect of shunt impedance on FRA measurement

component of the transfer admittance value. Consequently, the sensitivity of FRA measurement will be greatly reduced at higher frequencies, if the leads are longer in length. In other words, the lead length need to be as short as practicable for getting optimum sensitivity in FRA measurement for real-life measurements. Here it has to be remembered that for on-site measurement, depending upon the transformer and associated bushing size, there is a practical lower limit to the length of the measuring leads to be employed, which sets the sensitivity limit of FRA measurement, too. To get repeatability of FRA results the lead lengths must be kept same in all the measurements. The same is also true for making RC of FRA results.

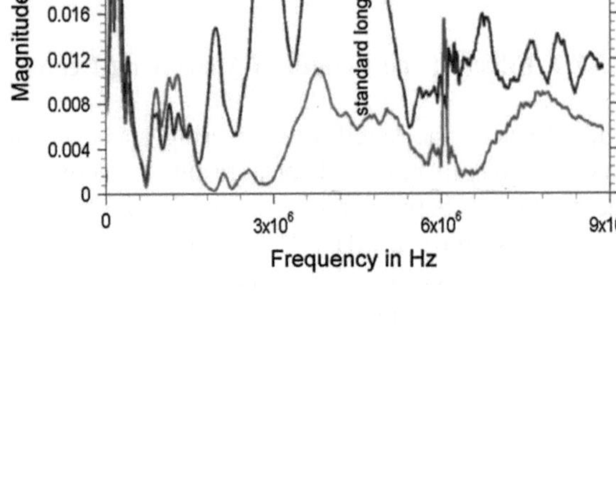

Fig. 7.26 Effect of the length of measuring leads on FRA measurement

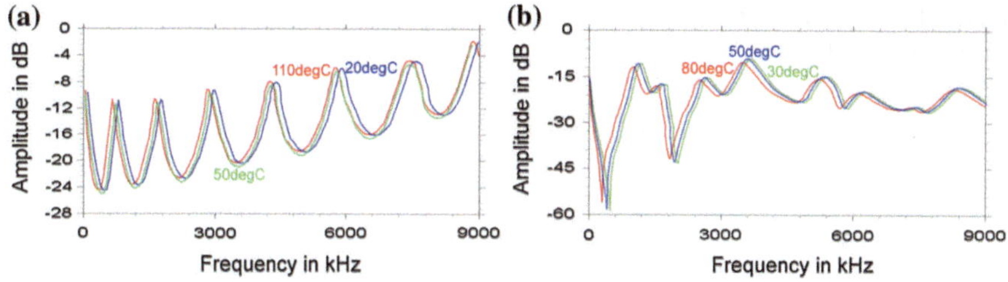

Fig. 7.27 Effect of temperature on FRA results

7.3.7.3 Effect of Temperature

FRA measurements are sensitive to temperature of transformer winding as it changes the impedance of the winding. In [27] studies have been conducted on the temperature dependence of FRA results for a disk type winding model and also a 50 kVA transformer having layer-type windings. In this study the FRA measurements were taken at three different temperatures, e.g., 20, 50, and 110 °C for the disk type model and 30, 50 and 80 °C for the 50 kVA transformer, after the sample under test was dried in a laboratory oven for about 24 h. Figure 7.27a shows that the FRA response shifts toward right as temperature is reduced from 110 to 20 °C. This right shift is due to the fact that a reduction in temperature causes a corresponding reduction in winding impedance. Figure 7.27b shows the effect of temperature on the FRA response for the 50 kVA transformer, which depicts similar trend of right shift of the resonant peaks with the reduction in temperature. Thus it is important to take note of the temperatures at which FRA measurements were taken while comparing FRA responses.

7.3.7.4 Effect of Moisture Content in Insulation

Moisture content in transformer insulation changes the real part of dielectric permittivity as well as the losses. As a result it causes a shift in the resonant frequencies along with damping of the peaks. A change in the real part of dielectric permittivity affects the entire frequency range and hence a change in the moisture content in transformer insulation affects all the resonant frequencies. On the other hand, radial or axial deformation of transformer winding do not cause change in the resonant peaks that appear in the low frequency change. As discussed earlier, such resonant peaks at low frequency range are due to changes in the magnetic flux distribution related to the core. Thus the frequency range where the core resonance appears, i.e., around 1 kHz, is the best region for observing the effect of moisture content in insulation on FRA results [27]. Figure 7.28 shows typical effects of moisture content in insulation on FRA measurements [28].

Dielectric response measurement in frequency domain (FDS) is used widely in practice for the estimation of moisture content in solid insulation of transformers.

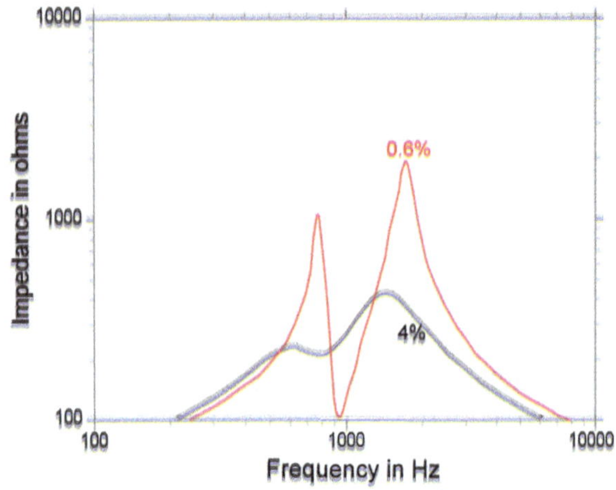

Fig. 7.28 Effect of moisture content in insulation on FRA results

But such measurements are carried out at very low frequencies, typically from 1 mHz to 1 kHz. Such diagnostic measurements at very low frequencies are time consuming. Shortening of time required for measurement could be done by shifting analyzed response zone to higher frequencies. This is possible by means of FRA by carrying out measurements in the range of 10 Hz to 10 MHZ.

Figure 7.29 shows the effect of moisture removal from the oil-paper insulation of a 15MVA transformer during different stages of drying of the transformer insulation [27]. Figure 7.29a shows that most of the moisture was removed from the insulation during the first day of drying, which is shown by a significant shift of the response toward higher frequencies after the first day of drying. The measurements after 6-days of drying was taken at an elevated temperature (95 °C) compared to that after the first day of drying (67 °C). It should have shifted the response toward left because of an increase in the impedance of the winding. But Fig. 7.29b shows that the response was shifted slightly toward right after 6 days of drying. It suggests that the effect of removal of moisture, which lowers the real part of dielectric permittivity, is more than that of impedance change due to temperature increase.

7.3.7.5 Effect of Insulation Aging

Aging of transformer solid insulation causes increased losses which increases imaginary part of complex permittivity. This in turn causes higher damping of the resonant peaks at frequencies below 100 kHz. Figure 7.30 shows the effect of aging on FRA response at a fixed temperature and constant moisture content of insulation [28]. Aging of solid insulation does not cause notable shift in resonant frequencies, as the influence of insulation aging on real part of dielectric permittivity is less [20].

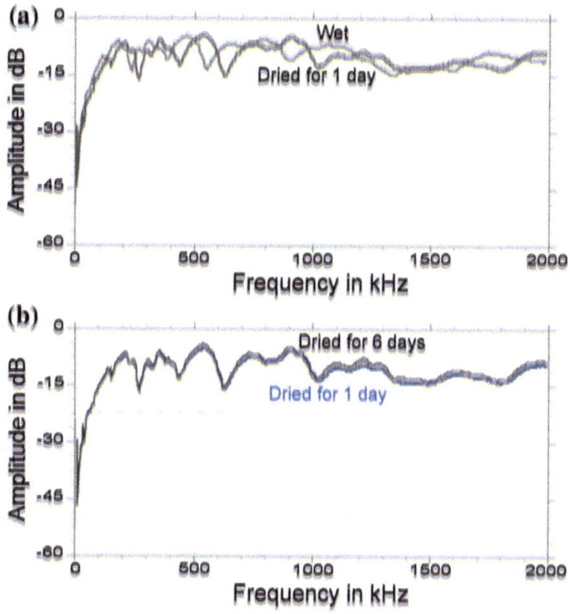

Fig. 7.29 Effect of moisture content in insulation on FRA results after different durations of drying

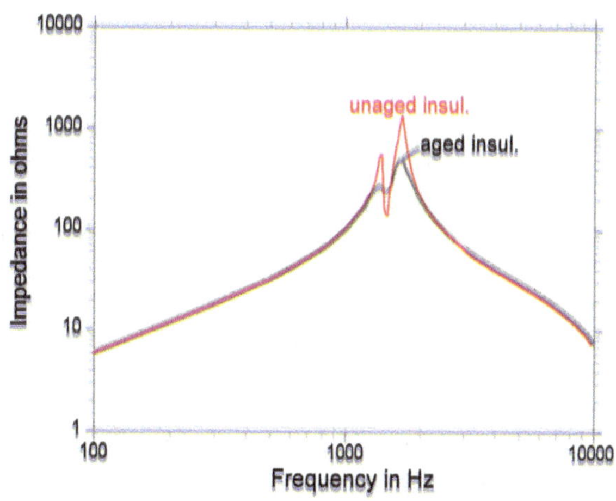

Fig. 7.30 Effect insulation aging on FRA response

7.3.7.6 FRA Measurements After Demagnetization of Transformer

A significant study on this aspect has been reported in [29]. In this study a 100 kVA transformer was supplied from the LV side from a three-phase supply to get the same core magnetization as under normal operating condition. Then the three-phase supply was switched off at different instants of time at different phases so that different residual magnetizations are left on the three limbs. Figure 7.31 shows the variation of impedance measured which started immediately after the

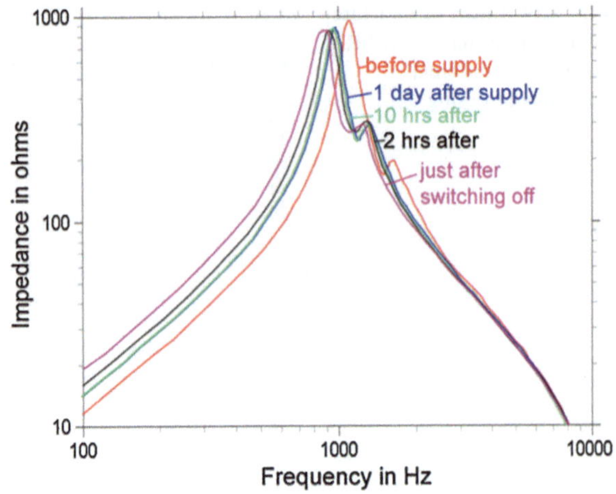

Fig. 7.31 Effect of core demagnetization after switching off three-phase supply to a transformer on its FRA response

switching off and continued for several days thereafter. The FRA response shown in Fig. 7.31 clearly depicts that the impedance increases immediately after the switching off and then it relaxes in the process of achieving the demagnetized state of minimum energy [29].

7.3.8 Online FRA Measurement

Detection of an inter-turn fault in a transformer at an early stage is of paramount importance to power utilities to prevent major damage to a transformer in service, which is an expensive equipment, and to avoid unwanted outage leading to considerable revenue loss. Such detection of minor faults at the earliest stage is possible through continuous online monitoring of the transformer. Frequency response analysis, which has evolved into a reliable tool for detecting winding faults, is an offline testing methodology and requires that the transformer be taken out of service for conducting the test.

Over the years researchers have given efforts to monitor transformers, while in service, by impulse response method. The main focus of these works [30–33] is to use the system transients, which are often present in the network, to detect mechanical displacement within transformers. In such attempts, the transfer functions are computed from the transients at the terminal of the transformers resulting from switching operations in the power system. The measured frequency response is, therefore, dependent on the wave-shape of the transients, which may vary from measurement to measurement. Consequently, a very significant drawback of this approach is poor reproducibility of results. Moreover, continuous monitoring of the transformer is also not possible by this methodology as the system transients are not present at all times.

Taking the limitations of online IFRA measurements into cognizance, an online monitoring system based on SFRA has been proposed in [34] for detecting inter-turn faults. In this system a signal of known frequency is injected into one end of the transformer winding through a specially designed sensor, which does not require any direct connection to the HV line, and measuring the response at the other side of the winding. In [34] it has been reported that accurate, reliable and reproducible measurements are possible using the proposed system.

For online monitoring, injection of a signal to the HV bus is not only complex but also expensive due the requirement of very high isolation. In [34] a non-invasive capacitive sensor has been designed for this purpose, which could be located on the surface of the transformer bushing. As the capacitance of the sensor is small compared to the bushing capacitance, it does not affect the voltage distribution across the layers of bushing insulation. As it is necessary that the 50 Hz HV will not have any effect on the measuring system, the signal to be injected is required to pass through a protective circuit, which will absorb the system over-voltages in addition to normal power frequency voltage. But, in the process the protective circuit itself contributes to the measured response and affects the winding transfer function, particularly in the low frequency range.

Figure 7.32a shows the transfer function as measured on one phase of a transformer in the online condition. Figure 7.32b shows the transfer function in the frequency range of 100 Hz to 1 MHz of the same phase of the same transformer as measured under offline condition [34]. Comparison of Fig.7.32a with 7.32b reveals that online measurements produce frequency responses having higher attenuation and fewer fluctuations. Different measurement methods and measurement set-up used in the online and offline methods contribute to the deviations in the two plots, as shown in Fig. 7.32. The most significant problems of the online method come from noise and environmental effects at lower frequencies, which make online measurements nearly impossible for frequencies below 1 kHz.

It has been reported in [34] that the apparatus used for online measurements masks some features of the transfer function of transformer winding. It also reduces the sensitivity of the method in detecting minor inter-turn faults in caparison with offline measurements. But on the whole the online system gives adequate information on fault occurrence and is reasonably sensitive to detect

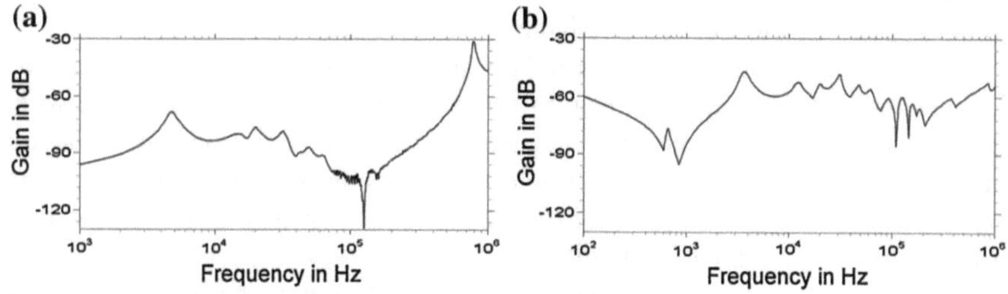

Fig. 7.32 Frequency response of one phase of a real-life transformer

minor inter-turn faults. But at the same it has been recommended that the online system should be used in conjunction with some other diagnostic method to take unambiguous decision regarding fault occurrence in transformer winding.

7.4 Future Scope of Work

Although FRA has been proved to be a useful tool for diagnosing mechanical faults in transformer windings, there is still no procedure for localization of mechanical deformations that is universally accepted. In fact when the FRA diagnostics problems are viewed in its entirety, there are several issues that need to be addressed. One such issue is to identify the interactions that cause the differences in FRA responses of different winding types such as continuous disk, interleaved disk, layer-type windings, etc. Another such issue is the interaction between the winding under test and the non-tested windings and also the interactions between the phases in the case of delta connection. A very significant practical issue is the identification of the terminal connection for FRA measurements that will be best suited for all types of winding structures. Measuring systems also need to be improved in terms of reliability and accuracy for conducting FRA tests in field conditions so that more precise evaluation could be made from the data obtained from on-site testing.

In FRA assessment process an expert considers all the measurement data that are available and perform cross-comparisons to arrive at a decision. Such a decision-making process is bound to generate differing opinions when the same data are evaluated by different experts. Thus it would be useful to have a formalized and meaningful framework to integrate all available information to make the overall evaluation. In this context, more experimental works need to be done to extract reliable interpretation features from FRA responses related to various winding failures that occur in real-life transformers.

Accurate simulation incorporating frequency dependent parameters of the core and winding of transformers aids in identifying diagnostic features for accurate fault diagnosis and reduces the number of experiments that need to be performed, which saves both time and cost. Sensitivity studies over a wide frequency range through simulations by changing the frequency dependent parameters are required to make FRA diagnostics more reliable and accurate.

It has been established by researchers that FRA technique can be used for judging the insulation quality in transformers. However, more work need to be done to study the frequency dependence of complex permittivity of dielectric media used in transformers. It will help in developing new approaches to characterization and assessment of transformer insulation quality in relation to temperature, moisture content, and aging.

Online monitoring of transformers using SFRA technique has been reported to be sensitive enough to identify inter-turn faults. But a lot of work are required in future so that different types of faults, such as axial movement, radial deformation, core defects, open-circuited winding to name a few, could be identified unambiguously with the help of the online monitoring system. Noise and environmental effects on online measuring system need to be reduced. The attenuation of the measured signal caused by the protective system used in online monitoring system should also be minimized by improving the design of the sensors and associated instrumentation.

References

1. Malewski R, Poulin B (1985) Digital monitoring techniques for high voltage impulse tests. IEEE Trans Power Apparatus Sys PAS-104:3108–3114
2. Malewski R, Poulin B (1988) Impulse testing of power transformers using the transfer function method. IEEE Trans Power Deliv 3(2):476–489
3. Ragavan K, Satish L (2007) Localization of changes of model winding based on terminal measurements: experimental study. IEEE Trans Power Deliv 22(3):1557–1565
4. Satish L, Sahoo SK (2009) Locating fault in a transformer winding: an experimental study. Electric Power Sys Res 79:89–97
5. Lech W, Tyminski L (1966) Detecting transformer winding damage – The low voltage impulse method. Electr Rev 179(21):768–772
6. Waters M, Stalewski A, Farr JC, Whitaker JD (1968) Short-circuit testing of power transformers and the detection and location of damage, CIGRE paper no. 12-05
7. Rogers EJ, Humbard LE, Gillies DA (1972) Instrumentation techniques for low voltage impulse testing of power transformers. IEEE Trans Power Apparatus Sys PAS-91:1281–1293
8. Ryder S (2003) Diagnosing transformer faults using frequency response analysis. IEEE Electr Insul Mag 19(2):16–22
9. Ryder S, Tenbohlen S (2003) Comparison of swept frequency and impulse response methods for making FRA measurements. In: Proceeding of conference of doble clients, Boston
10. Dick EP, Erven CC (1978) Transformer diagnostic testing by frequency response analysis. IEEE Trans Power Apparatus Sys PAS-97:2144–2153
11. Vaessen PTM, Hanique E (1992) A new frequency response analysis method for power transformers. IEEE Trans Power Deliv 7(1):384–391
12. IEEE PC57.149/D1 (2006) Draft trial use guide for the application and interpretation of frequency response analysis for oil immersed transformers
13. Ryder S (2001) Frequency response analysis for diagnostic testing of power transformers. Electr Today 13(6):14–19
14. Kim JW, Park BK, Jeong SC, Kim SW, Park PG (2005) Fault diagnosis of a power transformer using an improved frequency response analysis. IEEE Trans Power Deliv 20(1):169–178
15. Wang ZD, Li J, Sofian DM (2009) Interpretation of transformer FRA responses - Part I: influence of winding structure. IEEE Trans Power Deliv 24(2):703–710
16. Ragavan K, Satish L (2005) An efficient method to compute transfer function of a transformer from its equivalent circuit. IEEE Trans Power Deliv 20(2):780–788
17. Satish L, Jain A (2002) Structure of transfer function of transformers with special reference to interleaved windings. IEEE Trans Power Deliv 17(3):754–760
18. Jayasinghe JASB, Wang ZD, Jarman PN, Darwin AW (2006) Winding movement in power transformers: a comparison of FRA measurement connection methods. IEEE Trans Dielectr Electr Insul 13(6):1342–1349
19. Satish L, Sahoo SK (2005) An effort to understand what factors affect the transfer function of a two-winding transformer. IEEE Trans Power Deliv 20(2):1430–1440

20. Abeywickrama KGNB, Serdyuk YV, Gubanski SM (2006) Exploring possibilities for characterization of power transformer insulation by frequency response analysis (FRA). IEEE Trans Power Deliv 21(3):1375–1382
21. Sofian DM, Wang ZD, Li J (2010) Interpretation of transformer FRA responses-part II: influence of transformer structure. IEEE Trans Power Deliv 25(4):2582–2589
22. Shintemirov A, Tang WH, Wu QH (2010) Transformer winding condition assessment using frequency response analysis and evidential reasoning. IET Electr Power Appl 4(3):198–212
23. Pramanik S, Satish L (2011) Localisation of discrete change in a transformer winding: a network-function-loci approach. IET Electr Power Appl 5(6):540–548
24. Prout P, Lawrence M, McGrail T, Sweetser C (2004) Substation diagnostics with SFRA: transformers, line traps and synchronous compensators. In: Proceeding of EPRI substation diagnostics conference XIII, New Orleans, p 22
25. Kennedy GM, McGrail AJ, Lapworth JA (2007) Transformer sweep frequency response analysis (SFRA), Energize, pp 28–33
26. Wang M, Vandermaar AJ, Srivastava KD (2004) Transformer winding movement monitoring in service – Key factors affecting FRA measurements. IEEE Electr Insul Mag 20(5):5–12
27. Reykherdt AA, Davydov V (2011) Case studies of factors influencing frequency response analysis measurements and power transformer diagnostics. IEEE Electr Insul Mag 27(1):22–30
28. Abeywickrama N, Serdyuk YV, Gubanski SM (2008) High frequency modeling of power transformers for use in frequency response analysis (FRA). IEEE Trans Power Deliv 23(4):2042–2049
29. Abeywickrama N, Serdyuk YV, Gubanski SM (2008) Effect of core magnetization on frequency response analysis (FRA) of power transformers. IEEE Trans Power Deliv 23(3):1432–1438
30. Wang M (2003) Winding movement and condition monitoring of power transformers in service. Ph.D. Dissertation, University of British Columbia, Vancouver
31. Birlasekaran S, Fetherston F (1999) Off/on-line FRA condition monitoring technique for power transformer. IEEE Power Eng Rev 19(8):54–56
32. Leibfried T, Feser K (1999) Monitoring of power transformers using the transfer function method. IEEE Trans Power Deliv 14(4):1333–1341
33. Wimmer R, Feser K (2004) Calculation of the transfer function of a power transformer with online measuring data. In: Proceeding of advance in processing, testing and application of dielectric materials (APTADM 2004), Wroclaw, Poland, 15–17 Sept 2004
34. Behjat V, Vahedi A, Setayeshmehr A, Borsi H, Gockenbach E (2011) Diagnosing shorted turns on the windings of power transformers based upon online FRA using capacitive and inductive couplings. IEEE Trans Power Deliv 26(4):2123–2133

Chapter 8
Remaining Life Analysis

8.1 Introduction

The lifetime of an equipment may be conceptualized in three different ways: (a) *Physical Lifetime*: It corresponds to the state in which the equipment can no longer be used in its normal operating state, (b) *Technical Lifetime*: It refers to the state in which the equipment needs to be replaced due to technical reasons even though it is possible to use it physically. For example, the state at which maintenance becomes extremely difficult due to unavailability of spares, and (c) *Economic Lifetime*: The capital value of a piece of equipment depreciates every year. Moreover, operating and maintenance costs increase over time due to aging of the equipment and may even exceed the depreciated value of the equipment. In such cases, it is cost-effective to replace the equipment.

In the global sense, aging of electrical power infrastructure can be viewed as a process in which the whole system is getting old. In other words, if no new equipment is added, then a system continues to age. It is also possible for a system to get aged even when new equipments are added, if the addition of new equipments is slower than the aging of existing equipment. Whereas viewed from the angle of a piece of equipment, aging refers to an internal process through which gradual deterioration takes place, which brings the equipment closer to failure. Although in practice both the processes are active simultaneously, aging of electricity infrastructure refers mostly to the internal process, as increased redundancy can mitigate the problem arising out of the external process. The physical lifetime of important power system equipment such as transformers can be extended by means of suitable maintenance strategy, which slows down the deterioration mentioned in the internal process.

Age of power system equipment can be conceptualized in two different ways: (a) *Natural Age*: It is the difference between the date on which an equipment was brought to service for the first time and the present date and (b) *Functional Age*: If the focus is on a specific equipment, it is better to have an estimate of life in such a way that it depends on various factors such as deterioration status, usage history, operating and environmental conditions. The natural age of the majority of power

S. Chakravorti et al., *Recent Trends in the Condition Monitoring of Transformers*, Power Systems, DOI: 10.1007/978-1-4471-5550-8_8, © Springer-Verlag London 2013

transformers that are in service all over the globe varies between 25 to 50 years. Figure 8.1a shows the number transformers installed by year from 1950 onward for a UK power utility [1], while Fig. 8.1b shows the same for two Dutch utilities [2]. Both these figures indicate that depending on loading history, service duration and equipment aging, many power transformers will reach their end-of-life soon and replacements in near future will be required.

Remaining life analysis of transformer is made considering the following aspects: (a) Implementation of an improved maintenance strategy, e.g., condition-based or reliability-centered maintenance, (b) Maintenance of continuity of supply and thereby reduction of revenue loss due to unwanted interruption caused by transformer failure, (c) Minimization of remedial actions, and (d) Implementation of a comprehensive asset management program involving life extension and replacement of equipment.

Transformer failure on the whole is a probabilistic phenomenon which is controlled mainly by three parameters: design, aging, and external triggering events. Thus end-of-life modeling attempts to develop relationship between the aging conditions and transformer intrinsic dielectric properties as well as transformer thermal life, expressed as the service years to lose the mechanical strength of solid insulation under operational conditions.

8.2 Life Estimation Based on Thermal Modeling

A transformer is designed to have sufficient dielectric and mechanical strength, with some safety margin, to withstand the maximum operating stresses. Thermal stresses during the service life of the transformer will degrade the transformer insulation and reduce the safety margin until it is unable to withstand the stresses caused by external events and hence the transformer becomes prone to failure. However, it is now known from operational experience that transformers do not

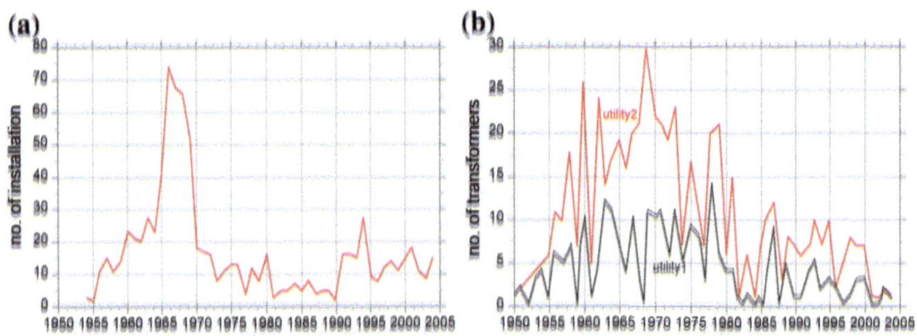

Fig. 8.1 **a** Transformers installed per year for a UK power utility. **b** Transformers installed per year for two Dutch power utilities

fail due to thermal stresses only. But thermal stress is still considered to be the most important factor in determining the useful life of a transformer.

Effect of temperature on the life of transformer having Class-A insulation was identified in 1930 by Montsinger [3], as shown in Fig. 8.2, which depicts a curve showing the probable life of Class-A insulation in terms of the time required for the insulation to become practically useless in oil for different temperatures.

The equation of the straight line shown in Fig. 8.2 is given by

$$Y = 7.15 \times 10^4 e^{-0.088x} \tag{8.1}$$

where, Y = time in years, x = temperature in °C.

The actual life of the insulation was determined in [3] from the solid line curve within the temperature limits of 30 and 120 °C. Montsinger's rule [3] states that the life of insulation is halved for every 10 °C temperature rise. But this rule is no longer valid now as the insulation quality is much superior at present than it was in 1930. But on the other hand, the factor of safety in transformer design is much lower now, which means that the transformer insulation is stressed to much higher level now even under normal operating conditions.

Figure 8.3 shows the results of several types of insulation deterioration tests as reported in 1947 [4]. In this study, the underlying consideration was that the reciprocal of the time to reach the point of failure is equal to a reaction rate constant as given below.

$$\frac{f_0'(p)}{t} = K_0 = A\,e^{-\frac{B}{T}} \tag{8.2}$$

where, A and B are constants, T = absolute temperature, K_0 = reaction rate constant and P could be a particular value of an insulation property, e.g., 20 % of the original tensile strength of paper as the limiting value. Then considering the value of $f_0'(p)$, the time to reach the limiting point could be determined.

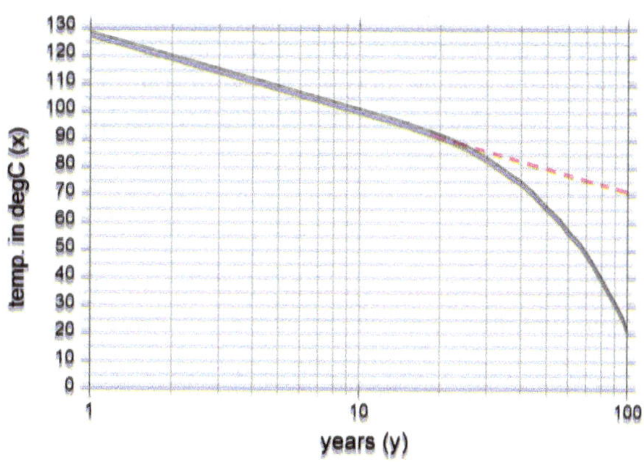

Fig. 8.2 Effect of temperature on the life of Class-A insulation

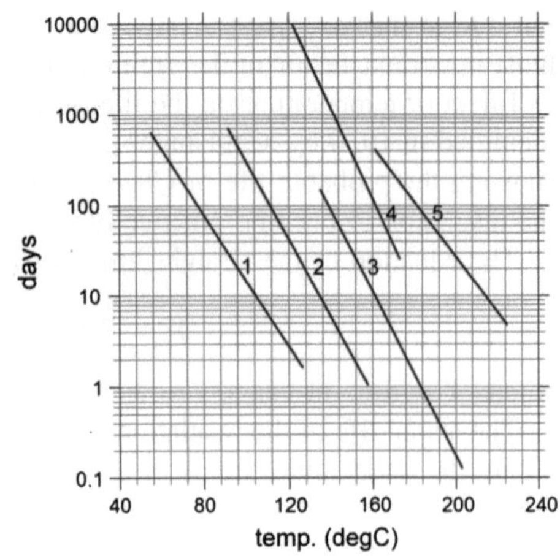

Fig. 8.3 Variation of life with temperature for different types of insulation

8.2.1 Aging Acceleration Factor

Experimental evidences obtained over the years have resulted in IEEE guide [5] and IEC standard [6] for evaluation of transformer insulation life, which are supplemented by CIGRE document [7]. Standard test procedures for thermal evaluation of liquid immersed transformers are given in IEEE guide [8], which also provides a method of estimating the hot-spot temperature of transformer.

Although aging of transformer insulation is a time-dependent function of temperature, moisture content, and oxygen content, IEEE guide [5] considers insulation temperature as the controlling parameter, with the premise that the moisture and oxygen contributions to insulation deterioration is minimized with the help of modern oil preservation systems. In aging studies, it is normal to consider the aging effects produced by the highest or hottest-spot temperature, as the part of insulation which operates at the highest temperature will usually undergo the greatest deterioration.

Figure 8.4 shows the curve that relates per unit transformer insulation life to winding hottest-spot temperature as per IEEE guide [5] considering temperature as the principal variable affecting thermal life. Figure 8.4 indicates that the aging is accelerated beyond normal for temperature above a reference temperature of 110 °C and is reduced below normal for temperature below 110 °C. The equation for the curve is as follows:

$$Per\,unit\,life = 9.80 \times 10^{-18}\, e^{\frac{1500}{\theta_H + 273}} \tag{8.3}$$

Based on Fig. 8.4, an aging acceleration factor (F_{AA}) can be calculated for a given load and temperature or for a varying load and temperature profile over a 24 h period. The equation for F_{AA} is given in [5], which may be used to calculate equivalent aging of the transformer.

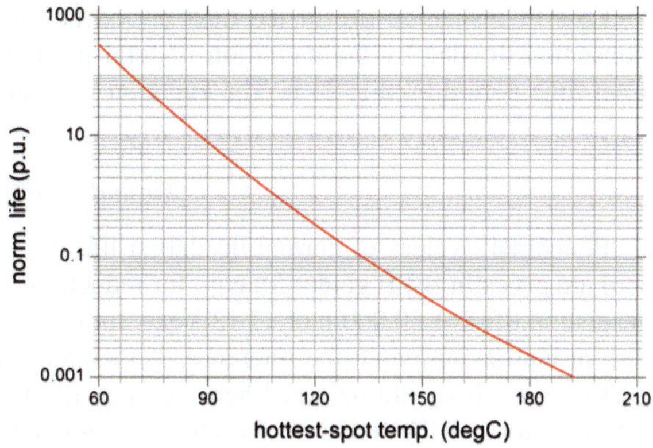

Fig. 8.4 Transformer insulation life versus hottest-spot temperature

$$F_{AA} = e^{\left(\frac{1500}{110+273} - \frac{1500}{\theta_H + 273}\right)} \tag{8.4}$$

For winding hottest-spot temperatures greater than the reference temperature 110 °C, F_{AA} will be greater than 1 and will be less than 1 for temperatures below 110 °C. A curve of F_{AA} versus hottest-spot temperature for a 65 °C rise insulation system is shown in Fig. 8.5.

The following equation gives the equivalent life at the reference temperature that will be consumed in a given time period for the given temperature cycle [5]:

$$F_{EQA} = \frac{\sum_{n=1}^{N} F_{AA,n}\,\Delta t_n}{\sum_{n=1}^{N} \Delta t_n} \tag{8.5}$$

where, F_{EQA} is the equivalent aging factor for the total time period, n is index of the time interval, t, N is total number of time intervals, Δt_n is the time interval, and $F_{AA,n}$ is aging acceleration factor for the temperature which exists during the time interval Δt_n.

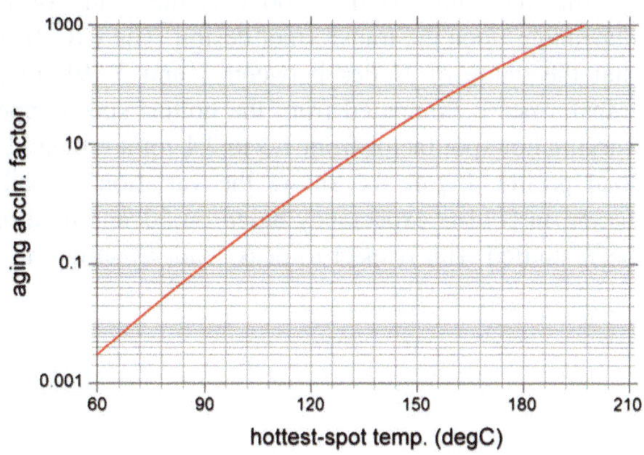

Fig. 8.5 Aging acceleration factor versus hottest-spot temperature

Loss of insulation life in the time period is computed as the equivalent life consumed divided by the definition of total normal insulation life. Usually, the total time period used is 24 h. The equation is given as follows [5]:

$$\%\,Loss\;of\;life = \frac{F_{EQA} \times t \times 100}{Normal\;insulation\;life} \tag{8.6}$$

Benchmark values of normal insulation life for a well-dried, oxygen-free system are given in IEEE guide [5], e.g., for reduction of DP value to 200 the normal insulation life is 1,50,000 h or 17.12 Years for 65 °C average winding temperature rise insulation system at the reference temperature of 110 °C.

However, it should be noted here that the uncertainty involved in the computation of loss of life based on the above-mentioned methodology is high. Moreover, as the deterioration of transformer insulation is caused by many factors that influence the cumulative effect of temperature over time, it is very difficult to predict with any great degree of accuracy the useful life of the insulation in a transformer, particularly under widely varying service conditions.

8.2.2 Hot-Spot Factor

The hottest-spot temperature (θ_H) could be computed according to IEC standard [6] as follows:

$$\theta_H = \theta_a + \Delta\theta_{or}\left(\frac{1 + rK^2}{1 + r}\right)^x + Hg_r K^y \tag{8.7}$$

where, K is the time-dependent relative load per unit, x = oil exponent, y = winding exponent, r = loss ratio, θ_a = ambient temperature, Hg_r = Hot-spot to top-oil gradient, and θ_{or} = top-of-tank oil rise. Details of these parameters are available in [6].

The hottest-spot temperature rise can also be estimated using a hot-spot factor, H [6]. According to IEC 60076-7, the value of H varies over a wide range depending upon the transformer design. The CIGRE working group 12.09 showed that the value of H lies between 1 and 3 [9]. In [1] the daily loss of life of transformer is computed from the hottest-spot temperature which is obtained by varying hot-spot factor from 1.3 to 2.1 for a typical daily ambient temperature and load profile. Figure 8.6 shows the variation of daily loss of life with hot-spot factor. In [1] life expectancy based on 50 % reduction of tensile strength has also been studied with respect to peak load, as shown in Fig. 8.7.

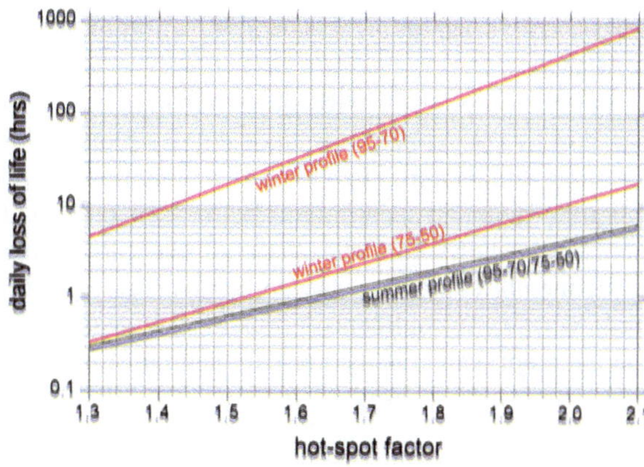

Fig. 8.6 Variation of daily loss of life with hot-spot factor

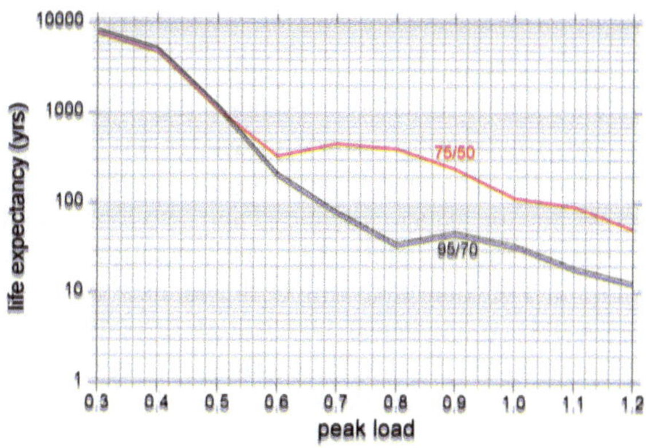

Fig. 8.7 Variation of life expectancy with peak load

8.3 Life Estimation Based on Cellulose Degradation Kinetics

Activation energy of cellulose degradation varies from a lower value of around 85 kJ/mole for degradation in oil [10, 11] to a higher value around 120 kJ/mole for hydrolytic degradation [12]. Above 140 °C, the rate of degradation apparently increases implying a further change in activation energy at that temperature. In [13] it has been shown that the mechanism of random scission of chain used to describe cellulose degradation in air and vacuum can be extended to describe degradation of cellulose in oil. The approach is similar to that used in [10, 11]. The reaction rate at any time is related to the number of unbroken polymer chains available at that time and the relationship can be expressed as follows:

$$\frac{1}{DP_t} - \frac{1}{DP_0} = kt \tag{8.8}$$

where, DP_t = DP at any time t, DP_0 = initial value of DP, and k = reaction rate constant.

For the purpose of prediction, the slope of the kinetic plot is related to temperature by the Arrhenius relationship by plotting log(k) against the reciprocal of absolute temperature. In [13, 14] a collation of relevant data available in the literature replotted according to Eq. (8.8) is presented, which is shown in Fig. 8.8. Figure 8.8 shows that the slope of the plot remains constant under different experimental conditions. Hence, it gives a constant activation energy of 111 kJ/mole with 95 % confidence limits on the mean of 105 and 117 kJ/mole [14]. But the intercept, which is the pre-exponential of the Arrhenius relationship, increases with increasing severity of reaction conditions and increasing susceptibility of paper to degradation [13]. The equation of the lines is given by [14]

$$k = A\exp\left(-\frac{111000 \pm 6000}{R(t+273)}\right) \tag{8.9}$$

where, t = Temperature in °C and R = Universal gas constant = 8.314 J/mole/K

Values of A including confidence limits are given in Table 8.1 [14].

Two of the most quoted measurements of degradation of paper in oil are reported by Fabre and Pichon in [15] and Shroff and Stannett in [16]. Figure 8.9 shows how the data reported by Fabre and Pichon as well as Shroff and Stannett fit to the collation of plots [13]. The activation energy that will be obtained from Fabre and Pichon data will be lower than the average value mentioned above below 100 °C and hence the resulting life estimates will be higher below 100 °C. But above 100 °C, it will be just the reverse. On the other hand, the activation energy obtained from Shroff and Stannett data will be higher than the mean value

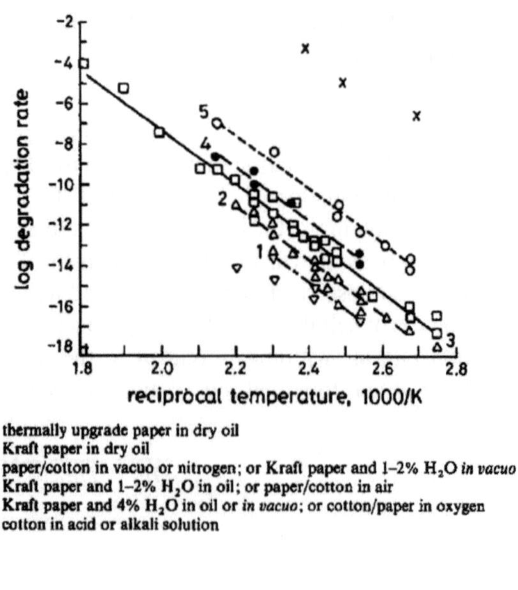

Fig. 8.8 Analysis of covariance of first-order rates of degradation of cellulose from DP measurements [14]. Reproduced with kind permission ©IET 2012

Table 8.1 Pre-exponential factors (*A*) as obtained from Fig. 8.8

Data	Pre-exponential	95 % confidence limits
Kraft paper +1 % moisture in oil	35.0×10^7	8.41×10^7 and 146×10^7
Kraft paper +2 % moisture in oil	77.8×10^7	0.18×10^7 and 330×10^7
Kraft paper +4 % moisture in oil	347×10^7	76.6×10^7 and 1570×10^7
Upgraded paper in oil	3.65×10^7	0.79×10^7 and 16.8×10^7
Dry Kraft paper in oil	10.7×10^7	2.41×10^7 and 47.1×10^7

and hence the resulting life estimation will be less for wet paper below 100 °C, but the opposite is true for dry paper below 100 °C [14].

Substituting the value of k from Eq. (8.9) and putting initial and final values of DP as 1,000 and 200, respectively, Eq. (8.8) can be simplified as [14].

$$Life = \frac{0.004}{A}\exp\left(\frac{13350}{t+273}\right) hours \tag{8.10}$$

Predicted insulation life as obtained from Eq. (8.10) for dry and wet insulation is shown in Fig. 8.10 [13], where the initial and final values of DP are chosen as 1,000 and 200, respectively. Figure 8.10 clearly shows the effect of moisture in reducing the insulation life. However, it should be noted here that the life estimation from Eq. (8.10) is based on an idealized accelerated reaction condition and hence is subject to certain limitations: (a) The degradation kinetics and mechanisms remain same in transformers in-service as in model experiments, (b) Operating conditions remain constant, (c) The initial DP value of paper insulation is known, and (d) A minimum acceptable DP value is assumed below which the failure probability of transformer becomes high. Such limitations bring in uncertainties in the estimates of life expectancy of transformers. Effect of different initial and final values of DP on life estimates are shown in Fig. 8.11 [13]. In this figure, the initial value of DP is varied from 1,300 as quoted by Fabre and Pichon to 900 as quoted by Shroff and Stannett and the final value of DP is varied in the range of 150–250.

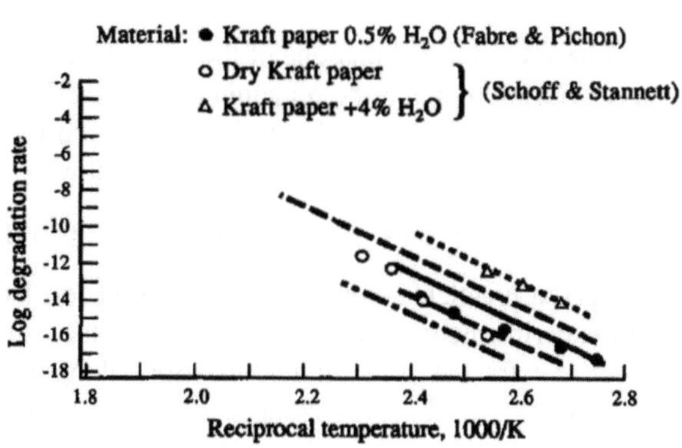

Fig. 8.9 Comparison of Fabre and Pichon and Shroff and Stannett data with the results of covariance analysis [13]. Reproduced with kind permission ©IET 2012

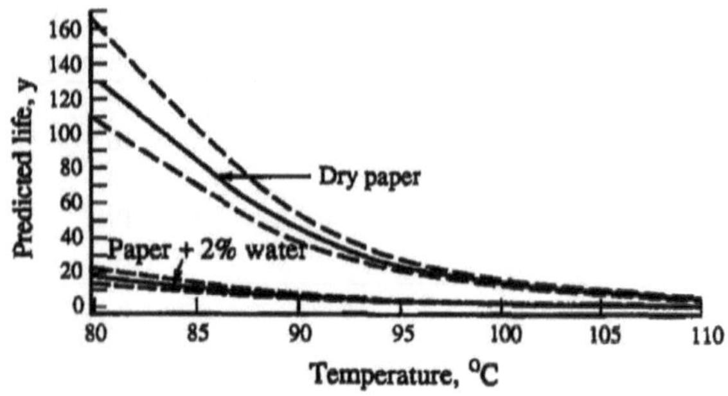

Fig. 8.10 Life prediction using Eq. (8.10) for dry and wet paper insulation [13]. Reproduced with kind permission ©IET 2012

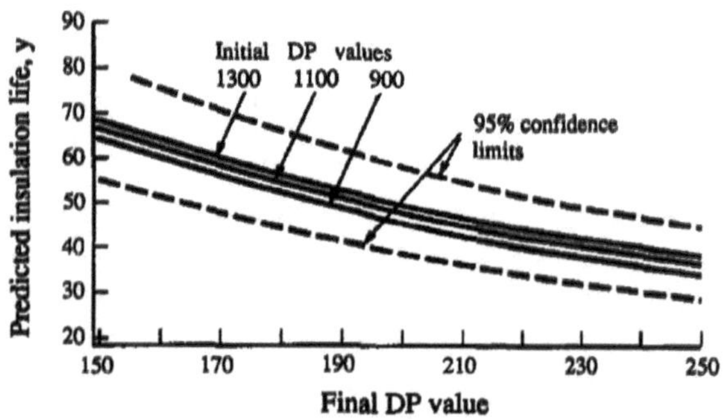

Fig. 8.11 Predicted life for different initial; and final values of DP [13]. Reproduced with kind permission ©IET 2012

To make the life estimation more rigorous and reliable, the uncertainties were incorporated in the computation in [2] as follows:

$$\frac{\Delta k}{k} = \sqrt{\left(\frac{\Delta A}{A}\right)^2 + \left(\frac{\Delta E_a}{RT}\right)^2 + \left(\frac{E_a \Delta t}{RT^2}\right)^2} \tag{8.11}$$

8.3.1 Life Estimation from DP Value

It has been established that one of the most dependable means of determination of paper deterioration and the remaining life is the DP value of cellulose insulation. In [17], an empirical formula has been proposed for percentage of remaining life (*%RL*) estimation from DP value as given below:

$$\%RL = -381.91574 + 72.053012\ \ln(DP) \tag{8.12}$$

In [1], a practical study has been reported for determination of remaining life from experimentally obtained DP values. In this study, cellulose insulation samples were extracted from different locations along the length of the windings of the transformers which were scrapped due to various reasons and the lowest DP

(DP_{min}) was obtained. Then the following steps were followed to determine the transformer life: (a) The reaction rate constant (k) of a scrapped unit is computed using Eq. (8.8) with the help of the DP_{min} and the operational life $0(t_l)$, assuming a suitable value of the initial DP value of cellulose insulation, which usually lies between 900 and 1,100, (b) Assuming that the scrapped unit would have aged following the same value of k, had it been in service ever after t_l, the residual year of operation could be computed from the obtained value of k and the final DP value of 200. This residual life would then have been the remaining life of the transformer under consideration. However, categorization of the transformer population of a utility and how well it is represented by the scrapped units needs to be studied in depth.

8.3.2 Life Estimation from Furan Value

Attempts have been made by researchers to estimate the DP without actually conducting the invasive test. In [17] one such relationship has been proposed as given below:

$$\begin{aligned} y &= 1303.5863 - 123.97843\ \ln(x_1) \\ y &= 1387.1026 - 149.26419\ \ln(x_2) \end{aligned} \tag{8.13}$$

where, y = estimated degree of polymerization, x_l = 2FAL in ppb for transformers having 55 °C temperature rise and x_2 = Total furans in ppb for transformers having 65 °C temperature rise.

The proposed empirical relationships between the percentage of remaining life ($\%RL$) and concentration of furans are as follows [17]:

$$\begin{aligned} \sqrt{\%RL} &= 10.598376 - 0.092978923\sqrt{x_1} \\ \sqrt{\%RL} &= 11.10373 - 0.15201543\sqrt{x_2} \end{aligned} \tag{8.14}$$

There are other attempts reported in literature toward determination of DP value from furan concentrations for life estimation studies. In [18], a correlation of the amount of furfural (2FAL), expressed in mg/kg of oil, with the DP of the cellulosic paper in the windings of a transformer in-service has been presented as given ion Eq. (8.15).

$$DP = \frac{7100}{8.88 + 2FAL} \tag{8.15}$$

Equation (8.15) is based on the results of a series of experiments carried out in a number of European laboratories represented in CIGRE Task Force 15.01.03. From these experiments, it was determined that every three cellulose chain scissions give one furfural molecule.

Equation (8.16) shows another empirical relationship for computing DP from total furan concentration in oil expressed in mg/liter [19].

$$DP = 418.95 - 233.48 \ln(furfural) \tag{8.16}$$

8.3.3 Life Estimation from Furan and DGA Results

A methodology has been proposed in [20] to relate DP with furans as well as DGA results and then to compute the elapsed life with this computed DP value. In this method, the DP is computed using the following equation:

$$DP = \sum_{i=1}^{6} g_i w_i \tag{8.17}$$

where,

$$g_1(CO) = a_0 + a_1 \log(CO),\ g_2(CO_2) = b_0 + b_1 \log(CO_2)$$
$$g_3(TF) = c_0 + c_1 \log(TF),\ g_4(FD) = d_0 + d_1 \log(FD)$$
$$g_5(AF) = e_0 + e_1 \log(AF),\ g_6(M2F) = f_0 + f_1 \log(M2F)$$

Such that CO = Carbon monoxide dissolved in oil in ppm, CO_2 = Carbon dioxide dissolved in oil in ppm, TF = Total Furan content dissolved in oil in μg/kg, FD = 2-furfuraldehyde dissolved in oil in μg/kg, AF = 2-Acetylefuran dissolved in oil in μg/kg, and $M2F$ = 5-Methyle-2-Furfural dissolved in oil in μg/kg.

When expressed as a function of all the six gases and furans, the DP is given by

$$DP = C_0 + a \log(CO) + b \log(CO_2) + c \log(TF) + d \log(FD) + e \log(AF) + f \log(M2F) \tag{8.18}$$

where, $C_0 = w_1 a_0 + w_2 b_0 + w_3 c_0 + w_4 d_0 + w_5 e_0 + w_6 f_0$, $a = w_1 a_1$, $b = w_2 b_1$, $c = w_3 c_1$, $d = w_4 d_1$, $e = w_5 e_1$ and $f = w_6 f_1$

The values of coefficients and the weights are available in [20]. The elapsed life is then given by the empirical formula as given below [20]:

$$Elapsed\ Life = 20.5 \times \ln(1100/DP)\ in\ yr \tag{8.19}$$

In Eq. (8.19), the initial value of mean DP is taken as 1,100 for new paper insulation.

Since the values of input quantities in this methodology could be obtained from testing oils samples, which can be collected even when the transformer is in-service, this method could also be used for online monitoring of transformer life.

8.4 Probabilistic Approach Towards Life Estimation

In [21] a probabilistic approach toward remaining life estimation of large power transformers has been reported. There are three main steps in this method:

(i) Statistical analysis of aging indicators from oil test results to derive an "insulation age" of the transformer. The aging indicators are 2-furfural, CO and CO_2 contents in oil.
(ii) Extraction of paper samples from in situ transformers for DP measurement.
(iii) Remaining life estimation by application of appropriate life models of typical power transformers.

In this work [21], the oil test results of about 200 large power transformers have been used as the database. A normal level of each of the three indicators, as mentioned above, has been determined by statistical analysis of these data, for power transformers subjected to the operating conditions. Comparison of the level of any particular age indicator of a transformer under investigation to this normal level makes it possible to estimate the effective insulation age of that transformer compared to other transformers having similar operating conditions. The average insulation age is determined from the individual insulation ages obtained from the three age indicators and is then corroborated with the help of DP measurement so that a more reliable estimation of the effective age could be made. Subsequently, hazard function best fitting failure statistics is used to derive a probability density function which fits the related hazard function. The computed insulation age is then applied to this model to determine specific failure probabilities over the time period required.

8.5 Avenues for Further Work

Over the years, several criteria have been proposed by researchers for prediction of end-of-life (EOL) of insulation for high voltage equipment. But so far there is no unanimity in the global technical community about the best criterion for EOL prediction. Hence, a lot of scope of further work lies in the area of accurate prediction of EOL based on technically feasible and economically viable considerations.

Since actual failure of transformers have been found to have a large random element that are related to system stresses, a complete knowledge of thermal EOL of a particular transformer would only give a statistical risk of failure. Authors in [1] have indicated that more work are needed to incorporate the DP data from scrapped transformers into the observed failure statistics to achieve a reliable EOL model.

Authors in [20] have suggested that for elapsed life estimation based on DGA and Furan test results, more rigorous formulations are necessary to perform an objective assessment of the variability of the coefficients, estimation of their confidence intervals, and determination of the bounds of the estimated life.

In [22] it has been shown that the ultraviolet-to-visible (UV–Vis) spectral response of transformer oil can be well correlated with furan content in oil. Consequently, estimation of aging and hence life estimation of transformer insulation using oil UV–Vis spectral response could be a promising area of future work.

Since the correlation between the furan concentration and the DP has nonlinearity and the application of conventional statistical method such as regression models results in errors, a more accurate estimation of the remaining life of oil-immersed transformer using artificial intelligence is an area of work which could be explored in future.

References

1. Jarman P, Wang Z, Zhong Q, Ishak T (2009) End-of-Life modeling for power transformers in aged power system networks. CIGRE 6th Southern Africa Regional Conference, Paper C105
2. van Schijndel A, Wetzer JM, Wouters PAAF (2008) Remaining lifetime modelling for replacement of power transformer populations. In: Proceedings of international conference on condition monitoring and diagnosis, Beijing, China, April 21–24
3. Montsinger VM (1930) Loading the transformers by temperature. AIEE Trans 49:776–792
4. Dakin TW (1947) Electrical insulation deterioration treated as a chemical reaction rate phenomenon. AIEE Trans 66:113–122
5. IEEE Loading Guide for Mineral Oil Immersed Transformer, C57.91-1995
6. IEC 60076 Part 7: Loading Guide for Oil-immersed Power Transformers, 2005
7. CIGRE Working Group 12.09 (1993) Lifetime evaluation of transformers. Electra, pp 39–51
8. IEEE Standard Test Procedure for Thermal Evaluation Liquid-Immersed Distribution and Power Transformers, C57.100-1999
9. CIGRE Working Group 12.09 (1995) Experimental determination of power transformer hot-spot factor. Electra, pp 35–39
10. Fallou B (1970) Synthesis of work carried out at LCIE on paper degradation. Rev Gen Elect 79:645
11. Bouvier B (1970) Criteria for characterizing the degradation of paper based insulation. Rev Gen Elec 79:489–496
12. Skabula W, Roganwicz J (1975) Effect of using inhibitor 2,6-di-tert-butyl-p-cresol on ageing of electrical insulation paper. Prz Papier 31:357–360
13. Emsley AM, Stevens GC (1992) A reassessment of the low temperature thermal degradation of cellulose. In: Proceedings of sixth international conference on dielectric materials, measurements and applications, Manchester, pp 229–232
14. Emsley AM, Stevens GC (1994) Review of chemical indicators of degradation of cellulosic electrical paper insulation in oil-filled transformers. IEE Proc-Sci Meas Technol 141(5):324–334
15. Fabre J, Pichon A (1960) Deteriorating processes and products of paper in oil—Application to transformers. In: Proceedings of international conference on large high voltage electric system, CIGRE, Paris, P 137
16. Shroff DH, Stannett AW (1985) A review of paper aging in power transformers. IEE Proc Part C 132:312–319

17. Bandyopadhyay MN (2007) Transformer diagnostics in the practical field. In: International conference on sustainable development: challenges and opportunities for GMS, Pattaya, Thailand, 12–14 Dec
18. de Pablo A (1999) Furfural and ageing: how are they related. In: IEE Colloquium on Insulating Liquids, pp 5/1–5/4
19. Yan Z, Dong M, Shang Y, Muhr M (2004) Ageing diagnosis and life estimation of paper insulation for operating power transformer. In: Proceedings of international conference on solid dielectrics, Toulouse, France, July 5–9
20. Pradhan MK, Ramu TS (2005) On the estimation of elapsed life of oil-immersed power transformers. IEEE Trans on Power Delivery 20(3):1962–1969
21. Allan D, Jones C, Sharp B (1991) Studies of the condition of insulation in aged power transformers. Part-1: insulation condition and remnant life assessments for in-service units. In: Proceedings of the 3rd international conference on properties and applications of dielectric materials, Tokyo, Japan, July 8–12
22. Abu-Siada A, Pin LS, Islam S (2009) Remnant life estimation of power transformer using oil UV-Vis spectral response. In: IEEE-PES power systems conference and exposition, Seattle, pp 1–5

Index

S. Chakravorti et al., *Recent Trends in the Condition Monitoring of Transformers*, Power Systems, DOI: 10.1007/978-1-4471-5550-8, © Springer-Verlag London 2013

MIX
Papier aus verantwortungsvollen Quellen
Paper from responsible sources
FSC® C105338

If you have any concerns about our products,
you can contact us on
ProductSafety@springernature.com

In case Publisher is established outside the EU,
the EU authorized representative is:
Springer Nature Customer Service Center GmbH
Europaplatz 3, 69115 Heidelberg, Germany

Printed by Libri Plureos GmbH
in Hamburg, Germany